W0230619

Fundamentals of Modern Statistical Genetics

Fundamentals of Modern Statistical Genetics

Contributors

Tao He, Jian Sa et al.

AURIS
Reference

www.aurisreference.com

Fundamentals of Modern Statistical Genetics

Contributors: Tao He, Jian Sa et al.

Published by Auris Reference Limited

www.aurisreference.com

United Kingdom

Copyright 2016
Printed in 2017 for Sale in the Indian Subcontinent

The information in this book has been obtained from highly regarded resources. The copyrights for individual articles remain with the authors, as indicated. All chapters are distributed under the terms of the Creative Commons Attribution License, which permit unrestricted use, distribution, and reproduction in any medium, provided the original author and source are credited.

Notice

Contributors, whose names have been given on the book cover, are not associated with the Publisher. The editors and the Publisher have attempted to trace the copyright holders of all material reproduced in this publication and apologise to copyright holders if permission has not been obtained. If any copyright holder has not been acknowledged, please write to us so we may rectify.

Reasonable efforts have been made to publish reliable data. The views articulated in the chapters are those of the individual contributors, and not necessarily those of the editors or the Publisher. Editors and/or the Publisher are not responsible for the accuracy of the information in the published chapters or consequences from their use. The Publisher accepts no responsibility for any damage or grievance to individual(s) or property arising out of the use of any material(s), instruction(s), methods or thoughts in the book.

Fundamentals of Modern Statistical Genetics

ISBN: 978-1-78154-775-5

British Library Cataloguing in Publication Data
A CIP record for this book is available from the British Library

Printed in the United Kingdom

Exclusively distributed by CBS Publishers & Distributors Pvt. Ltd.

Sales & Distribution Rights only for India, Pakistan, Bangladesh, Sri Lanka, Nepal and Bhutan.This book is not to be sold outside these territories.

Contents

List of Abbreviations

AFLP	Amplified Fragment Length Polymorphism
BWT	Burrows-Wheeler Transform
CCA	Canonical Correlation Analysis
CTS	Cecum total score
CHDWB	Center for Health Discovery and Well Being
COAG	Clarification of Optimal Anticoagulation through Genetics
CA	Cluster Analysis
CNV	copy number variation
DSMB	Data Safety and Monitoring Board
ED	Euclidian Distance
EM	Expectation maximization
eQTL	Expression quantitative trait loci (
FDR	False discovery rate
FBAT	Family-based association tests
FWER	Family-wise error rate
GWAS	genome wide association study
GWA	genome-wide association
HWE	Hardy–Weinberg equilibrium
HMM	Hidden Markov model
ITAMs	immunoreceptor tyrosine-based activation motifs
IME	Individual mutation error
ISE	Individual switch error
INR	International Normalized Ratio
IQR	Interquartile range
KIRs	Killer cell immunoglobin-like receptors
LDA	Linear Discriminant Analysis
LD	Linkage disequilibrium
LMN	Lymphocytes, monocytes, neutrophils
MHC	Major histocompatibility complex
DSM	Manual of mental disorders
MLEs	`Maximum likelihood estimates
MANOVA	Multivariate Analysis of Variance
NGS	Next generation sequencing (
ORs	odds ratios
PLS	Partial Least Squares
PADI4	Peptidyl arginine deiminase-4
PCR	Polymerase chain reaction
PCA	Principal Component Analysis
QNM	Quantile normalization
QTL	Quantitative trait locus

RFLPs	Restriction Fragment Length Polymorphic
RIN	RNA Integrity Number,
RMSE	Root of mean squared error
SSR	Simple Sequence Repeats
SMRT	Single Molecule Real Time
SNPs	single nucleotide polymorphism
SBWR	Spleen/body weight ratio
SD	standard deviations
TDT	Transmission/disequilibrium test
WGCNA	Weighted Correlation Network Analysis
WTCCC	Wellcome Trust Case Control Consortium

List of Contributors

Tao He
Department of Statistics and Probability, Michigan State University, East Lansing, Michigan, United States of America

Jian Sa
Department of Statistics and Probability, Michigan State University, East Lansing, Michigan, United States of America
Division of Medical Statistics, School of Public Health, Shanxi Medical University, Taiyuan, Shanxi, China

Ping-Shou Zhong
Department of Statistics and Probability, Michigan State University, East Lansing, Michigan, United States of America

Yuehua Cui
Department of Statistics and Probability, Michigan State University, East Lansing, Michigan, United States of America
Division of Medical Statistics, School of Public Health, Shanxi Medical University, Taiyuan, Shanxi, China

Bernd Genser
Department of Epidemiology and Population Health, London School of Hygiene and Tropical Medicine, London, UK
Instituto de Saúde Coletiva, Federal University of Bahia, Salvador, Brazil

Philip J Cooper
Centre for Infection, St George's University of London, London, UK
Instituto de Microbiologia, Universidad San Francisco de Quito, Quito, Ecuador

Maria Yazdanbakhsh
Department of Parasitology, Leiden University Medical Center, Leiden, The Netherlands

Mauricio L Barreto
Instituto de Saúde Coletiva, Federal University of Bahia, Salvador, Brazil

Laura C Rodrigues
Instituto de Saúde Coletiva, Federal University of Bahia, Salvador, Brazil

Mogens Fenger

Clinical Biochemistry, Molecular Biology, and Genetics, KBA339, Hvidovre, Denmark

Yao Li
Center for Computational Biology, Beijing Forestry University, Beijing, People's Republic of China
Department of Statistics, West Virginia University, Morgantown, West Virginia, United States of America

Yunqian Guo
Center for Computational Biology, Beijing Forestry University, Beijing, People's Republic of China

Jianxin Wang
Center for Computational Biology, Beijing Forestry University, Beijing, People's Republic of China

Wei Hou
Department of Biostatistics, University of Florida, Gainesville, Florida, United States of America

Myron N. Chang
Department of Biostatistics, University of Florida, Gainesville, Florida, United States of America

Duanping Liao
Department of Public Health Sciences, Penn State College of Medicine, Hershey, Pennsylvania, United States of America

Rongling Wu
Center for Computational Biology, Beijing Forestry University, Beijing, People's Republic of China
Department of Public Health Sciences, Penn State College of Medicine, Hershey, Pennsylvania, United States of America

Miaoqing Shen
Boyce Thompson Institute for Plant Research, Ithaca, New York, United States of America
United States Department of Agriculture, Agricultural Research Service, RW Holley Center for Agriculture and Health, Ithaca, New York, United States of America

Corey D. Broeckling
Colorado State University, Proteomics and Metabolomics Facility, Fort Collins, Colorado, United States of America

Elly Yiyi Chu
United States Department of Agriculture, Agricultural Research Service, RW Holley
Center for Agriculture and Health, Ithaca, New York, United States of America

Gregory Ziegler
United States Department of Agriculture, Agricultural Research Service, Plant Genet-
ics Research Unit, St. Louis, Missouri, United States of America
Donald Danforth Plant Science Center, St. Louis, Missouri, United States of America

Ivan R. Baxter
United States Department of Agriculture, Agricultural Research Service, Plant Genetics Research Unit, St. Louis, Missouri, United States of America

Jessica E. Prenni
Colorado State University, Proteomics and Metabolomics Facility, Fort Collins, Colorado, United States of America

Owen A. Hoekenga
United States Department of Agriculture, Agricultural Research Service, RW Holley
Center for Agriculture and Health, Ithaca, New York, United States of America

A. N. Diaz-Lacava
Institute for Medical Biometry, Informatics, and Epidemiology, University of Bonn,
53127 Bonn, Germany
Cologne Center for Genomics, University of Cologne, 50931 Cologne, Germany
DNA Analysis Unit, Official College of Pharmacists and Biochemists, C1184ABA
Buenos Aires, Argentina

M. Walier
Institute for Medical Biometry, Informatics, and Epidemiology, University of Bonn,
53127 Bonn, Germany

D. Holler
Institute for Medical Biometry, Informatics, and Epidemiology, University of Bonn,
53127 Bonn, Germany

M. Steffens
Institute for Medical Biometry, Informatics, and Epidemiology, University of Bonn,
53127 Bonn, Germany

C. Gieger
Research Unit of Molecular Epidemiology, Helmholtz Zentrum München, German
Research Center for Environmental Health, 85764 Neuherberg, Germany
Institute of Epidemiology II, Helmholtz Zentrum München, German Research Center
for Environmental Health, 85764 Neuherberg, Germany

C. Furlanello
FBK, 38122 Trento, Italy

C. Lamina
Division of Genetic Epidemiology, Department of Medical Genetics, Molecular and Clinical Pharmacology, Medical University of Innsbruck, 6020 Innsbruck, Austria

H. E. Wichmann
Institute of Medical Informatics, Biometry and Epidemiology, Chair of Epidemiology, Ludwig-Maximilians-University, 81377 Munich, Germany
Institute of Epidemiology I, Helmholtz Zentrum München, German Research Center for Environmental Health, 85764 Neuherberg, Germany
Institute of Medical Statistics and Epidemiology, Technical University Munich, 81675 Munich, Germany

T. Becker
Institute for Medical Biometry, Informatics, and Epidemiology, University of Bonn, 53127 Bonn, Germany
German Center for Neurodegenerative Diseases (DZNE), 53127 Bonn, Germany

C. O. Aremu
Department of Crop Production and Soil Science Ladoke Akintola University of Technology, Ogbomoso, Oyo state Landmark University, Omu-Aran Nigeria

Alex Clarke
Imperial College London, Faculty of Medicine, Section of Molecular Genetics and Rheumatology, Hammersmith Hospital

Timothy J Vyse
Imperial College London, Faculty of Medicine, Section of Molecular Genetics and Rheumatology, Hammersmith Hospital

Benjamin French
Department of Biostatistics and Epidemiology, University of Pennsylvania School of Medicine, 423 Guardian Drive, Philadelphia, Pennsylvania 19104 USA

Jungnam Joo
Office of Biostatistics Research, National Heart, Lung and Blood Institute, 6701 Rockledge Drive MSC 7913, Bethesda, Maryland 20892 USA

Nancy L Geller
Office of Biostatistics Research, National Heart, Lung and Blood Institute, 6701 Rockledge Drive MSC 7913, Bethesda, Maryland 20892 USA

Stephen E Kimmel
Department of Biostatistics and Epidemiology, University of Pennsylvania School of Medicine, 423 Guardian Drive, Philadelphia, Pennsylvania 19104 USA

Yves Rosenberg
Atherothrombosis and Coronary Artery Disease Branch, National Heart, Lung and Blood Institute, 6701 Rockledge Drive MSC 7956, Bethesda, Maryland 20892 USA

Jeffrey L Anderson
JL Sorenson Heart and Lung Center, Intermountain Medical Center, 5121 S Cottonwood St, Murray, Utah 84107 USA

Brian F Gage
Division of General Medical Sciences, Washington University School of Medicine, 660 S Euclid Ave, St. Louis, Missouri 63110 USA

Julie A Johnson
Department of Pharmacotherapy and Translational Research, University of Florida College of Pharmacy, Box 100486, Gainesville, Florida 32610 USA.

Jonas H Ellenberg
Department of Biostatistics and Epidemiology, University of Pennsylvania School of Medicine, 423 Guardian Drive, Philadelphia, Pennsylvania 19104 USA

Michael Lässig
Institut für Theoretische Physik, Universität zu Köln

Heping Zhang
Yale School of Public Health, Yale University, 60 College Street, New Heaven, Connecticut 06520-8034, USA

Lin Li
Department of Biostatistics, Harvard School of Public Health, Boston, MA, USA

Michael Kabesch
Department of Pediatric Pneumology and Allergy, KUNO University Children's Hospital Regensburg, Regensburg, Germany

Emmanuelle Bouzigon
INSERM, Genetic Variation and Human Diseases Unit, U946, Paris, France
Sorbonne Paris Cité, Institut Universitaire d'Hématologie, Université Paris Diderot, Paris, France

Florence Demenais
INSERM, Genetic Variation and Human Diseases Unit, U946, Paris, France
Sorbonne Paris Cité, Institut Universitaire d'Hématologie, Université Paris Diderot, Paris, France

Martin Farrall
Wellcome Trust Centre for Human Genetics, Oxford, UK

Miriam F. Moffatt
Molecular Genetics and Genomics Section, National Heart and Lung Institute, Imperial College London, London, UK

Xihong Lin
Department of Biostatistics, Harvard School of Public Health, Boston, MA, USA

Liming Liang
Department of Biostatistics, Harvard School of Public Health, Boston, MA, USA
Department of Epidemiology, Harvard School of Public Health, Boston, MA, USA

Jihua Wu
Section on Statistical Genetics, Department of Biostatistics, University of Alabama at Birmingham, Birmingham, AL, USA

Guo-Bo Chen
Section on Statistical Genetics, Department of Biostatistics, University of Alabama at Birmingham, Birmingham, AL, USA
Queensland Brain Institute, The University of Queensland, St. Lucia, QLD, Australia

Degui Zhi
Section on Statistical Genetics, Department of Biostatistics, University of Alabama at Birmingham, Birmingham, AL, USA

Nianjun Liu
Section on Statistical Genetics, Department of Biostatistics, University of Alabama at Birmingham, Birmingham, AL, USA

Kui Zhang
Section on Statistical Genetics, Department of Biostatistics, University of Alabama at Birmingham, Birmingham, AL, USA

Debashis Ghosh
Department of Statistics and Public Health Sciences, Penn State University, 514A Wartik Building, University Park, PA 16802, USA

Zhaohui S. Qin
Department of Biostatistics and Bioinformatics, Rollins School of Public Health, Center for Comprehensive Informatics, Emory University, 1518 Clifton Rd., N.E., Atlanta, GA 30322, USA

Shaopu Qin
School of Biology, Georgia Institute of Technology, Atlanta, GA, USA

Jinhee Kim
School of Biology, Georgia Institute of Technology, Atlanta, GA, USA

Dalia Arafat
School of Biology, Georgia Institute of Technology, Atlanta, GA, USA

Greg Gibson
School of Biology, Georgia Institute of Technology, Atlanta, GA, USA

Preface

Statistical genetics is a scientific field concerned with the development and application of statistical methods for drawing inferences from genetic data. The term is most commonly used in the context of human genetics. Rapid advances in genomic technologies have galvanized the development of new techniques to interpret genetics and genomics data sets. Fundamentals of Modern Statistical Genetics emphasizes on the statistical models and methods that are used to understand genetics, following the historical and recent developments of genetics. First chapter presents an efficient statistical model for genome-wide estimating and testing the cytoplasmic effect, nuclear DNA imprinting effect as well as the interaction between them under reciprocal backcross and F2 designs derived from inbred lines. Second chapter provides an overview of the range of statistical methods that can be used to answer different immunological study questions. We discuss specific aspects of immunological studies and give examples of typical scientific questions related to immunological data. We review classical bivariate and multivariate statistical techniques (factor analysis, cluster analysis, discriminant analysis) and more advanced methods aimed to explore causal relationships and illustrate their application to immunological data. Third chapter aims at identifying genes in biochemical and physiological processes to reveal genetic causes of rare and common diseases. In fourth chapter, we propose a statistical design for detecting imprinted loci that control quantitative traits based on a random set of three-generation families from a natural population in humans. This design provides a pathway for characterizing the effects of imprinted genes on a complex trait or disease at different generations and testing transgenerational changes of imprinted effects. In fifth chapter, we describe the mass spectrometry based profiling of maize kernels, a model system for genomic studies and a cornerstone of the agroeconomy. Sixth chapter aims to investigate fine-scale patterns of genetic heterogeneity in modern humans from a geographic perspective, a genetic geostatistical approach framed within a geographic information system is presented. Seventh chapter reveals the underlying importance of genetic diversity and reviews useable statistical techniques for identifying and grouping genotypes for intraspecies crop improvement. In eighth chapter, we focus on rheumatoid arthritis, systemic lupus erythematosus and ankylosing spondylitis and describe some of the recently described genes that underlie these conditions and the extent to which they overlap. Ninth chapter reveals on statistical design of personalized medicine interventions. Chapter ten provides an introductory review on how genes interact to produce biological functions. In eleventh chapter, we present the challenges and methods from a statistical perspective and focus on genetic association studies. In last chapter, we propose using eQTL weights as prior information in SNP based association tests to improve test power while maintaining control of the family-wise error rate (FWER) or the false discovery rate (FDR).

Chapter 1

STATISTICAL DISSECTION OF CYTO-NUCLEAR EPISTASIS SUBJECT TO GENOMIC IMPRINTING IN LINE CROSSES

Tao He[1], Jian Sa[1,2], Ping-Shou Zhong[1], Yuehua Cui[1,2]

[1]Department of Statistics and Probability, Michigan State University, East Lansing, Michigan, United States of America

[2]Division of Medical Statistics, School of Public Health, Shanxi Medical University, Taiyuan, Shanxi, China

ABSTRACT

Cytoplasm contains important metabolism reaction organelles such as mitochondria and chloroplast (in plant). In particular, mitochondria contains special DNA information which can be passed to offsprings through maternal gametes, and has been confirmed to play a pivotal role in nuclear activities. Experimental evidences have documented the importance of cyto-nuclear interactions in affecting important biological traits. While studies have also pointed out the role of interaction between imprinting nuclear DNA and cytoplasm, no statistical method has been developed to efficiently model such effect and further quantify its effect size. In this work, we developed an efficient statistical model for genome-wide estimating and testing the cytoplasmic effect, nuclear DNA imprinting effect as well as the interaction between them under reciprocal backcross and F_2 designs derived from inbred lines. Parameters are estimated under maximum likelihood framework implemented with the EM algorithm. Extensive simulations show good performance in a variety of scenarios. The utility of the method is demonstrated by analyzing a published data set in an F_2 family derived from C3H/HeJBir and C57BL/6 J mouse strains. Important cyto-nuclear interactions were identified. Our approach provides a quantitative framework for identifying and estimating cyto-nuclear interactions subject to genomic imprinting involved in the genetic control of complex traits.

INTRODUCTION

One of the central foci in biological study is to unravel the genetic secrets of complex traits of agricultural, evolutional and biomedical importance. Quantitative trait locus (QTL) mapping has been the major tool for this purpose over decades [1]–[3]. In QTL mapping, the identified QTLs are chromosome segments harboring potential genetic variants that could give rise to phenotypical manifestation. Large successes have been witnessed in the past [4]. However, there are still many phenomena that could not be explained by Mendelian genetics, leading to the new exploration of research focus on epigenetics [5].

Genomic imprinting, one of the major epigenetic phenomena, plays key roles in controlling embryonic growth and development [6], [7]. Let subscript letter M and F denote the parental origin of two alleles in a diploid organism, then a locus with two alleles A and a is thought to be imprinted if two heterozygotes $A_M a_F$ and $a_M A_F$ have different expressions [8]. The malfunction of imprinted genes could potentially lead to abnormal characters such as cancers or other genetic disorders [9].

Genomic imprinting effect is considered as one type of parent-of-origin effect due to allelic effect with specific parental origin. In contrast to this, maternal effect or cytoplasmic effect is also considered as one type of parent-of-origin effect in which the offspring›s expression is influenced by maternal parent. For example, a mother›s genotype, even if not transmitted to her offspring, may influence in utero conditions and increase risk and/or interact with genetic predisposition for particular diseases among those offspring [10]. For cytoplasm, it contains a wide variety of organelles such as mitochondria and chloroplast (in plant). Almost all the reactions of cellular metabolism take place in such an environment. It has been demonstrated that cytoplasm plays a central role in coordinating the activities of nuclear genetic materials[11]–[15]. Thus, the identification of cyto-nuclear interaction could shed novel insights into the genetic and epigenetic control of phenotypic variation. A number of empirical studies have documented the significant contribution of cyto-nuclear interaction to phenotypic variation in organisms such as wheat, rice, mice, yeast and Drosophila [16]–[19].

On the other hand, the existence of such parent-of-origin effects may lead to incorrect interpretation of the (marginal) effects of particular genes when performing genetic mapping studies, unless such effects are appropriately accounted for in the analysis [20]. Statistical methods for dissecting genomic imprinting effect has been extensively studied in literature (e.g., [21]–[26]). Tang et al. [27] developed a model to evaluate cyto-nuclear interaction effect based on experimental crosses. However, how the two types of parent-of-origin effects, one in nuclear level and one in cytoplasmic level, interact to influence phenotypic variation is largely unknown due to the lack of proper statistical models.

In this work, we discuss potential scenarios of parent-of-origin effects, and present a statistical method to dissect the cyto-nuclear interaction effects subject to genomic imprinting. The developed framework is based on experimental crosses which can be realized through two different designs, reciprocal backcross and F_2 design. When an F_2 design is considered, sex-specific difference in recombination fractions, which is initially discovered in [28], [29] and later observed in many species, is incorporated into our model to distinguish the genetic differences between two reciprocal heterozygote. Such information can be found in literature, such as the female-to-male recombination ratio of 1.6:1 for human [30], 1.4:1 for pig [31], 1.4:1 for dog [32], 1.25:1 for mouse [33] on average across the whole genome. A genome-wide scan for the identification of iQTL mapping cyto-nuclear epistasis is performed based on the interval mapping theory, and parameters are estimated based on the framework of maximum likelihood method implemented with the EM algorithm [34]. Extensive simulations are conducted to evaluate the performance of our model under different scenarios, such as different sample sizes, different heritability levels, and different gene effects. The utility of the method is illustrated by applying it to a genome-wide scan of four traits in an F_2 family derived from two inbred mouse strains.

STATISTICAL METHODS

Genetic Designs

Consider a design initiated with two inbreed lines with two segregating alleles A and a. Let subscript letter M and F represent the parental origin of offspring alleles inherited from the mother and father, respectively. A complete dissection of the cyto-nuclear interaction subject to imprinting needs experimental designs that can distinguish the quantitative variation between two heterozygotes $A_M a_F$ and $a_M A_F$, and also against the cytoplasmic effect. For this purpose, a reciprocal backcross or F_2 design is proposed so that variations and interactions can be fully introduced.

Figure 1 illustrates the reciprocal backcross design. Let the maternal parent carrying genotype AA (denoted as P2) has a cytoplasmic effect in contrast to the maternal line carrying genotype aa (denoted as P1) as the reference line. In the diagram, individuals that carry the maternal effect coming from the maternal parent with genotype AA are denoted by gray squares. Four possible backcrosses can be initiated as illustrated in Fig. 1. As shown in the figure, any backcross offsprings coming from the middle two designs carry the cytoplasmic maternal effect derived from the AA genotype.

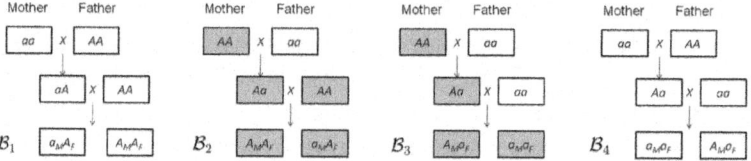

Figure 1: A reciprocal backcross design.

Figure 2 shows the F_2 design. Denote the left one as design S_1, and the right one as design S_2. The cross of two F_1's generates four possible allele-specific F_2 genotypes. Assuming there is a cytoplasmic effect, F_2 offsprings may show different phenotypes depending on the genotype of the maternal parental lines. For example, if maternal cytoplasmic effect exists, the offspring phenotypic value for AA genotype may be different depending on whether it comes from the S_1 or S_2 design. In the F_2 design, the two reciprocal heterozygotes $A_M a_F$ and $a_M A_F$ cannot be distinguished in general. Sex-specific recombination difference in male and female needs to be considered in order to distinguish the two (Cui et al. 2006) [25].

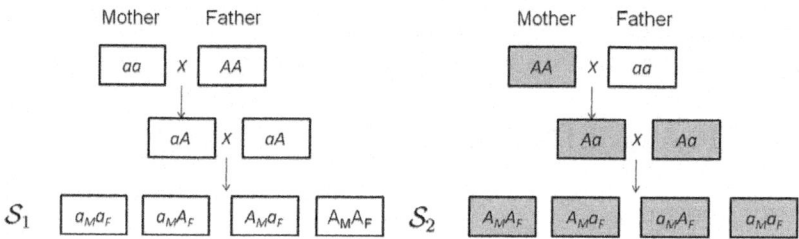

Figure 2: A reciprocal F_2 **design.**

Statistical Parameterization

For a particular cross, let y_j ($j = 1, \cdots, n$) denote the phenotypic value of interest. Following Tang et al. [27], the one-QTL genetic model can be expressed as,

$$y_j = \mu + c x_{1j} + a x_{2j} + d x_{3j} + i x_{4j} + i_{ca} x_{5j} + i_{cd} x_{6j} + i_{ci} x_{7j} + e_j \tag{1}$$

where μ is the overall mean; c is the cytoplasmic effect; a, d and i are the additive, dominance and imprinting effects of a QTL, respectively; and i_{ca}, i_{cd} and i_{ci} are the cytoplasm by additive, cytoplasm by dominance and cytoplasm by imprinting interactions, respectively; x_{1j} is an indicator variable, denoting $x_{1j} = 1$ for the AA maternal cytoplasm and $x_{1j} = -1$ for the aa maternal cytoplasm; $x_{2j}, x_{3j}, \cdots, x_{7j}$ are other indicators of various effects describing the additive, dominance and the interaction effect between the cytoplasm and genetic variables.

For the S_1 design in the F_2 population initiated with cross $AA \times aa$, the mean genotypic values of four possible genotypes formed by different allelic combination from the two F_1 parents can be expressed as,

$$\begin{cases} \mu_1 = \mu + c + a + i_{ca}, & \text{for } A_M A_F, \\ \mu_2 = \mu + c + d + i + i_{cd} + i_{ci}, & \text{for } A_M a_F, \\ \mu_3 = \mu + c + d - i + i_{cd} - i_{ci}, & \text{for } a_M A_F, \\ \mu_4 = \mu + c - a - i_{ca}, & \text{for } a_M a_F. \end{cases}$$
(2)

Similarly for the S_2 design initiated with cross $aa \times AA$, the genetic model can be expressed as,

$$\begin{cases} \mu_5 = \mu - c + a - i_{ca}, & \text{for } A_M A_F, \\ \mu_6 = \mu - c + d + i - i_{cd} - i_{ci}, & \text{for } A_M a_F, \\ \mu_7 = \mu - c + d - i - i_{cd} + i_{ci}, & \text{for } a_M A_F, \\ \mu_8 = \mu - c - a + i_{ca}, & \text{for } a_M a_F. \end{cases}$$
(3)

For the reciprocal backcross design which consists of B_1, B_2, B_3 and B_4, the indicator variables in Eq. (1) describing different QTL genotypes are defined in Table 1.

Table 1: QTL genotypes and corresponding genetic components under different backcross designs

Backcross design	BC QTL genotype	c	a	d	i	i_{ca}	i_{cd}	i_{ci}
B_1	$A_M A_F$	−1	1	0	0	−1	0	0
	$a_M A_F$	−1	0	1	−1	0	−1	1
B_2	$A_M A_F$	1	1	0	0	1	0	0
	$a_M A_F$	1	0	1	−1	0	1	−1
B_3	$A_M a_F$	1	0	1	1	0	1	1
	$a_M a_F$	1	−1	0	0	−1	0	0
B_4	$A_M a_F$	−1	0	1	1	0	−1	−1
	$a_M a_F$	−1	−1	0	0	1	0	0

doi:10.1371/journal.pone.0091702.t001

For simplicity, we will use matrix form to rewrite the models. Let us denote $\boldsymbol{\beta} = (\mu, c, a, d, i, i_{ca}, i_{cd}, i_{ci})^T$, $\boldsymbol{\mu} = (\mu_1, \mu_2, \mu_3, \mu_4, \mu_5, \mu_6, \mu_7, \mu_8)^T$. Then the relationship between the eight genetic means and the eight parameters can be written as

$$\boldsymbol{\mu} = \mathbf{D}\boldsymbol{\beta},$$
(4)

Where

$$\mathbf{D}=\mathbf{D}^{BC}=\begin{pmatrix} 1 & -1 & 1 & 0 & 0 & -1 & 0 & 0 \\ 1 & -1 & 0 & 1 & -1 & 0 & -1 & 1 \\ 1 & 1 & 1 & 0 & 0 & 1 & 0 & 0 \\ 1 & 1 & 0 & 1 & -1 & 0 & 1 & -1 \\ 1 & 1 & 0 & 1 & 1 & 0 & 1 & 1 \\ 1 & 1 & -1 & 0 & 0 & -1 & 0 & 0 \\ 1 & -1 & 0 & 1 & 1 & 0 & -1 & -1 \\ 1 & -1 & -1 & 0 & 0 & 1 & 0 & 0 \end{pmatrix},$$

(5)

for the reciprocal backcross design, and

$$\mathbf{D}=\mathbf{D}^{F2}=\begin{pmatrix} 1 & 1 & 1 & 0 & 0 & 1 & 0 & 0 \\ 1 & 1 & 0 & 1 & 1 & 0 & 1 & 1 \\ 1 & 1 & 0 & 1 & -1 & 0 & 1 & -1 \\ 1 & 1 & -1 & 0 & 0 & -1 & 0 & 0 \\ 1 & -1 & 1 & 0 & 0 & -1 & 0 & 0 \\ 1 & -1 & 0 & 1 & 1 & 0 & -1 & -1 \\ 1 & -1 & 0 & 1 & -1 & 0 & -1 & 1 \\ 1 & -1 & -1 & 0 & 0 & 1 & 0 & 0 \end{pmatrix},$$

(6)

for the F_2 design. For the purpose of illustration, the following estimation and inference is demonstrated through the F_2 design. The same procedure applies to BC design too.

The mixture Model and the Likelihood

Statistical methods for QTL interval mapping based on a mixture model traced back to the work by Lander and Botstein [1]. In the mixture model, each observation y is modeled as a weighted mixture of J (known and finite) components, and each component, which corresponds to a certain genotype category depending on the underlying genetic design, follows a certain distribution f_j with weight π_j. Conditional on the marker genotype $\mathbf{M}$ and unknown parameters ϕ and η, the density of the observed y has the following expression

$$y \sim p(y|\mathbf{M},\phi,\eta) = \pi_1 f_1(y|\mathbf{M},\phi_1,\eta) + \ldots + \pi_J f_J(y|\mathbf{M},\phi_J,\eta),$$

(7)

where $\pi = (\pi_1,...,\pi_J)^T$ refers to the mixture proportions which are constrained nonnegative and $\sum_{j=1}^{J} \pi_j = 1$; $\phi = (\phi_1,...,\phi_J)^T$ is a vector for the component-specific parameters, with ϕ_j being specific to jth component; and η consists of parameters (i.e., residual variance) that are common to all components.

For the F_2 design we described above, there are four genotypes at each locus $(A_M A_F, A_M a_F, a_M A_F,$ and $a_M a_F)$. The genotype of the QTL is generally unobservable, but can be inferred by using the two flanking markers' information. Given the flanking marker genotypes of the ith individual, the conditional probabilities $\pi_i = (\pi_{A_M A_F|i}, \pi_{A_M a_F|i}, \pi_{a_M A_F|i}, \pi_{a_M a_F|i})$ of the QTL genotype can be calculated. These conditional probabilities become the mixture proportions in the mixture model (7). Let us denote $\pi_i = (\pi_{1|i}, \pi_{2|i}, \pi_{3|i}, \pi_{4|i})$ to simplify the notation. These conditional probabilities are expressed in terms of sex-specific recombination rates in order to distinguish the two reciprocal heterozygotes. Please refer to Cui et al. [25] for the conditional probabilities of QTL genotypes given marker genotypes in terms of sex-specific recombination fractions for an F_2 design.

Assume the total number of F_2 offsprings for design S_1 and S_2 are n_1 and n_2 respectively, and let $n = n_1 + n_2$. Phenotype data for a certain quantitative trait can be observed and recorded as a vector $y = (y_1, ..., y_n)$, where $\mathbf{y}^{(1)} = (y_1, ..., y_{n_1})$ are from design S_1 and $\mathbf{y}^{(2)} = (y_{n_1+1}, ..., y_n)$ are from design S_2. Marker information can be reorganized as matrix $\mathbf{M} = (\mathbf{M}_1; \mathbf{M}_2)^T$, where the jth $(j = 1, ..., n_1)$ row of $\mathbf{M}_1$ (the jth row of matrix $\mathbf{M}$) contains all the marker information of the jth individual under design S_1 and the kth $(k = 1, ..., n_2)$ row of $\mathbf{M}_2$ (the $(n_1 + k)$th row of matrix $\mathbf{M}$) contains all the marker information of the kth individual under design S_2. Based on the mixture model (7), and with the independence assumption, the joint likelihood function for the F_2 family with total n individuals, constructed by combining design S_1 and S_2 together, can be formulated as,

$$L(\Theta|M, y) = \prod_{i=1}^{n_1} [\pi_{1|i} f_1(y_i) + \pi_{2|i} f_2(y_i) + \pi_{3|i} f_3(y_i) + \pi_{4|i} f_4(y_i)]$$

$$\times \prod_{i=n_1+1}^{n} [\pi_{1|i} f_5(y_i) + \pi_{2|i} f_6(y_i) + \pi_{3|i} f_7(y_i) + \pi_{4|i} f_8(y_i)],$$

$$(8)$$

where the unknown vector Θ contains the QTL position, QTL effects and residual variance, and the density function f_j $(j = 1, ..., 8)$ is assumed to follow a normal distribution with mean μ_j and common variance σ^2. More specifically, parameter vector Θ can be divided into two subsets, Θ_l and Θ_g, where Θ_l describes the location of QTL and Θ_g contains all the genetic parameters, including QTL-effects vector $\beta = (\mu, c, a, d, i, i_{ca}, i_{cd}, i_{ci})^T$ and residual variance σ^2 in our model.

Parameter Estimation

To estimate the unknown parameters $\Theta = (\Theta_l, \Theta_q)$, several algorithms could be implemented, such as Expectation-Maximization (EM), Newton Raphson and Fisher Scoring. Among all these methods, EM algorithm, which was initially developed by Dempster et al. [34], is most commonly used in QTL mapping study. In this paper, EM algorithm is applied to obtain the maximum likelihood estimates (MLEs). This procedure involves differentiating the log-likelihood function with respect to each unknown parameter, letting the derivatives equal to zero, and solving the log-likelihood equation for the corresponding parameter.

For the QTL position which is unknown in the model, we did not estimate parameters Θ_l directly. As commonly treated in QTL mapping studies, we applied a grid search approach to estimate the putative QTL position via scanning the entire linkage genome by 1 or 2 cM increment flanked by two markers and did a hypothesis testing at each putative position. A likelihood ratio or LOD profile plot can be generated to graphically display the LR or LOD test statistic for a putative QTL at each testing position. The genomic position which corresponds to a peak in the profile plot is the MLE of the QTL location, given that the peak passes the genome-wide threshold determined by the permutation tests detailed below. Bootstrap methods can be applied to assess the confidence interval of the estimated position [35].

Hypothesis Test

Testing the overall QTL effect on the quantitative trait is the first step toward a complete dissection of different genetic contributions to the trait. Once the MLEs of the parameters are obtained, the presence of QTL responsible for the variation of the quantitative phenotype can be tested by using the following hypotheses,

$$\begin{cases} H_0: & a = d = i = i_{ca} = i_{cd} = i_{ci} = 0 \\ H_1: & \text{not all equal zero} \end{cases}$$

The test statistic for testing the above hypotheses is calculated as the log-likelihood ratio test statistic (LR) of the full model (H_1) over the reduced model(H_0):

$$LR = -2\log[L(\tilde{\Omega}) - L(\hat{\Omega})] \tag{9}$$

where $\tilde{\Omega}$ and $\hat{\Omega}$ denote the MLEs of the unknown parameters under H_0 and H_1, respectively. Since a genome-wide scan involves multiple correlated tests, we use the permutation tests proposed by Churchill and Doerge [36] to find the threshold value.

If there is a QTL, a number of other hypothesis tests can also be performed to test the property of the detected QTL. To test the imprinting effect, we can simply formulate the hypothesis as $H_0 : i = 0$ vs $H_1 : i \neq 0$ to assess the mean difference of the two reciprocal heterozygotes. To test the cytoplasmic maternal effect, the hypothesis can be stated as $H_0 : c = 0$ vs $H_1 : c \neq 0$. The epistatic effects of all interaction terms can also be tested as

$$\begin{cases} H_0 : & i_{ca} = i_{cd} = i_{ci} = 0 \\ H_1 : & \text{not all equal zero} \end{cases}$$

Similarly, additive and dominance effects can be tested as

$$\begin{cases} H_0 : & a = d = 0 \\ H_1 : & \text{not all equal zero} \end{cases}$$

If specific interest is focused, for instance, the interaction of imprinting and maternal effect, the hypothesis can be formulated as $H_0 : i_{ci} = 0$ vs $H_1 : i_{ci} \neq 0$. All the above tests can be done by applying the likelihood ratio test in which the test statistic asymptotically follows a χ^2 distribution with degrees of freedom equal to the difference of the parameters under the null and the alternative hypotheses. For example, when testing $H_0 : a = d = 0$, the LR test statistics is compared with the χ_2^2 cutoff with 2 degrees of freedom.

RESULTS

Monte Carlo Simulation

Monte Carlo simulations were performed to investigate the statistical behavior of our model. We simulated an F_2 population, with one half of the population coming from design S_1 and the other half coming from design S_2. A genome with 100 cM long linkage group, composed of 6 equidistant markers, was constructed. The position of QTL was assumed to be located at 48 cM away from the first marker on the linkage group. The marker genotypes in the F_2 population were simulated by mimicking sex-specific recombination fractions in mice, i.e., $r_M = 1.25 r_P$ [33]. A series of simulation study with different sample size ($n = 400$ vs $n = 800$) and different heritability levels ($H^2 = 0.1, 0.25, 0.4$) was conducted to examine the impacts of parameter spaces on parameter estimation and testing power. These simulation designs, which were aimed to give a better understanding of model performance under different situations, can provide biologists some empirical evidences to design their experiments.

In the simulation study, the residual variance was calculated under different heritability levels. For the F_2 design, the genetic variances of various terms can be calculated as follows:

$$\sigma_a^2 = a^2/2,$$

$$\sigma_d^2 = d^2/4, \ \sigma_i^2 = i^2/2, \ \sigma_c^2 = c^2, \ \sigma_G^2 = \sigma_a^2 + \sigma_d^2 + \sigma_i^2 + i_{ca}^2/2 + i_{cd}^2/2 +$$

$$i_{ci}^2/2 + ci_{cd},$$

and the broad sense heritability level can be expressed as

$$H^2 = \sigma_G^2/(\sigma_G^2 + \sigma_c^2 + \sigma_e^2).$$

For given genetic parameters and the heritability level, the residual variance can be calculated as

$$\sigma_e^2 = (\sigma_Q^2 + \sigma_{cQ}^2)(1/H^2 - 1) - \sigma_c^2,$$

from which the phenotype data can be generated.

The MLEs of the QTL position and effect parameters, based on 200 simulation replicates under different heritability levels and sample sizes, are displayed in Table 2. The square root of mean squared error (RMSE) of parameter estimates are given in parenthesis to show the estimation accuracy. As we expected, the accuracy of parameter estimates increases as the sample size and heritability level increase. For instance, the RMSE of estimated QTL position decreases from 13.28 to 3.81, an 71% increase in accuracy when the sample size increases from 400 to 800 under a fixed heritability level 0.1. The other parameter estimates show the same pattern. If we increase the heritability level when the sample size is fixed, a clear reduction in RMSE can be observed. For example, with 400 samples, the RMSE of estimated QTL position decreases from 13.28 to 4.99, then to 2.77 as H^2 gradually increases from 0.1 to 0.4. From the decreasing RMSEs of the parameter estimates, we observed that simply increasing sample size is less efficient than increasing heritability level in order to increase the precision of parameter estimation. Since high heritability corresponds to small environmental variability [37], reducing environmental variation is of more practically important than just simply increasing sample size.

Table 2: The MLEs of the QTL position and effect parameters based on 200 simulation replicates under different heritabilities and sample sizes

H^2	n	Position 48 cM	$\mu=10$	$c=1$	$a=1$	$d=0.8$	$i=0.6$	$i_{ca}=0.6$	$i_{cd}=0.5$	$i_{ci}=0.4$	σ^2
0.1	400	47.93	10.00	0.99	0.96	0.78	·	0.59	0.49	·	3.679
		(13.28)	(0.33)	(0.35)	(0.35)	(0.55)	·	(0.32)	(0.53)	·	(0.22)
	800	47.35	10.01	1.02	0.98	0.77	·	0.60	0.45	·	3.73
		(3.81)	(0.23)	(0.21)	(0.24)	(0.39)	·	(0.22)	(0.33)	·	(0.15)
0.25	400	47.04	10.02	1.00	0.96	0.78	·	0.60	0.50	·	1.99
		(4.99)	(0.17)	(0.18)	(0.16)	(0.26)	·	(0.17)	(0.26)	·	(0.11)
	800	47.65	10.00	1.01	0.98	0.81	·	0.59	0.50	·	2.01
		(2.29)	(0.12)	(0.12)	(0.12)	(0.20)	·	(0.10)	(0.18)	·	(0.08)
0.4	400	48.02	9.99	1.00	0.99	0.81	·	0.60	0.50	·	1.23
		(2.77)	(0.10)	(0.11)	(0.09)	(0.16)	·	(0.10)	(0.17)	·	(0.06)
	800	47.95	9.99	0.99	0.99	0.81	·	0.60	0.51	·	1.24
		(1.78)	(0.07)	(0.07)	(0.08)	(0.12)	·	(0.07)	(0.11)	·	(0.04)

The squared roots of the mean squared errors (RMSE) of the MLEs are given in parentheses.
The locations of the QTL is described by the map distances (in cM) from the first marker of the linkage group. The hypothesized σ^2 value is 3.81 for $H^2=0.1$, 2.04 for $H^2=0.25$ and 1.26 for $H^2=0.4$.
doi:10.1371/journal.pone.0091702.t002

Note that we did not list the estimation of the imprinting effect i and cyto-imprinting interaction i_{ci}, which are not estimable under the F_2 design. The reason is that the imprinting direction cannot be inferred from the F_2 design [25]. However, we can still conduct hypothesis test to infer the imprinting effect as well as its interaction with cytoplasmic effect. To further investigate the testing performance of cytoplasmic and imprinting effects, we introduced two proportions, namely η_c and η_i, where $\eta_c=\sigma_c^2/\sigma_G^2$ and $\eta_i=\sigma_i^2/\sigma_G^2$. We can evaluate the test power under different cytoplasmic and imprinting effect sizes. Given all other genetic parameter values fixed (as shown in Table 2), simple algebra shows that the cytoplasmic effect c and imprinting effect i can be calculated for a given value of η_c or η_i, i.e.,

$$c=\frac{i_{cd}\eta_c+\sqrt{(i_{cd}\eta_c)^2+4\eta_c Q_1}}{2}, \text{ and } i=\sqrt{\frac{2\eta_i Q_2}{1-\eta_i}}$$

where

$$Q_1=\sigma_a^2+\sigma_d^2+\sigma_i^2+i_{ca}^2/2+i_{cd}^2/2+i_{ci}^2/2,$$

and

$$Q_2=\sigma_a^2+\sigma_d^2+i_{ca}^2/2+i_{cd}^2/2+i_{ci}^2/2+ci_{cd}.$$

Based on 1000 simulation runs under different heritabilities, sample sizes and variation proportions, the power of cytoplasmic effect test, imprinting effect test, interaction effects test and additive/dominance effects test are listed in Table 3. As we expect, the test power increases with the increasing of sample size and heritability level. For example, the cytoplasmic testing power increases from 0.663 to 0.905 as sample size increases from 400 to 800, a 36.5% increase in power for fixed $\eta_c = 0.1$ and $H^2 = 0.25$. The same pattern is observed for the imprinting test. As the proportion of variance explained by the cytoplasmic and imprinting effect increases, the power increases accordingly. Noted that when both η_c or η_i, are zeros, the testing power corresponds to the type I error rate for the corresponding factor. From the table we can see that the size of imprinting test is well controlled under different sample sizes and heritability levels. For the cytoplasmic effect, the size is a little inflated under $n = 400$, but it gets close to the nominal 5% level as sample size increases to 800, especially under large heritability (e.g., $H^2 = 0.4$). In sum, the simulation evidences show that the model performs reasonably well in both parameter estimation and testing.

Table 3: The power of four hypothesis tests based on 1000 samplings under different heritabilities, sample sizes and variation proportions

n	H^2	Power[1]			Power[2]			Power[3]	Power[4]
		$\eta_c = 0\%$	$\eta_c = 10\%$	$\eta_c = 20\%$	$\eta_i = 0\%$	$\eta_i = 10\%$	$\eta_i = 20\%$		
400	0.10	0.088	0.353	0.528	0.066	0.085	0.089	0.654	0.678
	0.25	0.081	0.663	0.928	0.051	0.263	0.432	0.990	0.992
	0.40	0.064	0.930	0.998	0.050	0.807	0.942	1.000	1.000
800	0.10	0.073	0.464	0.753	0.059	0.100	0.126	0.898	0.927
	0.25	0.066	0.905	0.998	0.047	0.473	0.713	1.000	1.000
	0.40	0.049	0.997	1.000	0.050	0.980	0.997	1.000	1.000

Power[k] ($k = 1,2,3,4$) refer to the powers for testing 1) $H_0 : c = 0$ vs $H_1 : c \neq 0$; 2) $H_0 : i = 0$ vs $H_1 : i \neq 0$; 3) $H_0 : i_{aa} = i_{ad} = i_{da} = 0$ vs H_1 : not all equal to 0; and 4) $H_0 : a = d = 0$ vs H_1 : not all equal to 0, respectively. For a given η_c, all other effect values are fixed as 0.8 except for c, which can be calculated in terms of η_c and other parameters. The hypothesized c value is 0 for $\eta_c = 0$, 0.461 for $\eta_c = 0.1$ and 0.679 for $\eta_c = 0.2$. Similarly, the value of i, which depends on imprinting effect variation proportion η_i, is 0 for $\eta_i = 0$, 0.680 for $\eta_i = 0.1$ and 1.020 for $\eta_i = 0.2$.
doi:10.1371/journal.pone.0091702.t003

A Case Study

We applied the model to a published F_2 cross data set based on design S_1 and design S_2 aimed to find QTLs that contribute to variation in quantitative traits related to colitis severity in 1L-10-deficient mice [38]. The data contain 411 F_2 mice derived from inbred strains, where 203 mice are from design S_1 and 208 mice are from design S_2. Ninety-one markers were obtained with an average length of ~15 cM spanning acroos the 19 autosome chromosomes. For more information about the data, the readers are referred to the original paper [38].

It has been reported that on average the female chromosome is 25% longer in genetic distance between homologous loci than the male in mice [38]. The sex-specific recombination fraction, expressed as $r_M = 1.25 r_P$, was reconstructed based on the marker information (see Cui et al. for details [25]). The method was applied to four phenotypes, cecum total score (CTS), spleen/body weight ratio (SBWR), mesenteric lymph node(MLN)/body weight ratio(MBWR), and secretory IgA (SIgA) level, where cecum total score was graded by using colitis-related criteria, including severity, hyperplasia, ulceration and the percentage of area involved. Box-Cox transformation was applied to all traits before fitting the Gaussian-mixture model. The genome-wide LOD profile plots for the four phenotypes are shown in Figure 3, where the solid blue curves correspond to the LOD values and the dashed red lines correspond to the 5% genome-wide threshold values out of 1000 permutations. The LOD score is calculated as $\log_{10}LR$, where LR is obtained from equation (9) to test the null hypothesis: $H_0 : a = d = i = i_{ca} = i_{cd} = i_{ci} = 0.$

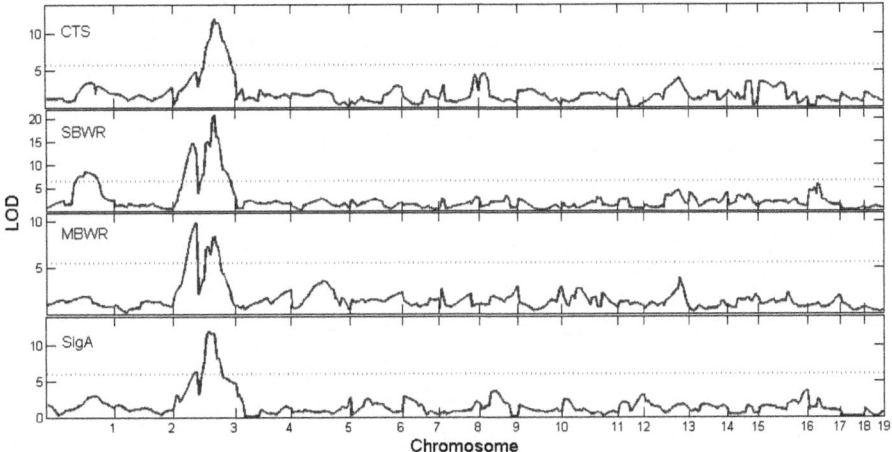

Figure 3: The LOD profiles of the four traits across the 19 chromosomes using the linkage map constructed from microsatellite markers [38]. The genomic positions corresponding to the peak of the curve are the MLEs of the QTL locations.

As shown in Figure 3, one QTL on chromosome 3 is detected for cecum total score trait, two QTLs on chromosomes 3 and one on chromosome 1 are detected for spleen/body weight ratio trait, three QTLs on chromosome 3 are detected for MLN/body weight ratio trait, and two QTLs located on chromosome 3 are detected for SIgA trait. The QTL located at 60.6 cM on chromosome 3 is common to three traits. The one located at 52.6 cM for SIgA trait is very close to it. It is highly possible that it is the same QTL that controls the four traits. Such a pleiotropic effect needs to be further evaluated. It should be mentioned that the QTL detected in the original paper for the four traits is located at 61.8 cM on chromosome 3 [38], which is 1.2 cM away from the one we found. Such a difference may arise because of the capitalization of sex-specific recombination rates and different models fitted.

In addition to the QTL identified in our analysis and the original paper, some other major QTLs which are not detected in Farmer et al. [14], such as those at 52.6 cM and 38.6 cM on chromosome 3, stand out in our model and therefore need further investigation. Almost all the QTLs on chromosome 3 are clustered together, whose local LOD profiles of four traits are shown in Figure 4.

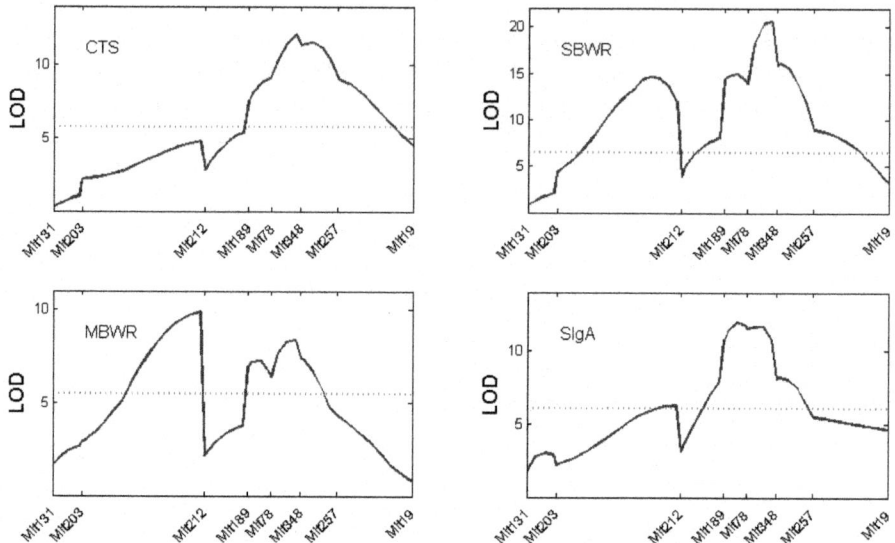

Figure 4: The local LOD profiles of the four traits across chromosomes 3.

The marker interval for each QTL is listed in Table 4, which also tabulates the p-values under four different tests for each estimated QTL using permutation tests. From the test results, it can be seen that most QTLs have strong additive and dominance genetic effect, except for the spleen/body weight ratio QTL located at D1Mit156+31.1 cM on chromosome 1. This QTL shows evidence of cytoplasmic, imprinting as well as cyto-nuclear interaction effects (p-values for the three tests are 0.0282, 0.0296 and 0.0073, respectively), but shows no sign of additive and dominance effect. In addition, the MLN/body weight ratio QTL located at D3Mit78+5.6 cM shows evidence of cytoplasmic effect. In summary, we identified one QTL on chromosome 1 with evidence of cyto-nuclear interaction effect and this QTL also shows evidence of cytoplasmic and imprinting effect. Further functional validation is needed to confirm the results.

Table 4: List of QTL positions, corresponding marker intervals and p-values under different tests for the four traits

Trait	Ch	QTL postion	Marker interval	p-value[1]	p-value[2]	p-value[3]	p-value[4]
CTS	3	60.6 cM	D3Mit78-D3Mit348	0.6680	0.3046	0.5005	$<10^{-11*}$
			(D3Mit78+5.6 cM)				
SBWR	3	60.6 cM	D3Mit78-D3Mit348	0.4322	0.6107	0.4502	$<10^{-16*}$
			(D3Mit78+5.6 cM)				
	3	32.6 cM	D3Mit203-D3Mit212	0.2003	0.9502	0.2700	$<10^{-13*}$
			(D3Mit203+11.4 cM)				
	1	63.9 cM	D3Mit156-D1Mit17	0.0282*	0.0296*	0.0073*	0.8064
			(D1Mit156+31.1 cM)				
MBWR	3	38.6 cM	D3Mit203-D3Mit212	0.3574	0.2651	0.1658	$<10^{-9*}$
			(D3Mit203+27.4 cM)				
	3	60.6 cM	D3Mit78-D3Mit348	0.0102*	0.9708	0.4831	$<10^{-8*}$
			(D3Mit78+5.6 cM)				
	3	52.6 cM	D3Mit189-D3Mit78	0.2120	0.6421	0.7095	$<10^{-3*}$
			(D3Mit189+2.9 cM)				
SIgA	3	52.6 cM	D3Mit189-D3Mit78	0.1016	0.6488	0.3195	$<10^{-9*}$
			(D3Mit189+2.9 cM)				
	3	38.6 cM	D3Mit203-D3Mit212	0.7174	0.4051	0.8020	$<10^{-7*}$
			(D3Mit203+27.4 cM)				

p-value[k] ($k = 1,2,3,4$) refer to the p-values obtained with the likelihood ratio tests for testing 1) $H_0 : c = 0$ vs $H_1 : c \neq 0$; 2) $H_0 : i = 0$ vs $H_1 : i \neq 0$; 3) $H_0 : i_{ca} = i_{cd} = i_{ci} = 0$ vs H_1 : not all equal to 0; and 4) $H_0 : a = d = 0$ vs H_1 : not all equal to 0, respectively. Significant test results are indicated with the "*" sign.
doi:10.1371/journal.pone.0091702.t004

DISCUSSION

The cytoplasmic environment influences the expression of nuclear information in a very complicated way, which is still an unravel mystery to many organisms. For example, Burgess and Husband have demonstrated great maternal contributions to the fitness of mulberry hybrids [39]. While it is an important parent-of-origin effect affecting offspring fitness, genomic imprinting, another source of parent-of-origin effect can also lead to phenotypic variation. Increasing evidence from cytoplasmic substation and cell fusion experiments suggests that weakness of hybrids may connect with the interactions between cytoplasm and nuclear [40],[41], and the evidences about interaction between cytoplasm and imprinting have also been observed [42], [43]. As the source of genetic variation for many traits, these two types of parent-of-origin effects are often confounded, making it difficult to distinguish without proper statistical dissection. Although the role of cross-talk between the two sets of factors on phenotypic variation has been recognized, which genes are involved in the process and in what form they respond to the cytoplasmic changes are still unclear.

In this paper, we developed a statistical model to evaluate the cytoplasmic environment and nuclear gene interaction subject to imprinting effect within the framework of QTL interval mapping. The model that considers eight genetic factors which measure the degree of imprinting, cytoplasmic, additive, dominance effects as well as the interaction effects among them, provides a complete dissection of cyto-nuclear epistasis subject to imprinting effect. A number of hypothesis tests can be performed not only to assess major genetic effect(s) responsible for phenotypic variation, but also to find the statistical evidence for the existence of imprinting, cytoplasmic effect as well as the cyto-nuclear interactions. Simulation study showed relative good performance of the model under the F_2 design, in which parameters are estimated efficiently with modest heritability and sample size. Low heritability level ($H^2 = 0.1$) and small sample size ($n = 400$) result in large mean squared error of parameter estimation. This result is valuable in practice as we need to be careful about the interpretation of genetic effects obtained in real data analysis when the proportion of variance explained by the QTL is small (i.e., low heritability). Although our model cannot estimate the imprinting effect (so the cyto-imprinting interaction effect) due to the nature of the F_2 design, existence test of imprinting (or cyto-imprinting interaction effect) can be achieved. Nevertheless, such imprinting estimation problem can be solved under the reciprocal backcross design illustrated in Figure 1.

In the real data analysis, one QTL located on chromosome 1 were found to have significant cytoplasmic, imprinting effect and cyto-nuclear interaction effects for spleen/body weight ratio. Other than that, no imprinting effect was found for all other QTLs, and only one on chromosome 3 that shows cytoplasmic effect for the MLN/body weigh ratio trait. It is worth mentioning that several QTL on chromosome 3 are detected by our model, but they are located relatively close to each other, as shown in Figure 3. For the sake of cautiousness, we reported all of them. However, these detected clustered QTLs may be due to the limitation of interval mapping, which can be overcome by fitting a composite interval mapping model as following,

$$y_j = \mu + cx_{1j} + ax_{2j} + dx_{3j} + ix_{4j} + i_{ca}x_{5j} + i_{cd}x_{6j} + i_{ci}x_{7j}$$

$$+ \sum_{k=1}^{K}(a_k x_{2jk} + d_k x_{3jk} + i_k x_{4jk} + i_{ca_k}x_{5jk} + i_{cd_k}x_{6jk} + i_{ci_k}x_{7jk}) + e_j$$

where $x_{2jk}, \cdots, x_{7jk}$ are corresponding variables for the kth marker, assuming total K markers are selected for controlling background genetic effect. Although more variables are introduced in the model, theoretically some dimension-reduction techniques such as LASSO, can be applied to implement the variable selection for each marker before fitting them into the final model [44]. The composite interval mapping is know for its improved resolution in QTL detection. Regardless of the potential limitations mentioned above, the integration of cyto-nuclear interactions into the QTL mapping framework provides a testable platform with feasible experimental design for biologists to test the existence of cytoplasmic and imprinting effects, as well as the interactions of interested. The proposed model will have important biological implications with potentials to lift a corner of the great veil of the genetic system.

AUTHOR CONTRIBUTIONS

Conceived and designed the experiments: YC. Analyzed the data: TH. Wrote the paper: TH YC. Participated in analysis: JS PSZ.

REFERENCES

1. Lander E, Bostein E (1989) Mapping Mendelian factors underlying quantitative traits using RFLP linkage maps. Genetics 121: 185–199.

2. Zeng Z (1994) Precision mapping of quantitative trait loci. Genetics 136: 228–235.

3. Wu R, Ma C, Gallo-Meagher M, Littell R, Casella G (2002) Statistical methods for dissectiong triploid endosperm traits using molecular markers: an autogamous model. Genetics 162: 875–892.

4. Mackay T (2001) Quantitative trait loci in Drosophila. Nat Rev Genet 2: 11–20. doi: 10.1038/nrg1206

5. Jirtle R, Skinner M (2007) Environmental epigenomics and disease susceptibility. Nat Rev Genet 8: 253–262. doi: 10.1038/nrg2045

6. Isles A, Holland A (2005) Imprinted genes and mother-offspring interactions. Early Hum Dev 81: 73–77. doi: 10.1016/j.earlhumdev.2004.10.006

7. Tycko B, Morison I (2002) Physiological functions of imprinted genes. J Cell Physiol 192: 245–258. doi: 10.1002/jcp.10129

8. Reik W, Walter J (2001) Genomic imprinting: parental influence on the genome. Nat Rev Genet 2: 21–32. doi: 10.1038/35047554

9. Falls J, Pulford D, Wylie A, Jirtle R (1999) Genomic imprinting: implications for human disease. Am J Pathol 154: 635–647. doi: 10.1016/s0002-9440(10)65309-6

10. Clerget-Darpoux F, Babron MC, Deschamps I, Hors J (1991) Complementation and maternal effect in insulin-dependent diabetes. Am J Hum Genet 49: 42–48.

11. Nanney D (1953) Nucleo-cytoplasmic interaction during conjugation in Tetrahymena. Biol Bull 105: 133–148. doi: 10.2307/1538562

12. Haga N (1995) Elucidation of nucleus-cytoplasm interaction: change in ability of the nucleus to express sexuality according to clonal age in Paramecium. J Cell Sci 108: 3671–3676.

13. Elfgang C, Rosorius O, Hofer L, Jaksche H, Hauber J, et al. (1999) Evidence for specific nucleocytoplasmic transport pathways used by leucine-rich nuclear export signals. Proc Natl Acad Sci USA 96: 6229–6234. doi: 10.1073/pnas.96.11.6229

14. Chinnery P (2003) Searching for nuclear-mitochondrial genes. Trends Genet 19: 60–62. doi: 10.1016/s0168-9525(02)00030-6

15. Rand D, Haney R, Fry A (2004) Cytonuclear coevolution: the genomics of cooperation. Trends Ecol Evol 19: 645–653. doi: 10.1016/j.tree.2004.10.003

16. Roubertoux P, Sluyter F, Carlier M, Marcet B, Maaroufveray F (2003) Mitochondrial DNA modifies cognition in interaction with the nuclear genome and age in mice. Nat Genet 35: 65–69. doi: 10.1038/ng1230

17. Tao D, Hu F, Yang J, Yang G, Yang Y, et al. (2004) Cytoplasm and cytoplasm-nucleus interactions affect afronomic traits in japonica rice. Euphytica 135: 129–134. doi: 10.1023/b:euph.0000009548.81467.73

18. Zeyl C, Andreson B, Weninck E (2005) Nuclear-mitochondrial epistasis for fitness in Saccharomyces cerevisiae. Evolution 59: 910–914. doi: 10.1554/04-487

19. Rand D, Fry A, Sheldahl L (2006) Nuclear-mitochondrial epistasis and Drosophila aging: introgression of Drosophila simulans mtDNA modifies longevity in D. melanogaster nuclear backgrounds. Genetics 172: 329–341. doi: 10.1534/genetics.105.046698

20. Li S, Lu Q, Fu W, Romero R, Cui Y (2009) A regularized regression approach for dissecting genetic conflicts that increase disease risk in pregnancy. Stat Appl Genet Mol Biol 8: 45. doi: 10.2202/1544-6115.1474

21. Hanson R, Kobes S, Lindsay R, Kmowler W (2001) Assessment of parent-of-origin effects in linkage analysis of quantitative traits. Am J Hum Genet 68: 951–962. doi: 10.1086/319508

22. de Koning DJ, Bovenhuis H, van Arendonk J (2002) On the detection of imprinted quantitative trait loci in experimental crosses of outbred species. Genetics 161: 931–938.

23. Cui Y (2007) A statistical framework for genome-wide scanning and testing imprinted quantitative trait loci. J Theo Biol 244: 115–126. doi: 10.1016/j.jtbi.2006.07.009

24. Cui Y, Cheverud J, Wu R (2007) A statistical model for dissecting genomic imprinting through genetic mapping. Genetica 130: 227–239. doi: 10.1007/s10709-006-9101-x

25. Cui Y, Lu Q, Cheverud J, Littell R, Wu R (2006) Model for mapping imprinted quantitative trait loci in an inbred F2 design. Genomics 87: 543–551. doi: 10.1016/j.ygeno.2005.11.021

26. Li G, Cui Y (2010) A general statistical framework for dissecting parent-of-origin effects underlying triploid endosperm traits in flowering plants. Ann Appl Stat 4: 1214–1233. doi: 10.1214/09-aoas323

27. Tang Z, Wang X, Hu Z, Yang Z, Xu C (2007) Genetic Dissection of Cytonuclear Epistasis in Line Crosses. Genetics 177: 669–672. doi: 10.1534/genetics.107.074302

28. Haldane J (1922) The part played by recurrent muation in evolution. Am Nat 67: 5–9.

29. Huxley J (1928) Sexual difference of linkage Grammarus chereuxi. J Genet 20: 145–156.

30. Dib C, Fauer S, Fizames C, Samson S, Drouot N, et al. (1996) A comprehensive genetiv map of the human genome based on 5,264 microsatellites. Nature 380: 152–154. doi: 10.1038/380152a0

31. Marklund L, Moller M, Hoyheim B, Davies W, Fredholm M, et al. (1996) A comprehensive linkage map of the pig based on a wild pig-Large White intercross. Anim Genet 27: 255–269. doi: 10.1111/j.1365-2052.1996. tb00487.x

32. Neff M, Broman K, Mellersh C, Ray K, Acland GM, et al. (1999) A second-generation genetic linkage map of the domestic dog, Canis familiaris. Genetics 151: 803–820.

33. Dietrich W, Miller J, Steen R, Merchant M, Damron-Boles D, et al. (1996) A comprehensive genetic map of the mouse genome. Nature 380: 149–152. doi: 10.1038/380149a0

34. Dempster A, Laird N, Rubin D (1977) Maximum likelihood from incomplete data via the EM algorthim. J Roy Stat Soc B 39: 1–38.

35. Manichaikul A, Dupuis J, Sen S, Broman K (2006) Poor Performance of Bootstrap Confidence Intervals for the Location of a Quantitative Trait Locus. Genetics 174. doi: 10.1534/genetics.106.061549

36. Churchill G, Doerge RW (1994) Empirical threshold values for quantitative trait mapping. Genetics 138: 963–971.

37. Lynch M, Walsh B (1998) Genetics and Analysis of Quantitative Traits. Sunderland, MA: Sinauer Associates, Inc.

38. Farmer M, Sundberg J, Bristol I, Churchill G, Li R, et al. (2001) A major quantitative trait locus on chromosome 3 controls colitis severity in 1L-10-deficient mice. PNAS 98: 13820–13825. doi: 10.1073/pnas.241258698

39. Burgess K, Husband B (2004) Maternal and paternal contributions to the fitness of hybrids between red and white mulberry. Am J Bot 91: 1802–1808. doi: 10.3732/ajb.91.11.1802

40. Levin D (2003) The cytoplasmic factor in plant speciation. Syst Bot 28: 5–11.

41. Rhode J, Cruzan M (2005) Contributions of heterosis and epistasis to hybrid fitness. Am Nat 166: E124–E139. doi: 10.1086/491798

42. Moore T, Haig D (1991) Genomic imprinting in mammalian development: a parental tug-of-war. Tren Genet 7: 45–49. doi: 10.1016/0168-9525(91)90230-n

43. Cardoso M, Leonhardt H (1999) DNA Methyltransferase Is Actively Retained in the Cytoplasm during Early Development. J Cell Bio 147: 25–32. doi: 10.1083/jcb.147.1.25

44. Li Z, Sillanpää M (2012) Overview of LASSO-related penalized regression methods for quantitative trait mapping and genomic selection. Theor Appl Genet 125: 419–435. doi: 10.1007/s00122-012-1892-9

Chapter 2

A GUIDE TO MODERN STATISTICAL ANALYSIS OF IMMUNOLOGICAL DATA

Bernd Genser[1, 2], Philip J Cooper[3, 4], Maria Yazdanbakhsh[5], Mauricio L Barreto[2] and Laura C Rodrigues[2]

[1]Department of Epidemiology and Population Health, London School of Hygiene and Tropical Medicine, London, UK

[2]Instituto de Saúde Coletiva, Federal University of Bahia, Salvador, Brazil

[3]Centre for Infection, St George's University of London, London, UK

[4]Instituto de Microbiologia, Universidad San Francisco de Quito, Quito, Ecuador

[5]Department of Parasitology, Leiden University Medical Center, Leiden, The Netherlands

ABSTRACT

Background

The number of subjects that can be recruited in immunological studies and the number of immunological parameters that can be measured has increased rapidly over the past decade and is likely to continue to expand. Large and complex immunological datasets can now be used to investigate complex scientific questions, but to make the most of the potential in such data and to get the right answers sophisticated statistical approaches are necessary. Such approaches are used in many other scientific disciplines, but immunological studies on the whole still use simple statistical techniques for data analysis.

Results

The paper provides an overview of the range of statistical methods that can be used to answer different immunological study questions. We discuss specific aspects of immunological studies and give examples of typical scientific questions related to immunological data. We review classical bivariate and multivariate statistical techniques (factor analysis, cluster analysis, discriminant analysis) and more advanced methods aimed to explore causal relationships (path analysis/structural equation modelling) and illustrate their application to immunological data. We show the main features of each method, the type of study question they can answer, the type of data they can be applied to, the assumptions required for each method and the software that can be used.

Conclusion

This paper will help the immunologist to choose the correct statistical approach for a particular research question.

BACKGROUND

The understanding of the importance of immunological mechanisms underlying human disease and the identification of associated immunological markers have grown enormously over the past ten years and the number of published immunological studies that investigate the relationships between human disease and cytokines and other immunological parameters has increased rapidly. Technical developments in sample processing and sophisticated immunological techniques permit the analysis of more immunological parameters in larger samples of human subjects, containing information that allows not only the measurement of simple associations between two parameters but also the exploration of the complex relationships between immunity, disease, environmental, social and genetic factors. The potential complexity of the possible relationships between large numbers of immunological parameters poses a special challenge for the applied immunologist: how to select the appropriate statistical techniques to extract the maximum relevant information from complex datasets and avoid spurious findings.

Immunologists tend to use simple statistical approaches even when multiple relationships between immunological parameters are expected, [1, 2] instead of multivariate statistical approaches that can analyse simultaneously multiple measurements on the same individual. Multivariate statistical analysis techniques are being widely applied in other scientific fields and numerous books and articles have been published that describe these techniques in detail. Unfortunately, this literature is not easily accessible to the applied immunologist without a detailed knowledge of statistics and few articles have been written demonstrating the application of statistical techniques to immunological data [3, 4].

This paper provides an overview of statistical analysis techniques that may be considered for the analysis of immunological data. We discuss specific aspects of immunological studies, give examples of typical scientific questions related to immunological data and present a statistical framework to help the immunologist to choose the correct statistical approach for a particular research questions. Although we have focused on cytokine data in the examples provided, the methods presented are applicable to most other immunological parameters.

SPECIFIC ASPECTS OF IMMUNOLOGICAL STUDIES RELEVANT FOR STATISTICAL ANALYSIS

In the following section we discuss specific aspects of immunological studies that are relevant for statistical analysis.

Structure of Immunological Data

Before analysing immunological data it is very important to examine the structure of the data because most statistical methods will only give the correct answer if the data has the characteristics required for the use of that method ("satisfy the data assumptions"). For example, common data assumptions are that the observations are approximately normally distributed or that the variances are similar across different subpopulations. Unfortunately, immunological data very frequently do not meet these assumptions and investigators are obliged to either apply data transformations (e.g. a logarithmic transformation to make skewed data approximately normally distributed), or to choose an alternative statistical techniques with less stringent data assumptions (e.g. using a non-parametric statistical approach that does not require the data to be normally distributed [5] instead of a parametric statistical technique that does). A further important aspect of immunological data is that different immunological parameters measured in the same study subject are frequently highly correlated ("multicollinearity"). Hence the application of statistical techniques that assume independence among the observations is often not valid and in such situations a method should be used that takes into account the fact that study variables may be the result of a common underlying biological mechanism. Examples of underlying biological mechanisms that can not be directly observed but will influence that value of more than one immunological variable are: "immune maturation," "down regulation" or "Th2 shift."

Complexity of the Relationships in Immunological Parameters

Immunological parameters are often involved in complex immunological mechanisms; and relationships between immunological parameters may be changeable. For example, a specific parameter (e.g. a cytokine) may have different effects in different cell populations, at different times and in the presence (or absence) of other immunological parameters. We often aim to explain the complete causal pathway from a non-immunological factor (e.g. exposure to an allergen) to an outcome (e.g. atopy or asthma). Clearly simple univariate statistical analysis would not be able to identify such inter-relationships among several study variables and the underlying immunological mechanisms that cannot be measured; multivariate statistical techniques are required that can examine multiple parameters simultaneously. A fundamental step to guide statistical analysis is to make the hypothesis explicit in a conceptual framework [6]. Conceptual frameworkspresent the proposed inter-relationships among the study variables and define any larger underlying immunological mechanisms assumed to influence their values. Conceptual frameworks should be detailed and explicit as they are used to guide the analysis.

Two further important aspects of immunological data that are not the focus of this paper but should be mentioned are:

Reproducibility of the Measurement of an Immunological Parameter

Reproducibility reflects how often we obtain the same result using the same laboratory test and sample. Some variation is expected for any measurement and statistical analysis must take into account the degree of variation. Although reproducibility of immunologic measurements is well defined for some immunological outcomes, particularly those used for diagnostic purposes, such as antibody levels associated with vaccine protection (e.g. levels >10 IU/mL for anti-HBS for vaccine response to hepatitis B vaccine), and for phenotypic characterization (CD4 counts or CD4/CD8 ratio for evaluation of immune status in HIV), reproducibility is not well defined for most immunological parameters. This is independent of the separate but important issue of repeatability between centres in multicentre studies that measure the same immunological parameters in different laboratories or even the measurement of the same parameter between different studies.

Multiple Testing

The problem of multiple testing is becoming increasingly relevant in immunological studies as the number of immunological parameters that can be measured increases and investigators conduct a large number of statistical tests on the same study data [7]. A specific concern in statistical analysis is to separate associations that occur by chance (because of «random variation» or «noise») from those reflecting true biological relationships («systematic variation,» often assumed to be a causal relationship). Most researchers use a statistical significance level («type I error,» for example $P = 0.05$) to decide whether the result of an analysis is likely to be due to chance. Conducting multiple hypothesis testing may result in substantial inflation of type I errors (depending on the degree of dependence between the tests). For example, if the value $p < 0.05$ is used, conducting twenty independent significance tests within a data set is likely to result in one comparison being significant just by chance. There are numerous multiple comparison procedures to adjust statistical analysis for type I error inflation, for example simultaneous test procedures, such as the approaches by Bonferroni, Tukey, Scheffé or Dunnet or more sophisticated step-wise procedures, such as the techniques by Newman-Keuls or Ryan. A good overview about the most important multiple comparison techniques is given by Toothaker [8].

RESEARCH OBJECTIVES OF IMMUNOLOGICAL STUDIES

In this section we list typical research questions from immunological studies.

Common objectives of immunological studies can be grouped into four overall categories:

i)   Those that investigate patterns of associations between several immunological parameters, without assuming any causal relationship (and therefore not classifying study variables as dependent variables (i.e. outcomes) and independent variables (i.e. explanatory variables or covariates).

For example, typical research questions of such studies are:

*   To assess the magnitude of the correlation between different cytokines or to quantify the balance between levels of cytokine expression. For example, the research question might be to measure the correlation between pro-inflammatory and anti-inflammatory cytokines (e.g. correlation between TNF-α and IL-10) or to quantify the «balance» between pro-inflammatory and anti-inflammatory cytokines (e.g. by calculating the ratio TNF-α/IL-10).

*   To identify highly correlated cytokines and to place them into groups which reflect an unobserved underlying mechanism. For example, Th1-related immune responses such as IFN-γ and TNF-α may mediate an inflammatory disease. Depending on the question being investigated, it may be more appropriate to first use a statistical analysis approach to «reduce the data», i.e., to aggregate the correlated Th1 related cytokines to form a «summarising variable» that reflects the underlying immunological mechanism (e.g. «degree of Th1 immune response») and use that summary variable in the analysis rather than using all the variables with the original cytokine levels.

- To identify individuals with similar profiles of immunological parameters and to place them into groups (so called "clusters"). For example, patients with a clinical outcome might be defined as "atopics» or «non-atopics» based on the values of skin prick tests; or subjects with a specific infection may be classified into groups (eg, active, chronic, or past) defined by the overall elevation in antibodies (e.g. IgE, IgA, IgM or IgG subclasses). However, within the same group of patients, clusters with distinct or overlapping profiles might be distinguishable and subsequent analyses might show associations between distinct clusters and disease (or some other outcome).

ii) The second group of research objectives investigates causal relationships between one or more immunological parameters (e.g. different cytokines, or summary measures) and other study variables (e.g. an outcome such as asthma). To guide the investigation of causation it is important to have developed an a priori causal pathway model. This will allow the appropriate definition of variables, i.e. defining which variables are dependent variables (outcomes), intervening variables (mediating the effect) or independent variables (exposures, confounding factors and effect modifiers) and will determine the choice of statistical approach.

Possible research objectives for causality include:

- Identification of determinants of immunological profiles. The objective may be to compare the expression of cytokines between two or more groups defined by an exposure, e.g. people infected or not infected with helminths, or people vaccinated and non-vaccinated in a vaccine trial where the vaccine exposure is assumed to influence the levels of the immunological parameter to be measured. For example, the question may be to determine if BCG vaccination influences the levels of IFN-γ secreted by mononuclear cells stimulated in vitro with a mycobacterial antigen. The immunological parameter is the outcome or dependent variable.

- Identification of clinical consequences of immunological profiles, (immunological parameter as the risk factor) or, in other words to identify associations between an immunological parameter and clinical (or other) outcomes. For example, immunologists are interested in predicting the probability of a disease occurrence by measurement of cytokine levels. For example, whether elevated TNF-α levels are associated with active disease in rheumatoid arthritis? The immunological parameter is the risk factor (often called «exposure») or independent variable.

iii)   The third group consists of more complex research questions that may include two or more of the objectives described above. Such questions may examine the role of cytokines in larger causal constructs, including more than one risk factor, intervening variables that mediate and modify an effect, and outcomes; and inter relationships between them. An example will be to investigate the causal inter relationships between early life infections, level of expression of pattern recognition receptors (such as Nods), activity of pro-inflammatory cytokines (IFN-γ and TNF-α) and the development of inflammatory bowel disease.

iv)   The field of in silico immunology (computer analysis generally in conjunction with informatics or immuno-informatics) is a rapidly developing and expanding field and has been used to address several types of study questions, such as:

•   The prediction of immunogenic sequences from microbial genomes to predict potential vaccine candidates [9].

•   The prediction of protein sequences in therapeutic antibodies that may be associated with adverse reactions [10].

•   Identification of regulatory molecules in the innate immune system [11].

These approaches are generally high-throughput analyses of large data sets (e.g. microbial genomes, human genome, etc) using available software (e.g. EpiMatrix) to either generate or test hypotheses and have been reviewed in detail elsewhere [12–15].

In silico statistical analyses use many of the multivariate statistical techniques discussed later in this review (e.g. cluster analysis). Because this is a highly specialised field for which there are many computational tools available [9], in silicoimmunology will not be discussed further in this review.

STATISTICAL METHODS FOR ANALYSIS OF IMMUNO-LOGICAL DATA

We conducted a systematic literature search in the database MEDLINE (1980–2005) to review statistical methods that have been previously applied to cytokine data. Because the objective was to get a crude overview rather than to reveal the exact number of papers published in this area we defined quite sensitive search criteria using the following key words:"cytokine$" or terms to identify specific cytokines (e.g. among others «IL$,» «interleukin$,» IF$, interferon$, TNF$, etc.) and common univariate and multivariate statistical techniques (e.g. among others "linear regression," «analysis of variance,»»cluster analysis,» «factor analysis» etc.).

Table 1 shows the results of our search. The most widely used methods found were simple statistical approaches that investigate the relationship between two variables (so called bivariate meth ods – also called univariate methods when variables are classified as dependent and independent variables). We frequently found standard methods to compare means of immunological parameters between independent groups (e.g. t-test, analysis of variance or their non-parametric equivalents), bivariate correlation analysis (Pearson's or Spearman's correlation coefficients) and univariate linear regression. By contrast, multivariate techniques (i.e. statistical approaches that consider three or more study variables simultaneously) were less frequently applied to cytokine data. Several studies used factor analysis (to identify groups of correlated immunological parameters) or cluster analysis (to identify groups of individuals with similar immunological profiles) or discrimination techniques such as logistic regression, discriminant analysis (to identify causes or consequences of immunological profiles). We also found a few examples of advanced modelling techniques (path analysis/structural equation modelling) that simultaneously model multiple relationships between the study variables.

Table 1: Results of a literature review conducted in the medical database MEDLINE (1980–2006) about statistical methods found in immunological studies investigating cytokine expressions

Statistical methods	**Number of references**
Univariate techniques	
Analysis of variance	2908
T-test	420
Mann-Whitney U-test	316
Wilcoxon/McNemar test	193
Univariate linear regression	163
Bivariate correlation analysis	157
Kruskal-Wallis H-test	95
Repeated measures analysis of variance	31
Friedman test	7
Non linear regression	5
Multivariate techniques	
Logistic regression	629
Cluster analysis	192
Multivariate analysis of variance	144

Multiple linear regression	91
Factor analysis/Principal components analysis	80
Analysis of covariance	56
Linear discriminant analysis	51
Partial correlation coefficient	24
Multinomial logistic regression	9
Multivariate analysis of covariance	7
Path analysis/Structural equation modelling	4

In the following section we provide an overview of statistical methods that can be considered for analysing immunological data that should help the applied immunologists without a detailed knowledge of statistics to select the appropriate statistical technique for each particular research question. The definition of which method is the most appropriate is strongly dependent on the research objective, the type of data collected, whether data assumptions are fulfilled and whether the sample size is sufficient. We begin with a short introduction to these topics.

Exploratory Data Analysis

An important first step in analysing immunological data is exploring and describing the data. Whatever the research question investigators should first explore the data in tables showing summary statistics (e.g. means, standard deviations, etc.) and apply graphical methods such as bar charts, histograms, Box and Whisker plots, or scatter plots. For multivariate data, scatter plot matrices are a powerful tool to examine associations among several immunological parameters [16]. A good overview of statistical methods for Exploratory Data Analysis is provided by Tukey [17].

Data Assumptions

The next thing investigators have to consider when selecting a statistical method is whether their data meet a number of data assumptions. The first assumption is the scale of measurement of the data, i.e. whether the data type is categorical (e.g. groups like male and female), ordinal (e.g. groups with a logical order, like order of birth) or continuous (also called metric [measured on a defined scale]). This restricts the statistical methods that can be used. When the investigator has to deal with continuous data the second assumption to test is whether the data follow a theoretical distribution (e.g. normal distribution). Distributional assumptions can be tested graphically by diagnostic plots or by applying a statistical test that compares the distribution of the data with a theoretical distribution. When the original data do not meet distributional assumptions required for a particular statistical technique (e.g. normal distribution to apply a t-test) a common approach is transforming the data to meet the data assumptions, for example to use the logarithm of the values [4]. However, immunological data frequently do not meet data assumptions even after applying different data transformations. In such cases it may be more appropriate to use an alternative statistical approach that requires fewer data assumptions (e.g. applying a non parametric Wilcoxon test instead of a t-test) [5]. Another approach for continuous variables that do not fulfil distributional assumptions is «categorising» the measurements by biological meaningful cut-off values (e.g. level >= 2.5: «positive», level < 2.5: «negative») or using centiles, and then apply a statistical test that is appropriate for categorical data. A less frequently applied strategy that allows for the use of a parametric approach even when data assumptions are violated is the «robust resampling variance estimator [18].» The concept behind this approach is to draw repeated random samples from the data, and then to estimate the parameter of interest (e.g. the mean) from each sample and finally to obtain an estimate of the variance by calculating the variance of the parameter across the samples. «Robust estimation» means the approach will provide valid estimations even when data assumptions (e.g. normal distribution) are violated [3]. For example, in a hypothetical immunological study IL-10 measurements have been obtained from a sample of 100 individuals. By drawing 500 random samples from these 100 individuals each including 50 subjects, a robust estimate of the variance of the parameter of interest (the population mean) can be calculated from the mean of each sample and from the variance of the mean (across the samples).

Sample Size Issues

A second major issue that immunologist frequently face and that also substantially affects the statistical approach to be used is to estimate the appropriate sample size for a statistical analysis. Whereas for univariate techniques (e.g. t-test, ANOVA etc.) sample size can be determined by conducting a power analysis [19], sample size formulae are available only for a couple of multivariate techniques. Cohen's Power analysis guide is a useful review for calculating sample sizes for common univariate and some multivariate techniques (analysis of covariance and multiple regression) [20]. Unfortunately, the theory underlying sample size estimation for other multivariate techniques is less developed. Recommendations for sample size for studies using factor analysis are that more subjects should be included than the number of unique correlations present in the correlation matrix [21, 22]. Authors suggest that for cluster analysis sample size calculations are dependent on how the investigator believes the study population is clustered. If some small clusters (say less than 10) are expected, then sufficient subjects will be required to sample at least 5 to 10 people in the smallest cluster [23]. There is a statistical framework and software available for sample size calculation for advanced techniques such as structural equation modelling and path analysis [24, 25].

SELECTING THE APPROPRIATE STATISTICAL METHOD FOR EACH PARTICULAR RESEARCH QUESTION

In this section we provide an overview of statistical methods that we consider useful for statistical analysis of immunological data. Our guide for data analysis was written with the "classical statistical approach" in mind, i.e. we provide statistical techniques to extract the maximum information from the present data. We do not discuss Bayesian statistical methods that also might be useful for application to immunological data seeking to combine a priori information with the information captured in the present data; a good overview is provided by Lee [26]. Moreover, statistical methods are important also for standardising immunological techniques and this is discussed elsewhere [27].

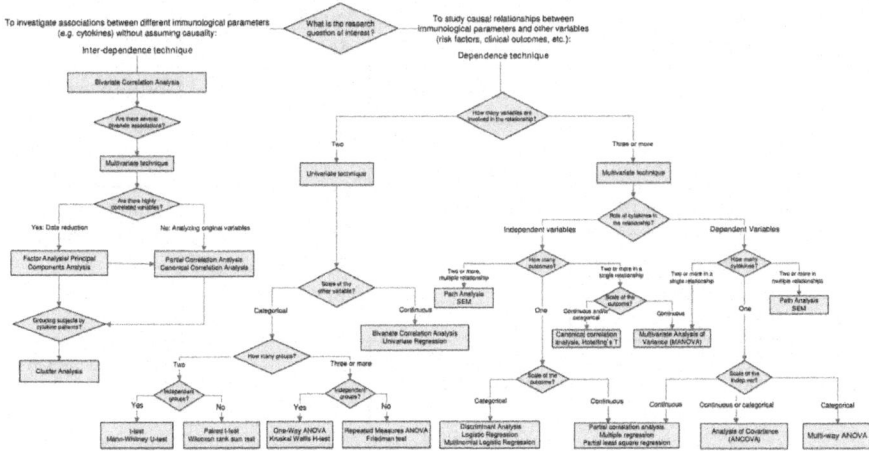

Figure 1: Selecting the appropriate statistical technique for analysis of immunological data.

We have prepared a summary table (Table 2) that lists the most important univariate and multivariate methods and links them to typical immunological research questions. In addition, we provide a flowchart that will help the immunologist to select the appropriate statistical analysis according to the research objective and the number and type of study variables to be analysed

(Figure 1). Most examples refer to cytokine data but the application to other continuous immunological data is straightforward.

Table 2: Selection of important statistical methods suitable for the analysis of immunological data

Example of research question	Type of data [D: dependent, I: independent]	Other data assumptions	Statistical method[1]
Univariate techniques			
Univariate group mean comparison techniques			
Compare expression of a cytokine between two independent groups (e.g. treatment vs. control)	D: continuous I: categorical	Normal distribution homogeneity of variances	t-test
	D: continuous or ordinal I: categorical		Mann Whitney-U test
Compare expression of a cytokine between two related groups (e.g. before and after treatment)	D: continuous I: categorical	Normal distribution, homogeneity of variances	Paired t-test
	D: continuous or ordinal I: categorical		Wilcoxon rank sum test
Compare expression of a cytokine between three or more independent groups defined by one factor (e.g. treatments A, B, C)	D: continuous I: categorical	Normal distribution, homogeneity of variances	One-way analysis of variance
	D: continuous or ordinal I: categorical		Kruskal Wallis – H test
Compare expression of a cytokine between three or more related groups (e.g. measurements 1, 2, and 3 weeks after treatment)	D: continuous I: categorical	Multivariate normal distribution, assumptions about covariance	Repeated measurements analysis of variance

	D: continuous or ordinal I: categorical		Friedman's ANOVA
Correlation and regression analysis			
Quantify association between two cytokines or a cytokine and another continuous variable	D: continuous I: continuous	Linear relationship, normality	Pearson correlation coefficient
	D: continuous or ordinal I: continuous or ordinal	Linear relationship	Spearman rank correlation coefficient
Predicting expression of a cytokine by a continuous independent variable	D: continuous I: continuous	Specified relationship (e.g. linearity for linear regression), normal distribution (for parametric regression)	Univariate regression
Multivariate techniques			
Multivariate correlation and regression techniques			
Quantify associations between two cytokines adjusted for the effect of a third continuous variable	All variables: continuous	Linear relationship, normality	Partial correlation coefficient
Predicting a continuous outcome (e.g. a cytokine) by several continuous or categorical independent variables	D: continuous I: continuous, ordinal or categorical	Specified relationship (e.g. linearity for linear regression), normal distribution for parametric regression, No multi-collinearity	Multiple regression
		Specified relationship, multi-collinearity	Partial least squares regression

Quantifying the magnitude of correlation between two groups of continuous variables (e.g. Th1 and Th2 related cytokines)	All variables: continuous		Canonical correlation analysis
Multivariate group mean comparison procedures			
Compare cytokine expressions between three or more independent groups defined by two or more factors (e.g. treatment and gender)	D: continuous I: categorical	Normal distribution, homogeneity of variances	Multi-way analysis of variance (ANOVA)
Simultaneously compare expressions of two or more cytokines between three or more independent groups defined by two or more factors	D: continuous I: categorical	Multivariate normal distribution, homogeneity of covariance matrices	Multivariate analysis of variance (MANOVA)
Compare cytokine expressions between three or more related groups defined by two or more factors (e.g. measurements at different time points during a study and treatment)	D: continuous I: categorical	Multivariate normality, homogeneity of covariance matrices	Multi-way repeated measurements analysis of variance
Grouping set of correlated cytokines to summary variables ("principal components")	All variables: continuous	High degree of multicollinearity	Factor analysis/ Principal components analysis
Grouping subjects in homogenous subgroups according to similar expression levels of two or more cytokines	All variables: continuous	Low degree of multicollinearity	Cluster analysis
Classification procedures			
Explaining or predicting group membership of two or more independent groups by cytokine levels	D: categorical I: continuous	Multivariate normal distribution, equal covariance matrices, low degree of multicollinearity	Linear discriminant analysis

Explaining or predicting group membership of two independent groups by cytokine levels	D: categorical I: continuous, ordinal or categorical		Logistic regression
Explaining or predicting group membership of three or more groups by cytokine levels	D: categorical I: continuous, ordinal or categorical		Multinomial logistic regression
Advanced techniques for multiple relationships			
Modelling multiple relationships between several immunological parameters and one or more outcome variables	All variables: categorical, ordinal or continuous data	Conceptual framework specifying the multiple relationships among the study variables	Path analysis/ Structural equation modelling

[1] All univariate and multivariate statistical approaches listed above can be implemented in general purpose statistical packages, e.g. among others S-PLUS® (Insightful Corporation, Seattle, WA), SAS® (SAS Institute Cary, NC, USA), SPSS® (Chicago: SPSS Inc.) or STATA® (StataCorp.Stata Statistical Software. College Station, TX: StataCorp LP). Path analysis/structural equation modelling can be implemented in STATA and SPSS that provide the extensions modules GLLAMM and AMOS, respectively, as well as in several special purpose software packages, e.g. among others LISREL® (Scientific Software International, Inc, IL, USA) or MX® (MCV, Department of Psychiatry, Richmond, VA, USA).

Inter-dependence Techniques

The first group embraces so called inter-dependence techniques, i.e. statistical methods aimed to explore relationships between study variables without assuming any causal relationship. These techniques are appropriate when the investigator cannot define (or may not wish to define) which variable is the independent variable (cause, exposure) and which is the dependent variable (effect, outcome). For example, inter-dependence techniques might be very useful in immunological studies to examine relationships between different cytokines measured in the same individual.

Correlation Analysis

One commonly applied inter-dependence approach is bivariate correlation analysis that aims to assess the magnitude of the linear relationship between two continuous variables (e.g. two cytokines, or a cytokine and another continuous variable. For example, Hartel et al conducted an observational, cross-sectional study in children aged between 1 and 96 months and in adults to investigate age-related changes in cytokine production [2]. The association between cytokine levels and age was analyzed using non-parametric rank correlation coefficients.

If there are more than two immunological parameters of interest and one finds several significant bivariate correlations, amultivariate correlation analysis should be conducted to examine the degree of multicollinearity in the data, i.e. whether multiple relationships are present between three or more study variables. A simple but useful approach to examine associations between three variables is to conduct stratified bivariate correlation analysis across strata defined by levels of a third variable. For example, to examine the association between a Th1-related and a Th2-related cytokine could be examined after stratifying by low and high levels of expression of IL-10.

Data Reduction Techniques

Datasets with several highly correlated immunological parameters can be simplified using "data reduction techniques."These methods are especially appropriate when it is assumed that many variables reflect aspects of an underlying process which is not directly measured. The most common technique is Principal Component Analysis (PCA) [28], which is a special type of Factor Analysis [29]. The idea behind this approach is to create summary variables (called «principal components»), that capture most of the information of the original data. For example, the technique can be used to derive two principal components from several correlated cytokines. After using this technique it is essential to consider whether the components identified are biologically plausible (e.g. it would be important to observe that the classification of cytokines grouped into two groups is consistent with the findings of published literature).

A useful feature of factor analysis and especially PCA is that the weights (the «factor loadings») for each variable within the components can be interpreted as correlation measures between the observed variable and the underlying unobservable component. Data assumptions in factor analysis are more conceptual than statistical. From a statistical point of view, normality is only necessary if a statistical test is applied to measure the significance of the factors, but these tests are rarely used in practice. The more important conceptual assumptions are that some underlying structure does exist in the set of selected variables and that results in some degree of multicollinearity.

In immunological studies factor analysis or PCA could be applied to extract information on the Th1- or Th2-related immune response from a set of cytokines or the ratio of a Th1- and Th2-score could be calculated to quantify the degree of «Th1/Th2 bias [30, 31].» For example, Turner et al [31] derived summary variables (called «principal components») from 11 different cytokines that were believed to better reflect the underlying mechanism and that were used in further analyses [32] instead of the original parameters. Data reduction techniques should be used when the objective of the study is not to investigate the role of each parameter but the role of the underlying mechanism.

Cluster Analysis

Cluster Analysis (CA) is the appropriate statistical approach when the researcher seeks to group individuals (not variables) according to their values of study variables (e.g. cytokine levels) [23]. CA groups individuals so that subjects in the same cluster have similar profiles of the parameters being studied (i.e. a high «within-cluster homogeneity») and subjects from different clusters have quite different immunological profiles (i.e. a high «between-cluster heterogeneity»). To perform CA the researcher has to define the variables on which the clustering is to be based and the type of cluster algorithm. Agglomerative algorithms treat each observation as a cluster and group similar individuals into clusters, while divisive algorithms start with the whole study population as a single cluster and divide the population by identifying homogeneous subgroups. The most appropriate clustering approach for a particular dataset depends on the type of data collected and the research question. Agglomerative clustering is preferable when there are extreme values in the data (outliers).

CA has strong mathematical properties but not statistical foundations. Data assumptions (e.g. normality and linearity) that are important in other multivariate techniques are of little importance in cluster analysis. However, the researcher is encouraged to examine the degree of multicollinearity in the data because each variable is weighted and variables that are multi-collinear are implicitly more heavily weighted in the clustering algorithm. For example, a cluster solution derived from a dataset with five highly correlated Th1-related cytokines and two correlated Th2-cytokines would substantially overestimate the importance of the Th1-component in the clustering. This can be avoided by first applying a data reduction technique (e.g. PCA) to derive the «principal components» that quantify the magnitude of Th1/Th2-immune response and afterwards clustering the individuals with respect to these immunological components.

In immunological studies cluster analysis may be useful to identify groups of individuals with similar immunological patterns (e.g. cytokine or antibody levels) that reflect an unknown common underlying immunological mechanism. For example, Mutapi et al [16] sought to group people infected with S. mansoni into clusters defined by levels of parasite specific IgE, IgA, IgM and the IgG subclasses. The authors identified two clusters, a cluster with high levels of IgM and low levels of all other antibodies and a cluster with high IgM and IgG1 and medium IgG4 and low levels of all other antibodies. In further analyses the authors investigated whether epidemiological features of schistosomiasis were associated with cluster membership and whether treatment changed this.

Another useful technique to study associations among immunological parameters without assuming any causal relationship is Canonical Correlation Analysis (CCA) [33] an approach that aims to quantify the correlation between two predefined sets of variables. In our MEDLINE search we did not find any previous applications of CCA to immunological data but we suggest that this statistical approach might be very useful to quantify the magnitude of correlation between two different sets of immunological parameters. A hypothetical example is where the investigators were interested to quantify the correlation between Th1- and Th2-related cytokines in individuals with and without helminths infections, or in atopic or non-atopic study subjects.

Dependence Techniques

The second group of techniques are statistical dependence techniques that are appropriate when the study investigates causation, i.e. variables can be classified as independent (cause, exposure) and dependent variables (effect, outcome), based on concrete a priori hypotheses about the underlying biological mechanisms. For example, an immunological parameter might be considered as an independent variable when it is the proposed cause (e.g. an autoantibody in an autoimmune disease) or the dependent variable when it is considered to be an effect (or outcome [e.g. the production of interferon gamma produced by lymphocytes stimulated with mycobacterial antigen following BCG vaccination]), or an intervening variable or intermediate factor (mediating variable) within a complex causal chain (e.g. the cytokine IL-13 would be an intermediate factor in studies that examine the relationship between exposure to aeroallergens and the development of atopic asthma). For example, Black et al [34] studied the IFN-γ response to Mycobacterium tuberculosis (as the outcome variable) before and after receiving BCG vaccination (the independent variable). Cooper et al [35] studied the effect of cholera vaccine (the independent variable) on the IL-2 response to recombinant cholera toxin B (as the outcome). In both studies, the authors pre-specified the classification of study variables before conducting statistical analysis. A more complex hypothetical example is the investigation of the effect of the impact of the intensity of infection with helminths on cytokine expressions (e.g. using Th2-related cytokines as outcomes) in a study of the relationship between helminths infections and atopy, and when the investigators may wish to consider the same cytokines as determinants, or risk factors, for atopy within the same study.

The choice of which statistical dependence technique is most appropriate for a particular research question will depend on the study design, the number and «scaling» (i.e. continuous, ordinal or categorical) of the study variables and other data assumptions (see Figure 1).

Univariate Dependence Techniques

These methods are appropriate when there is only one dependent and one independent variable. Common techniques areunivariate group mean comparison procedures [36] (aimed to compare the levels of a continuous variable (e.g. cytokine expressions) between groups of individuals pre-defined by an exposure that is considered to cause the immunological profile (e.g. vaccinated or not). The number of groups and whether the groups are independent or related will determine which approach is the most appropriate for each situation (see overview in Table 2).

By contrast, univariate regression analysis [37] is the best approach when the investigator seeks to model the relationship between two continuous variables and is able to decide which one is the outcome variable. The most common approach isLinear Regression a technique with stringent data assumptions (linearity of the relationship and normality of the error distribution). Robust alternatives when data assumptions are violated are non-parametric or non-linear regression techniques. Examples of applications for regression analysis in immunological studies are predicting the levels of expression of a continuous outcome variable (e.g. a cytokine) by a continuous variable (e.g. age or another continuous immunological parameter) when a causal relationship can be assumed.

Multivariate dependence techniques are needed when there are three or more variables involved and at least one variable can be considered the dependent variable.

Classification Techniques

These methods are required when there are several independent variables (e.g. cytokines) and one categorical outcome (e.g. atopy). One classical approach is Linear Discriminant Analysis (LDA), a method that derives linear combinations of the independent variables (called discriminant functions) that best discriminate between the two outcome groups (defined on the basis of the independent variable) [38]. LDA requires continuous normally distributed independent variables. A flexible approach which can be used with non normally distributed data and involves categorical independent variables is Logistic Regression [39].

In immunological studies classification techniques can be very useful to identify immunological profiles that best discriminate two or more pre-defined groups of interest (e.g. atopy vs. non atopy). For example, Gama et al [40] used LDA aimed to identify an immunological marker based on six cytokines (IL-2, IL-4, IL-10, IL-12, IFN-γ and TNF-α) to discriminate between clinical and asymptomatic forms of visceral leishmaniasis. The authors found that TNF-α, IL-10 and IL-4 were highly correlated with the clinical form, while IL-2, IL-12 and IFN-γ were correlated with the asymptomatic form. In another study, logistic regression was used to explore the role of parasite induced IL10 in decreasing the frequency of atopy: Van Biggelaar et al [41] sought to predict positive skin prick tests to house dust mite in children by mite-specific IgE, total IgE, IL-5 and IL-10 to worm and used logistic regression to show that positive skin prick test was positively associated with mite specific IgE but negatively associated with IL-10; and the probability of a skin test positivity was a result of the interaction between level of mite IgE and worm IL10.

Multivariate Group Mean Comparison Techniques

These techniques such as Multivariate Analysis of Variance (MANOVA) or Multi-way Analysis of Variance (Multi-way ANOVA), are used to compare the distributions of one or more continuous variables between groups defined by one ("one-way") or more ("multi-way") factors of interest (see overview Table 2) [21]. In contrast to LDA, where the groups are assumed to define a categorical outcome (e.g. asthmatic vs. non asthmatic) and the independent variables (e.g. cytokines) are used to discriminate between groups, in (M)ANOVA the groups are defined by the investigator considering one or more independent variables (e.g. treatments, vaccination status). A very useful application of MANOVA in immunological studies is to simultaneously compare the levels of two or more cytokines (e.g. IFN-γ, TNF-α) between two groups (asthmatic vs. non asthmatic). By applying MANOVA instead of repeated application of ANOVA the investigator can avoid the problem of type I error inflation for the whole experiment.

Multivariate Analysis of Covariance (MANCOVA) is an extension of MANOVA that additionally allows to control for the effect of an other continuous variable to be controlled (e.g. a confounder) [21]. An application of MANCOVA in immunological studies could be to simultaneously compare the expression levels of different cytokines (e.g. IF-γ, TNF-α, etc.) across groups defined by one or more experimental factors (e.g. vaccinated or control) and adjusted for age.

Multiple Regression Techniques

Multiple Regression is appropriate when the research question is to predict a single continuous dependent variable by a set of continuous and/or categorical independent variables [37]. The standard approach frequently used is multiple linear regression, however there are alternatives (e.g. non-parametric, non-linear multiple regression) when data assumptions are not met [42]. Regression analysis could be applied in immunological studies to predict the expression of a cytokine by explanatory variables (e.g. a set of other cytokines or other immunological parameters) or to predict the values of a continuous outcome variable (e.g. intensity of parasitic infection) by the expressions of one or more cytokines. For example, Dodoo et al [43] used multiple linear regression to predict malaria-related outcomes (fever, hemoglobin concentration) by levels of different cytokines (IFN-γ, TNF-α, IL-12, IL-10, TGF-β).

Partial Least Squares (PLS) Regression [44] is an extension of multiple linear regression for constructing predictive models when the factors are many and highly collinear. The approach could be very useful for analysis of immunological data, e.g. when the objective is to predict an outcome by a large set of highly correlated immunological parameters. Technically, the approach is a combination between principal components analysis and multiple linear regression, i.e. it produces factor scores as linear combinations of the original predictor variables so that there is no correlation between the factor score variables used in the predictive regression model.

Advanced Techniques

Path Analysis and Structural Equation Modelling

All the multivariate statistical methods that have been mentioned above have one common limitation: although they may include many variables, they all assume the presence of one single relationship between them. However, in modern immunological studies investigators often assume multiple relationships among immunological parameters and other study variables so that a simple multivariate approach might not be sufficient to reflect the complexity of the underlying immunological process. For example, in an immuno-epidemiological study conducted to study risk factors for asthma and allergy, investigators could define a conceptual framework that assumes multiple relationships between risk factors (e.g. allergens, vaccines, early life infections), immunological profiles (e.g. cytokine expression levels) and the occurrence of outcomes (e.g. asthma, atopy) (Fig 2). To model such complex immunological processes it will be important to simultaneously model all these multiple associations. Path Analysis and Structural Equation Modelling (SEM) are techniques developed by geneticists [45, 46] and economists [47] that can handle multiple relationships among study variables simultaneously and have been frequently applied in other scientific fields (e.g. economics, social sciences) [48]. SEM is an extension of path analysis that allows also for so called "latent variables" (a conceptual term for unobserved variables, see Appendix). A structural equation model consists of two components: "a measurement model" that defines how the observed measurements (called indicators) are related to the unobserved latent variables and a "structural model" that defines the assumed relationships between the observed study variables and one or more latent variables.

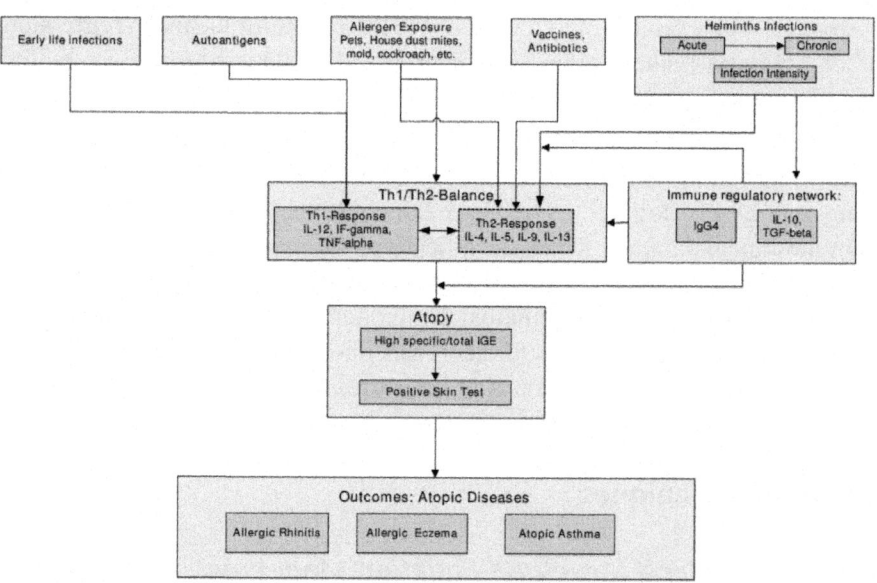

Figure 2: Conceptual framework that specifies multiple associations between potential risk factors, immunological parameters and outcomes (atopy and asthma).

The concept of latent variables and SEM is likely to be useful in immunological studies because immunologists frequently hypothesise that the measured immunological parameters are the result of unobservable underlying complex immunological processes. For example, imagine a hypothetical immunological study where different cytokines are measured to quantify two important immunological components. Figure 3 shows the path diagram of the study in which multiple relationships are assumed between the study variables. In this example we consider that the cytokines IL-12, IFN-γ and TNF-α represent an unobservable latent variable «Th1-related immune response» and the cytokines IL-4, IL-9 and IL-13 represent the latent variable «Th2-related immune response». Further, the structural model assumes relationships between the two latent variables (c1) and the «effects» (c2, c3) of these on an outcome variable (e.g. atopy, asthma). In SEM the concept of the «measurement model» is similar to factor analysis in which a linear relationship between the (observed) indicator variables and the (unobserved) latent variable is assumed. For example, in Figure 3, the «indicator loadings i1–i6» reflect the magnitude of association of each cytokine with the latent variable. However, the distinction between the two analyses is that in factor analysis the principal components are extracted to maximize the degree of variance explained by a specified number of factors while in SEM the investigator has to define a priory a path diagram that specifies which variables are the indicators of the underlying latent variables and the correlations of the measurements with the latent variables are derived that best reflect the whole conceptual framework e.g. maximizing the correlation between all latent and observed variables.

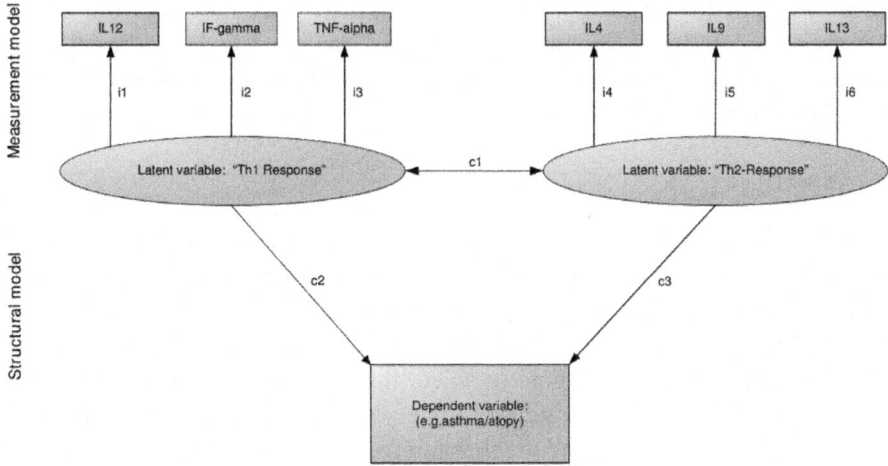

Figure 3: Example path diagram that specifies a structural equation model with two latent variables.

Application of these techniques to immunological data should allow the inference of complex immunological phenomena. For example, Chan et al [49] used SEM to study the role of obesity-associated dyslipidaemia, endothelial activation and two cytokines (IL-6 and TNF-α) in a complex causal chain of the metabolic syndrome. There are few applications of conceptual frameworks, path analysis and SEM to immunological data in the literature. However, because of increasing sample size and advancing knowledge about underlying complex immunological mechanisms, these approaches have a potentially important role in the analysis of data from modern immunological studies. The concept of latent variables in SEM will be especially useful because immunological mechanism can rarely be observed directly and must therefore be inferred through the measurement of different immunological indicators. Moreover, the unique feature of SEM that allows the analysis of multiple relationships between the study variables would be appropriate for the inference of causal chains of immunological processes.

Mixed Effects Models

The application of Random or Mixed Effects Models [50] for statistical analysis of clustered or longitudinal data is becoming a popular method in medicine. The basic idea of the approach is to adjust data analysis for an effect of the clustering by introducing a «random effect», i.e. an unobserved random variable that is specific to the clustering unit. In immunological studies the application of mixed effects models could be very useful when clustering must be assumed in the data, e.g. measurements of different cytokines from the same patients, repeated measurements of the same cytokine in a longitudinal study or in multi-centre studies where patients have been recruited from different populations.

CONCLUSION

The aim of this paper is to provide a modern overview for applied immunologists to explain and illustrate the statistical methods that can be employed for the analysis of immunological data. Our review should help immunologists without a detailed knowledge of statistics that are faced with the problem of statistical analysis of immunological data to select the appropriate statistical technique that will allow the valid extraction of the maximum information from the data collected. However, the statistical framework presented here should not be used as a substitute for an experienced biostatistician who should be involved from the beginning of the study for advice on study design, calculation of the sample size and planning of the statistical analysis. The systematic literature review illustrates the fact that most immunological studies still employ simple statistical approaches to immunological data even when multiple inter-relationships among several study variables are expected. We think it is important that more sophisticated statistical techniques are used for complex immunological data that will permit a better understanding of complex underlying immunological mechanism. Our focus has been on multivariate techniques that permit the analysis of multiple study variables simultaneously and hope that our examples from both real and hypothetical immunological studies will stimulate immunologists to make more use of these techniques. Moreover, bearing in mind the complexity of research questions addressed by modern immunological studies, we have introduced the idea of conceptual frameworks, latent variables and advanced statistical techniques (e.g. path analysis, SEM), providing a toolbox that should help the investigator to analyse multiple relationships among several study variables simultaneously. Finally, it should be pointed out that statistical techniques are a tool for the inference of underlying mechanisms and can never substitute for a priori hypotheses that are based on a sound knowledge of the scientific literature.

ACKNOWLEDGEMENTS

This research was funded by The Wellcome Trust, UK, HCPC Latin America Excellence Centre Program, Ref. 072405/Z/03/Z. PJC is supported also by the Wellcome Trust, grant no. 074679. BG is supported also by the Fundação de Amparo à Pesquisa do Estado da Bahia, Brazil (FAPESB), contract no. 1360/2006. The authors thank Dr Franca Hartgers and Dr Victoria Wright for their useful comments on the manuscript.

AUTHORS' CONTRIBUTIONS

BG and LCR had the idea for the paper, conducted the literature review, developed the guide for statistical analysis and wrote the manuscript. PJC and MY participated in writing the sections about research objectives of immunological studies and specific aspects of immunological data and evaluated the applicability of the proposed statistical framework for data from modern immunological studies. MLB participated in developing the statistical framework and helped to draft the manuscript. All authors have read and approved the final manuscript.

REFERENCES

1.   Geiger SM, Massara CL, Bethony J, Soboslay PT, Correa-Oliveira R: Cellular responses and cytokine production in post-treatment hookworm patients from an endemic area in Brazil. Clin Exp Immunol. 2004, 136: 334-40. 10.1111/j.1365-2249.2004.02449.x.

2.   Hartel C, Adam N, Strunk T, Temming P, Muller-Steinhardt M, Schultz C: Cytokine responses correlate differentially with age in infancy and early childhood. Clin Exp Immunol. 2005, 142: 446-53.

3. McGuinness D, Bennett S, Riley E: Statistical analysis of highly skewed immune response data. J Immunol Methods. 1997, 201: 99-114. 10.1016/S0022-1759(96)00216-5.

4. Bennett S, Riley EM: The statistical analysis of data from immunoepidemiological studies. J Immunol Methods. 1992, 146: 229-39. 10.1016/0022-1759(92)90232-I.

5. Hollander M, Wolfe DA: Nonparametric Statistical Methods. 1999, Wiley, 2

6. CG Victora, Huttly SR, Fuchs SC, Olinto MT: The role of conceptual frameworks in epidemiological analysis: a hierarchical approach. Int J Epidemiol. 1997, 26: 224-7. 10.1093/ije/26.1.224.

7. Hsu J: Multiple Comparisons. 1996, Chapman & Hall/CRC

8. Toothaker L: Multiple Comparison Procedures. 1992, London: SAGE University Paper

9. De Groot AS, Sbai H, Aubin CS, McMurry J, Martin W: Immuno-informatics: Mining genomes for vaccine components. Immunol Cell Biol. 2002, 80: 255-69. 10.1046/j.1440-1711.2002.01092.x.

10. Koren E, De Groot AS, Jawa V, Beck KD, Boone T, Rivera D, Li L, Mytych D, Koscec M, Weeraratne D, et al.,: Clinical validation of the "in silico" prediction of immunogenicity of a human recombinant therapeutic protein. Clin Immunol. 2007, 124: 26-32. 10.1016/j.clim.2007.03.544.

11. Gilchrist M, Thorsson V, Li B, Rust AG, Korb M, Kennedy K, Hai T, Bolouri H, Aderem A: Systems biology approaches identify ATF3 as a negative regulator of Toll-like receptor 4. Nature. 2006, 441: 173-8. 10.1038/nature04768.

12. Chakraborty AK, Dustin ML, Shaw AS: In silico models for cellular and molecular immunology: successes, promises and challenges. Nat Immunol. 2003, 4: 933-6. 10.1038/ni1003-933.

13. De Groot AS, Moise L: Prediction of immunogenicity for therapeutic proteins: state of the art. Curr Opin Drug Discov Devel. 2007, 10: 332-40.

14. De Groot AS, Moise L: New tools, new approaches and new ideas for vaccine development. Expert Rev Vaccines. 2007, 6: 125-7. 10.1586/14760584.6.2.125.

15. Hedeler C, Paton NW, Behnke JM, Bradley JE, Hamshere MG, Else KJ: A classification of tasks for the systematic study of immune response using functional genomics data. Parasitology. 2006, 132: 157-67. 10.1017/S0031182005008796.

16. Mutapi F, Mduluza T, Roddam AW: Cluster analysis of schistosome-specific antibody responses partitions the population into distinct epidemiological groups. Immunol Lett. 2005, 96: 231-40. 10.1016/j. imlet.2004.08.017.

17. Tukey JW: Exporatory Data Analysis. 1977, Boston: Addison-Wesley

18. Good PI: Resampling Methods: A Practical Guide to Data Analysis. 2005, Birkhauser

19. Chmura Kraemer H, Thiemann S: How many subjects? Statistical Power Analysis in Research. 1987, Sage Publications

20. Cohen J: Statistical Power Analysis for the Behavioral Sciences. 1988, Hillside, New Jersey: Lawrence Erlbaum Associates

21. Hair JF, Tatham RL, Anderson RE, Black W: Multivariate Data Analysis. 1998, Prentice Hall, 5

22. MacCallum RC, Widaman KF, Zhang S, Hong S: Sample size in factor analysis. Psychological Methods. 84-99.

23. Everitt BS, Landau S, Leese M: Cluster Analysis. 1993, London: Arnold

24. MacCallum RC, Browne MW, Sugawara HM: Power analysis and determination of sample size for covariance structure modeling. Psychological Methods. 1996, 130-149. 10.1037/1082-989X.1.2.130.

25. MacCallum RC, Hong S: Power analysis in covariance structure modeling using GFI and AGFI. Multivariate Behavioral Research. 1997, 193-210. 10.1207/s15327906mbr3202_5.

26. Lee PM: Bayesian Statistics: An Introduction. 2004, London: Arnold Publisher

27. Plikaytis BD, Carlone GM: Statistical considerations for vaccine immunogenicity trials. Part 1: Introduction and bioassay design and analysis. Vaccine. 2005, 23: 1596-605. 10.1016/j.vaccine.2004.06.046.

28. Jolliffe IT: Principal Component Analysis. 2002, Springer, 2

29. Gorsuch RL: Factor Analysis. 1983, Lawrence Erlbaum

30. Jackson JA, Turner JD, Kamal M, Wright V, Bickle Q, Else KJ, Ramsan M, Bradley JE: Gastrointestinal nematode infection is associated with variation in innate immune responsiveness. Microbes Infect. 2005

31. Turner JD, Faulkner H, Kamgno J, Cormont F, Van Snick J, Else KJ, Grencis RK, Behnke JM, Boussinesq M, Bradley JE: Th2 cytokines are associated with reduced worm burdens in a human intestinal helminth infection. J Infect Dis. 2003, 188: 1768-75. 10.1086/379370.

32. Jackson JA, Turner JD, Rentoul L, Faulkner H, Behnke JM, Hoyle M, Grencis RK, Else KJ, Kamgno J, Bradley JE, et al.,: Cytokine response

profiles predict species-specific infection patterns in human GI nematodes. Int J Parasitol. 2004, 34: 1237-44. 10.1016/j.ijpara.2004.07.009.

33. Thompson B: Canonical Correlation Analysis: Uses and Interpretation (Quantitative Applications in the Social Sciences). 1984, Sage Publications

34. Black GF, Weir RE, Floyd S, Bliss L, Warndorff DK, Crampin AC, Ngwira B, Sichali L, Nazareth B, Blackwell JM, et al.,: BCG-induced increase in interferon-gamma response to mycobacterial antigens and efficacy of BCG vaccination in Malawi and the UK: two randomised controlled studies. Lancet. 2002, 359: 1393-401. 10.1016/S0140-6736(02)08353-8.

35. Cooper PJ, Chico M, Sandoval C, Espinel I, Guevara A, Levine MM, Griffin GE, Nutman TB: Human infection with Ascaris lumbricoides is associated with suppression of the interleukin-2 response to recombinant cholera toxin B subunit following vaccination with the live oral cholera vaccine CVD 103-HgR. Infect Immun. 2001, 69: 1574-80. 10.1128/IAI.69.3.1574-1580.2001.

36. Montgomery DC: Design and Analysis of Experiments. 2000, Wiley, 5

37. Kleinbaum D, Kupper L, Muller KE, Nizam A: Applied Regression Analysis and Multivariable Methods. 1997, Duxbury Press

38. Huberty CJ: Applied Discriminant Analysis. 1994, Wiley

39. Hosmer DW, Lemeshow S: Applied Logistic Regression. 2000, Wiley, 2

40. Gama ME, Costa JM, Pereira JC, Gomes CM, Corbett CE: Serum cytokine profile in the subclinical form of visceral leishmaniasis. Braz J Med Biol Res. 2004, 37: 129-36. 10.1590/S0100-879X2004000100018.

41. van den Biggelaar AH, van Ree R, Rodrigues LC, Lell B, Deelder AM, Kremsner PG, Yazdanbakhsh M: Decreased atopy in children infected with Schistosoma haematobium: a role for parasite-induced interleukin-10. Lancet. 2000, 356: 1723-7. 10.1016/S0140-6736(00)03206-2.

42. Takezawa K: Introduction to nonparametric regression. 2005, Wiley

43. Dodoo D, Omer FM, Todd J, Akanmori BD, Koram KA, Riley EM: Absolute levels and ratios of proinflammatory and anti-inflammatory cytokine production in vitro predict clinical immunity to Plasmodium falciparum malaria. J Infect Dis. 2002, 185: 971-9. 10.1086/339408.

44. Geladi P, Kowalski B: Partial least squares regression: A tutorial. Analytica Chimica Acta. 1986, 185: 1-17. 10.1016/0003-2670(86)80028-9.

45. Wright S: Correlation and causation. Journal of Agricultural Research. 1921, 557-85.

46. MacCallum RC, Austin JT: Applications of structural equation modeling in psychological research. Annu Rev Psychol. 2000, 51: 201-26. 10.1146/annurev.psych.51.1.201.

47. Haavelmo T: The statistical implications of a system of simultaneous equations. Econometrica. 1943, 1-12. 10.2307/1905714.

48. Skrondal A, Rabe-Hesketh S: Generalized Latent Variable Modeling. Multilevel, Longitudinal, and Structural Equation Models. 2004, Chapman & Hall/CRC

49. Chan JC, Cheung JC, Stehouwer CD, Emeis JJ, Tong PC, Ko GT, Yudkin JS: The roles of obesity-associated dyslipidaemia, endothelial activation and cytokines in the Metabolic Syndrome – an analysis by structural equation modelling. Int J Obes Relat Metab Disord. 2002, 26: 994-1008. 10.1038/sj.ijo.0802017.

50. Brown H, Prescott R: Applied Mixed Models in Medicine. 1999, Chichester: John Wiley & Sons, Ltd

Chapter 3

NEXT GENERATION GENETICS

Mogens Fenger

Clinical Biochemistry, Molecular Biology, and Genetics, KBA339, Hvidovre, Denmark

ABSTRACT

One of the goals in genetic research aims at identifying genes in biochemical and physiological processes to reveal genetic causes of rare and common diseases. Previous obstacles such as costly genotyping or sequencing have been reduced with the chip-based genome wide association studies (GWAS), now culminating with the latest toy—next generation sequencing methodologies (NGS). Concomitantly, computer technologies have evolved to an increasing use of multicore processors and distributed computing on large networks or grids. Although the technologies are not perfect, we now have unprecedented opportunities to perform genetic studies not possible just 10 years ago. The hype about these new technologies have been large, but all the promises have however not been fulfilled entirely as hoped for. Maybe because the hype has been more about the technologies as such, and less about their intended use. Millions of genetic variations have been detected by GWAS and NGS, but only a few have been linked to diseases—with almost no practical clinical significance. A major reason for this apparent deadlock is the inadequacy of the models used, which are based on the traditional "Mendelian" approach, in which one gene is supposed to have a main effect on a trait or a disease. However, most genes claimed to be associated with a disease have small effects and only a tiny fraction of the genetic variance has been captured.

In this short notice it is argued, why this traditional approach should be supplemented or even replaced by modeling approaches in accordance with the complexity of biological systems, if we shall have any reasonable hope to understand the genetics behind any trait and bring genetics into practical use in medicine for common diseases (Costanzo et al., 2010; Ramanan et al., 2012).

EVOLUTION, FITNESS, AND EPISTASIS

Evolution of phyla is a complex process governed by genomic as well as environmental factors (Marshall, 2006). Much theoretic and practical work about evolution are based on theories of adaptive landscapes of fitness and natural selection, as advocated by Fisher in his geometrical model of adaption (Fisher, 1930;Martin and Lenormand, 2006; Chevin et al., 2010; Weinreich and Knies, 2013). In this model fitness is determined in a multidimensional landscape of phenotypes or traits, on which a selection pressure is imposed that limits the number of viable phenotypes. Although the space of theoretically phenotypes increases with the complexity of an organism, this may come with a cost of decreasing adaptability (Fisher, 1930; Orr, 2000; Martin and Lenormand, 2006; Borenstein and Krakauer, 2008). The Fisher model(s) is not explicitly rooted in genetic models but rather considers the phenotypic pleiotropic effect of mutations, i.e., particular genes and loci are not formulated in the model. In contrast, Wrights formulation of evolution (Wright, 1920) can be described as a multidimensional mutational or genetic landscape in which each dimension corresponds to a specific locus. In "modern" terms these ideas can be formulated as the occurrence of stabilizing selection acting on the increasing mutational load possibly involving pleiotropic behavior of a given mutation, that is a mutation may affect several endophenotypes (Weinreich et al., 2006; Masel and Siegal, 2009). Pleiotropicity also means that a selection pressure imposed on one endophenotype may not only constrict the number of viable genotypes but also inflict a collateral selection on other endophenotypes and genes (Gavrilets and De Jong, 1993; Snitkin and Segre, 2011). The latter may be regarded as "innocent" bystanders and the preserved genotypes may just be those that happen to be around at the moment of selection.

Fitness is a measure of the capability of survival and reproduction of a species as the result of integrated action of many subprocesses conditional on the imposed selection pressure. However, less fitness may not necessarily result in an entire loss of a phenotype or trait, but may prevail and in fact increase fitness if the selection pressure changes. This scenario is supported by the long known fact that a mutation may have major effect in one genotypic background, but may only have a minor influence in another and hence escape purging by selection (Nijhout and Paulsen, 1997; Kouyos et al., 2006). The fitness landscape (or any other trait landscape) may thus be roughed with several local optima. This is clearly obvious in the landscape of species, but is also present within a species (Marshall, 2006).

For long the question has been if a mutation impose a pure additive effect on fitness or if epistasis (the effect of mutations in a gene or regulatory structure on the effects of other genes) is the prevalent driving force in evolution. The effect of any mutation (genic as well as exgenic including possible changes in epigenetic processes) may be increased, buffered, or ameliorated in particular genetic backgrounds, while having negative (even lethal) effect in other genetic backgrounds. Buffering is the essence in evolutionary theory of canalization and organismal robustness, in which the phenotype appears robust to mutational perturbations. Mutations may accumulate and appear as cryptic or neutral variations as long as they are not selected against (Masel and Siegal, 2009). These cryptic genetic variations may be revealed if some genetic or environmental changes happen affecting the fitness and then contribute to evolvability (McGuigan and Sgró, 2009).

Canalization (or buffering) implies that the phenotypic mean tends to be preserved when a mutation occurs, but the cost is diminished variation of the phenotype, as new (deleterious) mutations are buffered leaving less degrees of freedom of variation compared to the pre-mutational genotype. Thus, a particular phenotype representing a local maximum in the phenotypic landscape, is generated by an ensemble of genotypes, each depicting a path or trajectory of the genetic network. Generally, the probability of a given genetic path being accessible to generate a phenotype decreases with the number of mutations. However, as the number of paths increases exponentially with the number of mutations a large and increasing number of paths may eventually generate a phenotype. This hypothesis has been confirmed empirically (Dowell et al., 2010; Franke et al., 2011). These and many other studies have firmly established epistasis as a primary driving force in evolution and as a fundamental principle in governing biological processes (Rice, 1998; Segre et al., 2005; Weinreich et al., 2005, 2013; Bershtein et al., 2006; Borenstein and Krakauer, 2008; Pavlicev et al., 2008; Chevin et al., 2010; Lunzer et al., 2010; Breen et al., 2012; Huang et al., 2012; Hemani et al., 2013; Weinreich and Knies, 2013).

The mechanism behind interactions and epistasis has been extensively studied and includes concepts as sign epistasis (Weinreich et al., 2005), reciprocal epistasis (Poelwijk et al., 2011), and the expansion of the protein universe (Povolotskaya and Kondrashov, 2010) to mention just a few outstanding contributions. The reader is referred to the cited work and to the vast literature appearing now.

THE GENETIC AND PHENOTYPIC SPACES

Complex species like humans are organized in interacting and interdependent functional units called organs or multicellular tissues. This extends the complexity of the genetics to another level. Despite the constrains this impose, the phenotypic space is vast.

Suppose that a diploid organism like *Homo sapiens* with 23 sets of chromosomes only harbor one mutation in each chromosome. The theoretically number of gametes emerging by random segregation amounts to approximately 8.4 million. If all gametes are viable then the number of possible zygotes will be more than $7*10^{13}$ or more than 11.000 fold the number human beings ever lived on planet Earth. Most probably a vast amount of the gametes or zygotes are not viable, but nevertheless, genetic variations so far discovered runs in the millions. This maps to as many phenotypes and hence, two human beings will never be genetically identical.

Similarly, in a physiological process like blood pressure, which are regulated by say 100 interacting genes, more than 10^{30} networks with exactly the same topology can be constructed if just one mutation is present in each gene. This would map to as many physiological states and dynamics. Adding to the number of genes their alternative spliced forms, the vast number of posttranslational modifications of proteins, non-protein regulatory elements (metabolites, small regulatory RNAs), epigenetic modifications, non-genic regulatory and genome-organizing structures, and not the least interactions and communications between cells in a multicellular organisms like humans, the combinatorial space of interactions and hence phenotypes is (almost) infinite.

POPULATION STRUCTURE AND GENETIC NETWORKS

Two basic aspects must be addressed in population genetics: (1) biological processes even in its most simple forms are blue-printed in the genome of interacting networks of genes; (2) the expression of the biological processes and phenotypes are conditional on the genetic variations and their inherent epistasis.

The genetic networks coding a trait can be mapped as a "continuum" reflecting the physiological states they define (see the Figure 1). Neighboring networks can be distinguished by variations in one or several genes or non-genetic regulatory structures, but may appear physiological similar as most genetic variations have small effects. The sensitivity of a network to external factors is encoded in the genome, and it is the variations in the process-specific genes and regulatory structures that determines the range of the response to an external perturbation. Identifying genetic networks are neither simple, nor transparent: functional networks are multipartite structures and are not secluded entities but rather interconnected with other networks (e.g., the glucose and the fat metabolisms are highly intermingled processes). Nevertheless, it may be possible to define a reasonable number of sub-ensembles of networks to be interpretable.

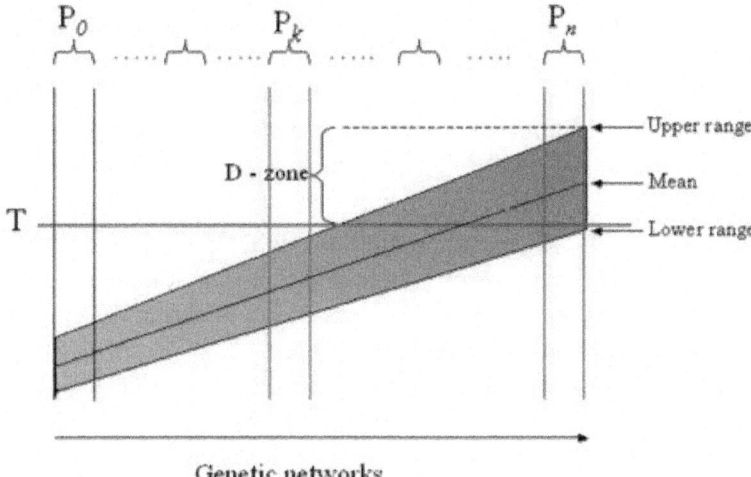

Figure 1: Risk profile in genetic networks. The population in this example is partitioned into more homogeneous subpopulations (P_i) by the LCA-SEM procedure indicated by the sets of two vertical lines. Each subpopulation is genetically defined by an ensemble of networks with exactly the same topology but differs in genetic variations. The networks in each subpopulation arise by successive mutational incidences that are balanced (buffered) to generate a phenotype similar for all subjects in the subpopulation (see also the text). In reality the subpopulations represent different local maxima in a miltidimensional phenotypic landscape, but are for illustrative purposes collapsed to flat, two-dimensional presentation. The range of the phenotype (e.g., diastolic blood pressure) depends on extra-genetic factors, but can never exceed the range defined by the genotype. Thus, some subjects (genotypes) will never exceed the threshold (T), while others will experience the extreme phenotype (e.g., diastolic hypertension) regardless of extra-genetic factors. The D(anger) - zone indicates the subjects or subpopulations which may be classified in this examples as diastolic hypertensive. However, this may depend on the circumstances under which the blood pressure is measured, i.e., subjects may be classified as normotensive although they have a massive propensity to develop hypertension. Dichotomizing the trait in a population is thus a dubious affaire and compromise most association studies to the point that information of the genetics of, in this case diastolic blood pressure, is entirely lost.

HETEROGENEITY

Population heterogeneity refers to the mixture of phenotypically homogeneous subpopulations, although in the extreme no subpopulation is truly homogenous as each subject harbors a unique genomewide genotype. A more or less well-defined phenotype thus comprise an ensemble of genotypes in the population.

The initial task is then to cluster subjects into more physiological homogeneous subpopulations to increase the accuracy and power of the genetic analysis (Fenger et al., 2008, 2011). The application of appropriate cluster algorithms is generally ill posed, as no universal formal criteria for the best clustering is available. Many of the well-known classification procedures implement some data-reduction or feature selection (Saeys et al., 2007), but any manipulations of the data space are likely to result in loss of information and should be avoided.

Allocating subjects to subpopulations is an art of modeling hidden or latent variables as the number of subpopulations is not known *a priori*. A way to resolve this is by applying the concept of latent class (LCA) in a structural equation modeling framework (SEM) (Bollen, 1989; Muthen, 2002; Skrondal and Rabe-Hesketh, 2007; Fenger et al., 2008, 2011). The idea of the LCA-SEM approach is to model a physiological process, and therefore the most appropriate study population would be a random selection of subjects as each subject provide information of the physiological process. Genetic structures and variations are not necessarily modeled directly, but are embedded as latent variables in the SEM structure and are mapped or reflected by the measured variables. Modeling in this framework addresses two pivotal issues in complex data: resolving the heterogeneity in the population, and simultaneously evaluating the data structure within the sub-populations. This approach outperforms most other classification methods in almost all aspects (Magidson and Vermunt, 2002).

An emerging line of methods implement ensembles of classification functions (Polikar, 2006). These approaches are particularly attractive when features in multi-source or distributed data sets are partly or completely disjoint, or if access to data in data set is limited to a subset of objects. Thus, the problem of missing data and hence reduced power may be circumvented to some extent and represent a potential alternative to imputing missing genetic data.

Undoubtedly, new and promising methods will merge, in particular as theoretical ideas mentioned below are integrated in future developments.

INHERITANCE: GENES OR INFORMATION?

Understanding and integrating the wealth of genetic data in a biological and medical context requires new approaches and techniques. Fortunately many new approaches are emerging increasingly embracing the nature of biological systems. In particularly the recent developments in network theory (Dorogovtsev and Mendes, 2002; Newman, 2009) including concepts of modularity (Newman and Girvan, 2004), stochastic block modeling (Karrer and Newman, 2011), statistical mechanics (Reichardt and Bornholdt, 2006;Ronhovde and Nussinov, 2009), and information theory (Anand and Bianconi, 2010) are promising.

All these methods comprise passing of information that have its corollary in genetics. Stabilizing selection may give rise to prevailing linkage disequilibrium of genes within and across chromosomes (Fenger et al., 2011). Such linkage disequilibrium arise as a consequence of preserving physiological processes regardless of the physical structure of the genome. Thus, inheritance is not a simple matter of passing on genetic material, but rather to combine information harbored in the genome into a viable organism. Information theoretic approaches in genetics may therefore be more promising (how abstract it may be) than traditional association methods, although transformation of these theories to biological structures may not be always straightforward.

IS VALIDATION IN GENETICS ACTUALLY POSSIBLE?

Validation is a standard requirements in genetic association studies today. However, validation of an association of a genetic variant to a trait or disease is often not or only sporadically obtained and for that reason a gene may be dismissed as disease related (Ioannidis, 2007; Shriner and Vaughan, 2011), or even to be pivotal in a physiological process (Fenger et al., 2011; Spijkers et al., 2011). It should hopefully be clear from the discussions above that validation should be expected to be an exception. A non-validated association may simply indicate a local optimum in the phenotypic landscape that happens to be detected because the genotypes in a population are permissive for expressing an apparent main effect. It has indeed been demonstrated that the lack of validation actually suggest more complex genetic structures governs a trait (Greene et al., 2009; Fenger et al., 2011), and including epistasis in the analysis may eventual confirm the importance of non-validated associations.

In the end the importance of genetic variations should be confirmed by cellular experiments. If possible, studies should be done in the cells where the genes in a network has its effect (which we often do not know). A gargantuan endeavor and at the moment maybe wild-fetched, but eventually any genetic variation has to be substantiated in a real biological context - not just as a statistical phenomenon.

CONFLICT OF INTEREST STATEMENT

The author declares that the research was conducted in the absence of any commercial or financial relationships that could be construed as a potential conflict of interest.

ACKNOWLEDGMENT

Elin R. Carlsson is thanked for helpful suggestions to this Opinion.

REFERENCES

1. Anand, K., and Bianconi, G. (2010). Gibbs entropy of network ensembles by cavity methods. *Phys. Rev. E Stat. Nonlin. Soft Matter Phys.* 82:011116. doi: 10.1103/PhysRevE.82.011116

2. Bershtein, S., Segal, M., Bekerman, R., Tokuriki, N., and Tawfik, D. S. (2006). Robustness-epistasis link shapes the fitness landscape of a randomly drifting protein. *Nature* 444, 929–932. doi: 10.1038/nature05385

3. Bollen, K. A. (1989). *Structural Equations with Latent Variables*. Hobroken, NJ: John Wiley and Sons.

4. Borenstein, E., and Krakauer, D. C. (2008). An end to endless forms: epistasis, phenotype distribution bias, and nonuniform evolution.*PLoS Comput. Biol.* 4:e1000202. doi: 10.1371/journal.pcbi.1000202

5. Breen, M. S., Kemena, C., Vlasov, P. K., Notredame, C., and Kondrashov, F. A. (2012). Epistasis as the primary factor in molecular evolution. *Nature* 490, 535–538. doi: 10.1038/nature11510

6. Chevin, L. M., Martin, G., and Lenormand, T. (2010). Fisher›s model and the genomics of adaption: restricted pleiotropy, heterogenous mutation, and parallel evolution. *Evolution* 64, 3213–3231. doi: 10.1111/j.1558-5646.2010.01058.x

7. Costanzo, M., Baryshnikova, A., Bellay, J., Kim, Y., Spear, E. D., Sevier, C. S., et al. (2010). The genetic landscape of a cell. *Science* 327, 425–431. doi: 10.1126/science.1180823

8. Dorogovtsev, S. N., and Mendes, J. F. (2002). Evolution of networks. *Adv. Phys.* 51, 1079–1187. doi: 10.1080/00018730110112519

9. Dowell, R. D., Ryan, O., Jansen, A., Cheung, D., Agarwala, S., Danford, T., et al. (2010). Genotype to phenotype: a complex problem.*Science* 328, 469. doi: 10.1126/science.1189015

10. Fenger, M., Linneberg, A., Jorgensen, T., Madsbad, S., Sobye, K., Eugen-Olsen, J., et al. (2011). Genetics of the ceramide/sphingosine-1-phosphate rheostat in blood pressure regulation and hypertension. *BMC Genet.* 12:44. doi: 10.1186/1471-2156-12-44

11. Fenger, M., Linneberg, A., Werge, T., and Jorgensen, T. (2008). Analysis of heterogeneity and epistasis in physiological mixed populations by combined structural equation modelling and latent class analysis. *BMC*

Genet. 9:43. doi: 10.1186/1471-2156-9-43

12. Fisher, R. A. (1930). *The Genetical Theory of Natural Selection.* London: Nature Publishing Group.

13. Franke, J., Klözer, A., de Visser, J. A., and Krug, J. (2011). Evolutionary accessibility of mutational pathways. *PLoS Comput. Biol.* 7:e1002134. doi: 10.1371/journal.pcbi.1002134

14. Gavrilets, S., and De Jong, G. (1993). Pleiotropic models of polygenic variation, stabilizing selection, and epistasis. *Genetics* 134, 609–625.

15. Greene, C. S., Penrod, N. M., Williams, S. M., and Moore, J. H. (2009). Failure to replicate a genetic association may provide important clues about genetic architecture. *PLoS ONE* 4:e5639. doi: 10.1371/journal. pone.0005639

16. Hemani, G., Knott, S., and Haley, C. (2013). An Evolutionary Perspective on epistasis and the missing heritability. *PLoS Genet.* 9:e1003295. doi: 10.1371/journal.pgen.1003295

17. Huang, W., Richards, S., Carbone, M. A., Zhu, D., Anholt, R. R. H., Ayroles, J. F., et al. (2012). Epistasis dominates the genetic architecture of Drosophila quantitative traits. *Proc. Natl. Acad. Sci. U.S.A.* 109, 15553–15559. doi: 10.1073/pnas.1213423109

18. Ioannidis, J. P. (2007). Non-replication and inconsistency in the genome-wide association setting. *Hum. Hered.* 64, 203–213. doi: 10.1159/000103512

19. Karrer, B., and Newman, M. E. (2011). Stochastic blockmodels and community structure in networks. *Phys. Rev. E Stat. Nonlin. Soft Matter Phys.* 83:016107. doi: 10.1103/PhysRevE.83.016107

20. Kouyos, R. D., Otto, S. P., and Bonhoeffer, S. (2006). Effect of varying epistasis on the evolution of recombination. *Genetics* 173, 589–597. doi: 10.1534/genetics.105.053108

21. Lunzer, M., Golding, G. B., and Dean, A. M. (2010). Pervasive cryptic epistasis in molecular evolution. *PLoS Genet.* 6:e1001162. doi: 10.1371/ journal.pgen.1001162

22. Magidson, J., and Vermunt, J. K. (2002). Latent class models for clustering: a comparison with K-means. *Can. J. Market. Res.* 20, 37–44. Available online at: http://statisticalinnovations.com/technicalsupport/ cjmr.pdf

23. Marshall, C. R. (2006). Explaining the Cambrian "Explosion" of animals. *Annu. Rev. Earth Planet. Sci.* 34, 355–384. doi: 10.1146/ annurev.earth.33.031504.103001

24. Martin, G., and Lenormand, T. (2006). A general multivariate extension of Fisher's geometrical model and the distribution of mutation fitness effects across species. *Evolution* 60, 893–907. doi: 10.1111/j.0014-3820.2006.tb01169.x

25. Masel, J., and Siegal, M. L. (2009). Robustness: mechanisms and consequences. *Trends Genet.* 25, 395–403. doi: 10.1016/j.tig.2009.07.005

26. McGuigan, K., and Sgró C. M. (2009). Evolutionary consequences of cryptic genetic variation. *Trends Ecol. Evol.* 24, 305–311. doi: 10.1016/j.tree.2009.02.001

27. Muthen, B. O. (2002). Beyond SEM: general latent variabel modeling. *Behaviormetrika* 29, 81–117. doi: 10.2333/bhmk.29.81

28. Newman, M. E. J. (2009). The structure and functions of complex networks. *SIAM Rev.* 45, 167–256. doi: 10.1137/S003614450342480

29. Newman, M. E. J., and Girvan, M. (2004). Finding and evaluating community structure in networks. *Phys. Rev. E Stat. Nonlin. Soft Matter Phys.* 69:026113. doi: 10.1103/PhysRevE.69.026113

30. Nijhout, H. F., and Paulsen, S. M. (1997). Develomental models and polygenic characters. *Am. Nat.* 149, 394–405. doi: 10.1086/285996

31. Orr, H. A. (2000). Adaptation and the cost of complexity. *Evolution* 54, 13–20. doi: 10.1111/j.0014-3820.2000.tb00002.x

32. Pavlicev, M., Kenney-Hunt, J. P., Norgard, E. A., Roseman, C. C., Wolf, J. B., and Cheverud, J. M. (2008). Genetic variation in pleiotropy: differential epistasis as a source of variation in the allometric relationship between long bone lengths and body weight.*Evolution* 62, 199–213. doi: 10.1111/j.1558-5646.2007.00255.x

33. Poelwijk, F. J., T-ânase-Nicola, S., Kiviet, D. J., and Tans, S. J. (2011). Reciprocal sign epistasis is a necessary condition for multi-peaked fitness landscapes. *J. Theor. Biol.* 272, 141–144. doi: 10.1016/j.jtbi.2010.12.015

34. Polikar, R. (2006). Ensemble based systems in decision making. *IEEE Circuits Syst. Mag.* 21–45. doi: 10.1109/MCAS.2006.1688199

35. Povolotskaya, I. S., and Kondrashov, F. A. (2010). Sequence space and the ongoing expansion of the protein universe. *Nature* 465, 922–926. doi: 10.1038/nature09105

36. Ramanan, V. K., Shen, L., Moore, J. H., and Saykin, A. J. (2012). Pathway analysis of genomic data: concepts, methods, and prospects for future development. *Trends Genet.* 28, 323–332. doi: 10.1016/j.tig.2012.03.004

37. Reichardt, J., and Bornholdt, S. (2006). Statistical mechanics of community detection. *Phys. Rev. E Stat. Nonlin. Soft Matter Phys.* 74:016110. doi: 10.1103/PhysRevE.74.016110

38. Rice, S. H. (1998). The evolution of canalization and the breaking of von Baer's laws: modeling the evolution of development with epistasis. *Evolution* 52, 647–656. doi: 10.2307/2411260

39. Ronhovde, P., and Nussinov, Z. (2009). Multiresolution community detection for megascale networks by information-based replica correlations. *Phys. Rev. E Stat. Nonlin. Soft Matter Phys.* 80:016109. doi: 10.1103/PhysRevE.80.016109

40. Saeys, Y., Inza, I., and Larranaga, P. (2007). A review of feature selection techniques in bioinformatics. *Bioinformatics* 23, 2507–2517. doi: 10.1093/bioinformatics/btm344

41. Segre, D., Deluna, A., Church, G. M., and Kishony, R. (2005). Modular epistasis in yeast metabolism. *Nat. Genet.* 37, 77–83. doi: 10.1038/ng1489

42. Shriner, D., and Vaughan, L. K. (2011). A unified framework for multi-locus association analysis of both common and rare variants. *BMC Genomics* 12:89. doi: 10.1186/1471-2164-12-89

43. Skrondal, A., and Rabe-Hesketh, S. (2007). Latent variable modelling: a survey. *Scand. J. Stat.* 34, 712–745. doi: 10.1111/j.1467-9469.2007.00573.x

44. Snitkin, E. S., and Segre, D. (2011). Epistatic interaction maps relative to multiple metabolic phenotypes. *PLoS Genet.* 7:e1001294. doi: 10.1371/journal.pgen.1001294

45. Spijkers, L. J., van den Akker, R. F., Janssen, B. J., Debets, J. J., De Mey, J. G., Stroes, E. S., et al. (2011). Hypertension is associated with marked alterations in sphingolipid biology: a potential role for ceramide. *PLoS ONE* 6:e21817. doi: 10.1371/journal.pone.0021817

46. Weinreich, D. M., Delaney, N. F., Depristo, M. A., and Hartl, D. L. (2006). Darwinian evolution can follow only very few mutational paths to fitter proteins. *Science* 312, 111–114. doi: 10.1126/science.1123539

47. Weinreich, D. M., and Knies, J. L. (2013). Fishers›s geometric model of adaption meets the functional synthesis: data on pairwise epistasis fro fitness yields insights into the shape and size of phenotypic space. *Evolution* 67, 2957–2972. doi: 10.1111/evo.12156

48. Weinreich, D. M., Lan, Y., Wylie, C. S., and Heckendorn, R. B. (2013). Should evolutionary geneticists worry about higher-order epistasis? *Curr. Opin. Genet. Dev.* 23, 700–707. doi: 10.1016/j.gde.2013.10.007

49. Weinreich, D. M., Watson, R. A., and Chao, L. (2005). Perspective: sign epistasis and genetic constraint on evolutionary trajectories.*Evolution* 59, 1165–1174. doi: 10.1554/04-272

50. Wright, S. (1920). The relative importance of heredity and environment in determining the piebald pattern of guinea-pigs. *Proc. Natl. Acad. Sci. U.S.A.* 6, 320–332. doi: 10.1073/pnas.6.6.320

Chapter 4

A STATISTICAL DESIGN FOR TESTING TRANSGENERATIONAL GENOMIC IMPRINTING IN NATURAL HUMAN POPULATIONS

Yao Li[1, 2], Yunqian Guo[1], Jianxin Wang[1], Wei Hou[3], Myron N. Chang[3], Duanping Liao[4], Rongling Wu[1, 4]

[1]Center for Computational Biology, Beijing Forestry University, Beijing, People's Republic of China

[2] Department of Statistics, West Virginia University, Morgantown, West Virginia, United States of America

[3] Department of Biostatistics, University of Florida, Gainesville, Florida, United States of America

[4]Department of Public Health Sciences, Penn State College of Medicine, Hershey, Pennsylvania, United States of America

ABSTRACT

Genomic imprinting is a phenomenon in which the same allele is expressed differently, depending on its parental origin. Such a phenomenon, also called the parent-of-origin effect, has been recognized to play a pivotal role in embryological development and pathogenesis in many species. Here we propose a statistical design for detecting imprinted loci that control quantitative traits based on a random set of three-generation families from a natural population in humans. This design provides a pathway for characterizing the effects of imprinted genes on a complex trait or disease at different generations and testing transgenerational changes of imprinted effects. The design is integrated with population and cytogenetic principles of gene segregation and transmission from a previous generation to next. The implementation of the EM algorithm within the design framework leads to the estimation of genetic parameters that define imprinted effects. A simulation study is used to investigate the statistical properties of the model and validate its utilization. This new design, coupled with increasingly used genome-wide association studies, should have an immediate implication for studying the genetic architecture of complex traits in humans.

INTRODUCTION

Genomic imprinting arises from a gene when either the maternally or paternally derived copy of it is expressed while the other copy is silenced [1], [2]. Caused by epigenetic modifications such as DNA methylation established during gametogenesis and maintained throughout somatic development in the offspring, genetic imprinting has been shown to play a pivotal role in regulating the formation, development, function, and evolution of complex traits and diseases [3], [4], [5], [6], [7], [8], [9], [10]. While most studies of genetic imprinting focus on the epigenetic and molecular mechanisms of this phenomenon [7], [11], the number and distribution of imprinted genes and their epistatic interactions for quantitative traits are poorly understood, limiting the scope of our inference about the effects of imprinting genes on the diversity of biological traits or processes. Several authors have started to use genome-wide association and linkage studies to identify the regions of the genome that contain imprinted sequence variants and further understand the epigenetic variation of complex traits [12], [13],[14], [15].

In a series of recent studies, Cheverud, Wolf, and colleagues categorized genetic imprinting into different types based on the pattern of its expression, i.e., maternal expression, paternal expression, bipolar dominance, polar overdominance, and polar underdominance [14], [15]. With a three-generation F2 design, they identified these types of imprinted quantitative trait loci (iQTL) affecting body weight and growth in mice, displaying much more complex and diverse effect patterns than previously assumed. A different design based on reciprocal backcrosses was proposed to test and estimate the distribution of iQTL responsible for physiological traits related to endosperm development in maize [16]. By modeling identical-by-descent relationships in multiple related families of canines, Liu et al. [13] derived a random effect model based on linkage analysis to genome-wide scan for the existence of iQTL that affect canine hip dysplasia. In a recent study, Wang et al. [9] used reciprocal F2 designs to identify the additive and dominant effects of iQTLs and their interactions with imprinting effects for hyperoxic acute lung injury survival time in mice. These authors also explore the transgenerational inheritance of iQTLs.

While epigenetic marks resulting in genomic imprinting can be generally stable in an organism›s lifetime, they may undergo reprogramming, i.e., a faithful clearing of the epigenetic state established in the previous generation, in the new generation during gametogenesis and early embryogenesis [17], [18],

. However, a growing body of evidence since the early 1980s indicates that genes may escape such reprogramming and, thus, inherit their imprinting effects into next generations [20], [21], [22], [23], [24], [25]. Two fundamental questions will naturally arise from this discovery: how common are imprinted genes of this type and how strong is the evidence for their existence in humans and other organisms? If epigenetic changes through imprinted genes can be inherited across generations, this would significantly alter the way we think about the inheritance of phenotype [26], [27]. Such transgenerational epigenetic inheritance, i.e., modifications of the chromosomes that pass to the next generation through gametes, may be related with health and diseases with a mechanism for transmitting environmental exposure information that alters gene expression in the next generation(s) [28]. The identification of imprinted loci displaying transgenerational epigenetic inheritance will be greatly helpful for addressing the two questions mentioned above, in a quest to elucidate the detailed genetic architecture of complex traits and diseases.

The motivation of this study is to develop a novel strategy for identifying imprinted genes for a quantitative trait and understanding the transgenerational changes of their effects with a three-generation family design by sampling multiple unrelated nuclear families, each composed of the grandfather, grandmother, father, mother, and grandchildren, from a natural population. This transgenerational design contains information about how alleles at different loci co-transmit during meiosis from one generation to next and, thus, has been widely used for genetic linkage analysis [29], [30]. By tracing the inheritance of alleles at a gene(s) from a paternal or maternal parent, this design allows the characterization of parent-of-origin of alleles and provides a powerful way to estimate genetic imprinting effects. Because only genotypes can be observed, we formulate a mixture model to specify allelic configurations in terms of parental origins of the alleles. The EM algorithm is implemented to estimate the effects of imprinted genes and their changes across generations. A testing procedure is proposed to study the pattern of transgenerational epigenetic inheritance. The statistical behavior of the model is examined through simulation studies.

RESULTS

Simulation studies were performed to examine the statistical behavior of the model. A three-generation design is simulated which include a certain number of first-generation families sorted into 9 mating types according to the genotype frequencies. Assume that the allele frequencies of a gene are 0.6 and 0.4 in a natural population at Hardy-Weinberg equilibrium. Our simulation will focus on the investigation of the impacts of different sampling strategies and heritabilities on parameter estimation and model power. For a given sample size, two sampling strategies are simulated, (1) a large family number and small family size, and (2) a small family number and large family size.

The first strategy samples 200 unrelated grandfathers and 200 unrelated grandmothers, who marry to form 200 the first-generation families. Each first-generation family is assumed to have one son who, as the father, form a second-generation family with the mother from the natural population. There is one child for each second-generation family. This allocation results in a total of 1000 subjects. All members in the design are typed for the gene, but only the fathers and offspring of the third generation are phenotyped for a normally distributed trait. The second strategy samples 50 unrelated grandfathers and 50 unrelated grandmothers. In each first-generation family, 3 sons are simulated, forming 150 second-generation families in which 4 children are assumed. This strategy also results in 1000 subjects.

Different genetic effects of the gene, additive, dominant, and imprinting, are simulated for the second- and third-generations using the designed. Two different heritability levels, 0.1 and 0.4, are simulated for each generation, from which variances are determined. Table 1 tabulates the estimates of population and quantitative genetic parameters from the three-generation design. As expected, allele frequency can be very well estimated. The model provides reasonable estimation accuracy and precision for all genetic parameters under different sampling strategies, even for a modest heritability level. Under both strategies, the model has great power (0.85 or higher) to test the significance of individual genetic effects, additive, dominant, and imprinting, expressed in different generations. The model is also powerful to detect differences of genetic effects between two consecutive generations. More interesting, the difference of imprinting effect between different generations, i.e., transgenerational inheritance of genetic imprinting, can be discerned with power 0.80 using our statistical design.

Table 1: The maximum likelihood estimates (MLEs) of additive (a), dominant (d), and imprinting effects (i) of a functional SNP on a complex trait in parental (F) and offspring (O) generations under two different strategies

Genetic Parameter	True Value	Strategy 1		Strategy 2	
		$H^2 = 0.1$	$H^2 = 0.4$	$H^2 = 0.1$	$H^2 = 0.4$
a_F	1.0	1.0198(0.0236)	1.0028(0.0112)	1.0387(0.0333)	0.9975(0.0125)
d_F	0.6	0.5702(0.0340)	0.6184(0.0144)	0.6272(0.0392)	0.6046(0.0185)
i_F	0.6	0.6024(0.0306)	0.5991(0.0093)	0.5897(0.0328)	0.6037(0.012)
a_O	1.0	0.9779(0.0429)	0.9889(0.0167)	1.0136(0.0243)	1.0082(0.0108)
d_O	1.5	1.5218(0.0462)	1.4397(0.0242)	1.5465(0.0389)	1.4962(0.0138)
i_O	1.0	0.9853(0.0393)	0.9885(0.0138)	1.0272(0.0271)	1.0212(0.0119)

The esimates are the means of MLEs obtained from 200 simulation replicates, with standard errors given in parentheses.
doi:10.1371/journal.pone.0016858.t001

One major aim of this study is to estimate the change of genetic effects over generation. Although our model has great power to detect the transgenerational change of genetic effects, its false positive rates should also be assessed. We conducted an additional simulation study to address this issue by simulating a SNP that has the same genetic effects between the two generations. The model detects a small proportion of simulation replicates ($<6\%$) which displays transgenerational differences in all types of genetic effects including additive, dominant, and imprinting. This suggests that the model has a small type I error rate for detecting the transgenerational difference of overall genetic effects. We particularly tested the type I error rate for the transgenerational difference of genetic imprinting, which is reasonably small ($<8\%$).

The haplotype model is also examined through simulation studies. We simulated two SNPs with a recombination fraction of $r = 0.05$ that are segregating in a human population. Of the four haplotypes, one is assumed to function as a risk haplotype. The remaining is collectively called the non-risk haplotype. The genetic values of composite diplotypes constituted by risk and non-risk haplotypes include the additive (a), dominant (d), and imprinting (i) genetic effects. We assume that some of these effects are different, and the others are the same between the parental and offspring generations. Combinations of different heritabilities between the two generations are simulated.

Table 2: Simulation results for transgeneration imprinting effects comparisons

			First Generation Parameters			Second Generation Parameters		
N	H²	r = 0.05	$a_F = 1$	$d_F = 0.5$	$i_F = 0.4$	$a_O = 1$	$d_O = 1.5$	$i_O = 0.6$
400	0.1	0.1	0.0561(0.0049)	0.9936(0.0185)	0.5228(0.0263)	0.3824(0.0195)	1.0182(0.0266)	1.4761(0.0405)	0.6592(0.0269)
	0.1	0.4	0.0450(0.0049)	0.9975(0.0162)	0.4690(0.0279)	0.4437(0.0184)	0.9860(0.0106)	1.4936(0.0174)	0.5937(0.0115)
	0.4	0.1	0.0639(0.0055)	1.0011(0.0082)	0.5032(0.0107)	0.4013(0.0082)	1.0210(0.0277)	1.5275(0.0383)	0.5631(0.0285)
	0.4	0.4	0.0681(0.0054)	1.0137(0.0075)	0.5072(0.0115)	0.3916(0.0083)	0.9939(0.0108)	0.4920(0.0144)	0.5975(0.0110)
800	0.1	0.1	0.0486(0.0044)	1.0003(0.0134)	0.4610(0.0183)	0.4162(0.0126)	1.0079(0.0194)	1.5219(0.0284)	0.6322(0.0175)
	0.1	0.4	0.0461(0.0040)	0.9956(0.0135)	0.4933(0.0180)	0.3796(0.0114)	1.0049(0.0075)	1.4978(0.0112)	0.5874(0.0072)
	0.4	0.1	0.0516(0.0041)	1.0047(0.0046)	0.5074(0.0077)	0.3915(0.0055)	0.9916(0.0083)	1.5027(0.0091)	0.6002(0.0073)
	0.4	0.4	0.0567(0.0036)	1.0011(0.0056)	0.5023(0.0080)	0.3976(0.0061)	0.9773(0.0077)	1.5069(0.0104)	0.5926(0.0083)
2000	0.1	0.1	0.0516(0.0032)	1.0109(0.0079)	0.4951(0.0116)	0.4059(0.0089)	1.0053(0.0122)	1.5095(0.0150)	0.5878(0.0119)
	0.1	0.4	0.0536(0.0029)	1.0078(0.0094)	0.5283(0.0107)	0.4038(0.0099)	1.0017(0.0042)	1.5011(0.0064)	0.5912(0.0053)
	0.4	0.1	0.0488(0.0027)	0.9996(0.0034)	0.5076(0.0044)	0.4064(0.0036)	0.9830(0.0115)	1.4997(0.0152)	0.6001(0.0138)
	0.4	0.4	0.0545(0.0028)	1.0009(0.0033)	0.5043(0.0047)	0.3986(0.0033)	0.9993(0.0050)	1.5012(0.0064)	0.5996(0.0048)

The genetic design scenarios are chosen as the combination of different heritabilities and sample sizes. They are: $H_1^2 = 0.1/0.4$, $H_2^2 = 0.1/0.4$, $n = 400, 800, 2000$.
doi:10.1371/journal.pone.0016858.t002

Table 2 gives the results of simulation for different heritabilities and sample sizes (all subjects used). Overall, all parameters can be estimated reasonably well. As expected, the precision of parameter estimation increases with heritability and sample size. The additive genetic effects in both generations can well be estimated with a modest sample size (say 400) for a small heritability (0.1). More sample sizes (say 800) are needed to provide a good estimate for genetic imprinting effects for a small heritability. To well estimate dominant genetic effects, an even larger sample size (say 2000) is required for the same level of heritability.

DISCUSSION

The traditional view of quantitative trait expression analysis assumes that the maternally and paternally derived alleles of each gene are expressed simultaneously at a similar level. However, this view is violated by a growing body of evidence that alleles are expressed from only one of the two parental chromosomes [1], [2]. This so-called genetic imprinting or parent-of-origin effect has been thought to play a pivotal role in regulating the phenotypic variation of a complex trait [3], [4], [6], [8], [9], [12], [13], [14], [15]. With the discovery of more imprinting genes involved in trait control through molecular and bioinformatics approaches, we will be in a position to elucidate the genetic architecture of quantitative variation for various organisms including humans.

Recent evidence shows that epigenetic inheritance in humans may experience a transgenerational change. This would represent a significant shift in our current understanding of inheritance and disease aetiology. Despite the development of new technologies that are reducing the time and cost of genotyping by several orders of magnitude [31], [32], the understanding of the underlying genetic events will be challenging. In this article, we present a computational model for identifying the genomic imprinting effect of genes on quantitative phenotypes and transgenerational change of genomic imprinting using a multigenerational sampling design for human families. The model formulates a general framework for testing the difference of genetic effects between different generations. By including multiple SNPs, the model was extended to estimate genomic imprinting and its transgenerational change expressed at the haplotype level. Although several models have been developed to estimate genomic imprinting for binary disease traits [33], [34], our model is among the first for estimating genetic imprinting operational in regulating the variation of quantitative traits and is certainly the first of its kind that can discern the transgenerational change of genetic imprinting.

Although no real data were analyzed for the moment, this model presents a conceptual design by which new data can be collected according to the sampling strategy proposed and then analyzed by the computational algorithm derived. Based on computer simulation, the model should display convincing statistical properties in parameter estimation and test and can be applied to a practical data set. However, several issues need to be addressed when the model is attempted to solve broader genetic questions. First, the maternal effects that cause parent-of-origin effects of alleles may be confounded with imprinting effects [35], which should be separated by developing a proper design in order to better study the patterns of gene expression and evolutionary dynamics.

Second, this study assumes the unisex (sons) produced from the first-generation family. One can also assume daughters with no change of the model, allowing the test of genomic imprinting between mother and offspring. In fact, our model can involve both sexes so that in the second generation sex-specific genetic effects can be characterized. If the sexes in the third generation are considered, the model can be extended to study the transgenerational changes of gene-sex interactions. Third, it is possible that part of parental genotypes are missing in practice. To infer genomic imprinting using such data sets, a multi-hierachical mixture model can be derived to estimate the missing parental genotypes based on observed offspring genotypes. Fourth, although a basic premise of epigenetic processes was that, once established, these marks were maintained through rounds of mitotic cell division and stable for the life of the organism, several recent studies have shown that at some loci the epigenetic state can be altered by the environment [36]. The questions are how common are genes of this type and how strong is the evidence for their existence in humans? The development of our design and model will help to address these biological questions of fundamental importance in elucidating the genetic architecture of complex traits.

METHODS

Sampling Strategies

Suppose there is a natural human population at Hardy–Weinberg equilibrium (HWE) from which a panel of three-generation families, each composed of the grandfather, grandmother, father, mother, and grandchildren, are sampled. Each member in a family is typed for single nucleotide polymorphisms (SNPs) from the human genome. Consider a quantitative trait affected by a SNP with two alleles A in a frequency of p and a in a frequency of q, leading to three genotypes AA, Aa, and aa with the frequencies of p^2, $2pq$, and q^2, respectively. In the grandparent generation, these three genotypes are mating randomly to produce nine cross types. Given a cross type, the genotypes of sons or daughters can be inferred. Here we first assume one sex (say son) in the second generation, although both sexes can be considered. The sons from a family serve as the father to mate with the females as the mother derived from a natural population, with genotypes, AA, Aa, and aa, characterized by frequencies p^2, $2pq$, and q^2, respectively. Each of such second-generation families produces a certain number of grandchildren. The genotype frequencies in the third generation are derived according to Mendel's first law.

According to this design, the grandfathers and grandmothers are founders whose parents are unknown. Alleles of sons from a first-generation family can be traced directly or indirectly, but the females used to generate the second-generation family are the founders with the unknown origin of alleles. For this reason, we will measure the phenotype for sons from the first-generation families and grandchildren from the second-generation families. This design will allow us to characterize imprinting effects of a gene in the second- and third-generations.

Genetic Models

There are three genotypes AA, Aa, and aa, for a biallelic gene according to Mendelian segregation pattern. Considering the parent-of-origin of alleles, these genotypes are described by four configurations, $A|A$ (coded as 2), $A|a$ (coded as 1), $a|A$ (coded as 1_t), and $a|a$ (coded as 0), where symbol $|$ is used to separate the maternally- (left) and paternally-derived alleles (right). The genotypic values of the four configurations in two different generations are defined as follows:

Configuration	Paternal	Offspring
$A|A$	$\mu_2^F = \mu_F + a_F$	$\mu_2^O = \mu_O + a_O,$
$A|a$	$\mu_1^F = \mu_F + d_F + i_F$	$\mu_1^O = \mu_O + d_O + i_O,$
$a|A$	$\mu_{1'}^F = \mu_F + d_F - i_F$	$\mu_{1'}^O = \mu_O + d_O - i_O,$
$a|a$	$\mu_0^F = \mu_F - a_F$	$\mu_0^O = \mu_O - a_O,$

$$(1)$$

where μ_F and μ_O are the overall means of the paternal and offspring generations, a_F, d_F, and i_F are the additive, dominant and imprinting genetic effects of the gene in the parental generation, and a_O, d_O, and i_O are the additive, dominant and imprinting genetic effects of the gene in the offspring generation.

The difference in the genetic architecture of a complex trait between two different generations is described as

$$\Delta_a = a_F - a_O,$$
$$(2)$$

$$\Delta_d = d_F - d_O,$$
$$(3)$$

$$\Delta_i = i_F - i_O.$$
$$(4)$$

By testing whether these differences are equal to zero jointly or individually, we can determine the transgenerational changes of the pattern of genetic control. If a significant imprinting effect is detected, we can test the type of genetic imprinting, i.e., parental or maternal dominance, by incorporating the imprinting models of Cheverud et al. [14].

Estimation

The grandfather and grandmother in the first generation from a natural population constitutes $3 \times 3 = 9$ mating types for three genotypes. For the j th first-generation mating type listed in Table $(j = 1,...,9)$, let N_j denote the family number of this mating type. Each first-generation family may have one or multiple sons who serve the father of the second generation. Those families in the second generation with the father derived from the jth first-generation mating type and the mother of a particular genotype from the natural population are summed together, denoted by N_{jl}^M, for mother genotype l ($l=2$ for AA, 1 for Aa, and 0 for aa). Thus, we have a total of $N_{.l}^M = \sum_{j=1}^9 N_{jl}^M$ second-generation mothers who carry genotype l.

It is not difficult to derive the maximum likelihood estimate of allele frequency from the three-generation family design as

$$p = \frac{4N_1 + 3(N_2 + N_4) + 2(N_3 + N_5 + N_7) + (N_6 + N_8) + 2N_2^M + N_1^M}{\sum_{j=1}^9 4N_j + 2(N_2^M + N_1^M + N_0^M)}$$

$$q = \frac{4N_8 + 3(N_5 + N_7) + 2(N_2 + N_4 + N_6) + (N_1 + N_3) + 2N_0^M + N_1^M}{\sum_{j=1}^9 4N_j + 2(N_2^M + N_1^M + N_0^M)}.$$

The male individuals from the first generation are typed for the marker, with four distinct configurations, $A|A$ (2), $A|a$ (1), $a|A$ (1'), and $a|a$ (0). Let n_{jk}^F denote the cumulative number of male individuals (as the father for the second generation) bearing configuration k ($k = 2,1,1',0$) from n_j first-generation families. In the third generation, only genotypes rather than configurations can be observed. We use N_{jkls}^O to denote the number of children who carry genotype s ($s = 2,1,0$) from a second-generation family with father k (from the jth first-generation mating type) and mother l from a natural population. The phenotypic values measured are expressed as y_{jki}^F ($i = 1,...,n_{jk}^F$) for the second-generation fathers and y_{jklsi}^O ($i = 1,...,N_{jkls}^O$) for the third-generation children. Both y_{jki}^F and y_{jklsi}^O are assumed to follow a normal distribution with mean depending on genotypes and residual variances σ_F^2 and σ_O^2, respectively.

Since offspring genotypes depend on parental genotypes, the log-likelihood of paternal and offspring parameters given marker (**M**) and phenotypic (**y**) data from the three generations is decomposed into two components, one related to the paternal parameters and the second related to the offspring parameters given the paternal parameters, expressed as

$$L(\Omega_F,\Omega_O|y_F,y_O,\mathbf{M}_F^{\{1\}},\mathbf{M}_M^{\{1\}},\mathbf{M}_F^{\{2\}},\mathbf{M}_M^{\{2\}},\mathbf{M}_O^{\{3\}})$$

$$=L(\Omega_F|y_F,\mathbf{M}_F^{\{1\}},\mathbf{M}_M^{\{1\}},\mathbf{M}_F^{\{2\}})+$$

$$L(\Omega_O|y_O,\mathbf{M}_F^{\{1\}},\mathbf{M}_M^{\{1\}},\mathbf{M}_F^{\{2\}},\mathbf{M}_M^{\{2\}},\mathbf{M}_O^{\{3\}}),$$

(5)

where $\Omega_F=(\mu_F,a_F,d_F,i_F,\sigma_F^2)$ are the paternal parameters and $\Omega_O=(\mu_O,a_O,d_O,i_O,\sigma_O^2)$ are the offspring parameters. Maximizing joint likelihood (5) is equivalent to maximizing its two likelihood components independently. The estimates of parameters Ω_F that maximize the first component can be obtained with the EM algorithm. In the E step, the posterior probability with which the double heterozygote father of the second generation from the 5th first-generation mating type in Table has a particular configuration is calculated by

$$\Phi_{51i}^F=\frac{\frac{1}{2}f_1(y_{51i}^F)}{\frac{1}{2}f_1(y_{51i}^F)+\frac{1}{2}f_{1'}(y_{51'i}^F)} \quad \text{and} \quad \Phi_{51'i}^F=\frac{\frac{1}{2}f_{1'}(y_{51'i}^F)}{\frac{1}{2}f_1(y_{51i}^F)+\frac{1}{2}f_{1'}(y_{51'i}^F)}$$

(6)

In the M step, the genotypic values of configurations and variance are calculated by

$$\mu_2^F=\frac{\sum_{i=1}^{N_{12}^F}y_{12i}^F+\sum_{i=1}^{N_{22}^F}y_{22i}^F+\sum_{i=1}^{N_{42}^F}y_{42i}^F+\sum_{i=1}^{N_{52}^F}y_{52i}^F}{N_{12}^F+N_{22}^F+N_{42}^F+N_{52}^F},$$

$$\mu_1^F=\frac{\sum_{i=1}^{N_{21}^F}y_{21i}^F+\sum_{i=1}^{N_{31}^F}y_{31i}^F+\sum_{i=1}^{N_{51}^F}\Phi_{51i}^F y_{51i}^F+\sum_{i=1}^{N_{61}^F}y_{61i}^F}{N_{21}^F+N_{31}^F+\sum_{i=1}^{N_{51}^F}\Phi_{51i}^F+N_{61}^F},$$

$$\mu_{1'}^F=\frac{\sum_{i=1}^{N_{41'}^F}y_{41pi}^F+\sum_{i=1}^{N_{51'}^F}\Phi_{51'i}^F y_{51'i}^F+\sum_{i=1}^{N_{71'}^F}y_{71'i}^F+\sum_{i=1}^{N_{81'}^F}y_{81'i}^F}{N_{41'}^F+\sum_{i=1}^{N_{51'}^F}\Phi_{51'i}^F+N_{71'}^F+N_{81'}^F},$$

$$\mu_0^F=\frac{\sum_{i=1}^{N_{50}^F}y_{50i}^F+\sum_{i=1}^{N_{60}^F}y_{60i}^F+\sum_{i=1}^{N_{80}^F}y_{80i}^F+\sum_{i=1}^{N_{90}^F}y_{90i}^F}{N_{50}^F+N_{60}^F+N_{80}^F+N_{90}^F},$$

$$\sigma_F^2 = \frac{1}{\sum_{j=1}^{9} N_j} \left\{ \sum_{i=1}^{N_{12}^F} (y_{12i}^F - \mu_2^F)^2 + \sum_{i=1}^{N_{22}^F} (y_{22i}^F - \mu_2^F)^2 \right.$$

$$+ \sum_{i=1}^{N_{21}^F} (y_{21i}^F - \mu_1^F)^2 + \sum_{i=1}^{N_{31}^F} (y_{31i}^F - \mu_1^F)^2$$

$$+ \sum_{i=1}^{N_{42}^F} (y_{42i}^F - \mu_2^F)^2 + \sum_{i=1}^{N_{41'}^F} [(y_{41'i}^F - \mu_{1'}^F)^2$$

$$+ \sum_{i=1}^{N_{52}^F} (y_{52i}^F - \mu_2^F)^2 + \sum_{i=1}^{N_{51}^F} (y_{51i}^F - \mu_1^F)^2$$

$$+ \sum_{i=1}^{N_{51'}^F} (y_{51'i}^F - \mu_{1'}^F)^2 + \sum_{i=1}^{N_{50}^F} (y_{50i}^F - \mu_0^F)^2$$

$$+ \sum_{i=1}^{N_{61}^F} (y_{61i}^F - \mu_1^F)^2 + \sum_{i=1}^{N_{60}^F} (y_{60i}^F - \mu_0^F)^2$$

$$+ \sum_{i=1}^{N_{71'}^F} (y_{71'i}^F - \mu_{1'}^F)^2 + \sum_{i=1}^{N_{81'}^F} (y_{81'i}^F - \mu_{1'}^F)^2$$

$$+ \sum_{i=1}^{N_{80}^F} (y_{80i}^F - \mu_0^F)^2 + \sum_{i=1}^{N_{90}^F} (y_{90i}^F - \mu_0^F)^2 \left. \right\}.$$

$$(7)$$

The EM algorithm can also be implemented to estimate genetic parameters Ω_O in the third generation that maximize the second component in (5). In the E step, the posterior probability with which the double heterozygote offspring of the third generation derived from the combination of two double heterozygote parents in the second generation has a particular configuration is calculated by

$$\Phi_{jkl1i}^O = \frac{\frac{1}{4}f_1(y_{jkli}^O)}{\frac{1}{4}f_1(y_{jkli}^O) + \frac{1}{4}f_{1'}(y_{jkli}^O)} \quad \text{and} \quad \Phi_{jkl1'i}^O = \frac{\frac{1}{4}f_{1'}(y_{jkli}^O)}{\frac{1}{4}f_1(y_{jkli}^O) + \frac{1}{4}f_{1'}(y_{jkli}^O)}$$

$$(8)$$

In the M step, the genotypic values of configurations and variance are calculated by

$$\mu_2^O = \frac{\sum_{j=1}^{9}\sum_{k=0}^{2}\sum_{l=0}^{2}\sum_{i=1}^{N_{jkl2}^O} y_{jkl2i}^O \xi_{jkl2i}}{\sum_{j=1}^{9}\sum_{k=0}^{2}\sum_{l=0}^{2} N_{jkl2}^O}, k=0,1',1,2; l=0,1,2,$$

$$\mu_1^O = \frac{\sum_{j=1}^{9}\sum_{k=0}^{2}\sum_{l=0}^{2}\sum_{i=1}^{N_{jkl}^O} \Phi_{jkl1i}^O y_{jkl1i}^O \xi_{jkl1i}}{\sum_{j=1}^{9}\sum_{k=0}^{2}\sum_{l=0}^{2}\sum_{i=1}^{N_{jkl}^O} \Phi_{jkl1i}^O \xi_{jkl1i}}, k=0,1',1,2; l=0,1,2,$$

$$\mu_{1'}^O = \frac{\sum_{j=1}^{9}\sum_{k=0}^{2}\sum_{l=0}^{2}\sum_{i=1}^{N_{jkl}^O} \Phi_{jkl1'i}^O y_{jkl1i}^O \xi_{jkl1i}}{\sum_{j=1}^{9}\sum_{k=0}^{2}\sum_{l=0}^{2}\sum_{i=1}^{N_{jkl}^O} \Phi_{jkl1'i}^O \xi_{jkl1i}}, k=0,1',1,2; l=0,1,2,$$

$$\mu_0^O = \frac{\sum_{j=1}^{9}\sum_{k=0}^{2}\sum_{l=0}^{2}\sum_{i=1}^{N_{jkl0}^O} y_{jkl0i}^O \xi_{jkl0i}}{\sum_{j=1}^{9}\sum_{k=0}^{2}\sum_{l=0}^{2} N_{jkl0}^O}, k=0,1',1,2; l=0,1,2,$$

$$\sigma_O^2 = \frac{1}{\sum_{j=1}^{9}\sum_{k=0}^{2}\sum_{l=0}^{2} (N_{jkl2}^O + N_{jkl1}^O + N_{jkl0}^O)}$$

$$\times \sum_{j=1}^{9}\sum_{k=0}^{2}\sum_{l=0}^{2}\left\{ \sum_{i=1}^{N_{jkl2}^O} \xi_{jkl2i}(y_{jkl2i}^O - \mu_2^O)^2 + \sum_{i=1}^{N_{jkl1}^O} \xi_{jkl1i}\left[\Phi_{jkl1i}^O(y_{jkl1i}^O - \mu_1^O)^2\right]\right.$$

$$\left. + \Phi_{jkl1'i}^O(y_{jkl1i}^O - \mu_{1'}^O)^2 + \sum_{i=1}^{N_{jkl0}^O} \xi_{jkl0i}(y_{jkl0i}^O - \mu_0^O)^2 \right\}, k=0,1',1,2; l=0,1,2,$$

where ξ_{jkl2i}, ξ_{jkl1i}, and ξ_{jkl0i} are the indicator variables that are defined as 1 if offspring i in the third generation from the combination of father k from the jth first-generation mating type and mother l from the natural population has genotype AA, Aa, and aa, respectively, and 0 otherwise. The EM steps are iterated between equations (6) and (7) to obtain the MLEs of Ω_F and between equations (8) and (9) to obtain the MLEs of Ω_O.

Hypothesis Tests

It is imperative to know whether there exists a significant association between a specific SNP and a complex trait and how a significant SNP triggers an additive, dominant, or imprinting effect on the trait. To test for the overall significant association of SNP genotype and trait phenotype, we generate the following hypotheses:

$$H_0 : a_F = d_F = i_F = a_O = d_O = i_O = 0$$

H_1 : At least one of these equalities above does not hold.

The log-likelihood ratio under the null and alternative hypotheses is calculated. Since the null hypothesis contains a nuisance parameter, allele frequency, this log-likelihood ratio test statistic may have an unclear distribution. For this reason, the critical threshold for claiming the existence of a significant SNP is determined from permutation tests [37]. If our interest is in testing whether there is an additive, dominant, or imprinting effect, the null hypothesis should be $H_0 : a_F = a_O = 0$, $H_0 : d_F = d_O = 0$, and $H_0 : i_F = i_O = 0$, respectively. Because each of these null hypotheses is nested within its alternative, the log-likelihood ratio test statistic can be thought to asymptotically follow a χ^2-distribution for a large sample size.

The transgenerational changes of different genetic effects can also be tested. The null hypotheses used to test whether the additive, dominant, and imprinting effects display significant changes from one generation to next are expressed as $H_0 : \Delta_a = 0$, $H_0 : \Delta_d = 0$, and $H_0 : \Delta_i = 0$, respectively. These null hypotheses can be considered singly or jointly, in order to better study the transgenerational changes of the genetic architecture of a trait.

Haplotyping Model

Recent molecular surveys suggest that the human genome contains many discrete haplotype blocks that are sites of closely located SNPs [38], [39], [40]. Each block may have a few common haplotypes which account for a large proportion of chromosomal variation. Between adjacent blocks are there large regions, called hotspots, in which recombination events occur with high frequencies. Several algorithms have been developed to identify a minimal subset of SNPs, i.e., tagging SNPs, that can characterize the most common haplotypes [41]. The number and type of tagging SNPs within each haplotype block can be determined prior to association studies. In this section, we will derive a model for detecting the association between haplotypes constructed by alleles at a set of SNPs and complex traits.

For the simplicity of our description, consider two SNPs **A** (with two alleles A and a) and **B**(with two alleles B and b). They form four haplotypes AB, Ab, aB, and ab, of which one that is distinct from the rest three is defined as a risk haplotype W and all the others are defined as a non-risk haplotype w [42]. Risk and non-risk haplotypes from the maternal and paternal parents generate four composite diplotypes, $W|W$, $W|w$, $w|W$, and $w|w$, whose genotypic values are described by the additive (a), dominant (d), and imprinting genetic effects (i). Cheng et al. [43] and Wang et al. [44] proposed a two- and three-SNP model for estimating and testing genetic imprinting effects in a natural population, respectively. Wu et al.'s procedure[45] allows the choice of an optimal number and combination of risk haplotypes within a multiallelic model framework. Here, we adopted Cheng et al.›s two-SNP model to estimate haplotype imprinting genetic effects and their transgenerational change.

In this example, four haplotypes AB, Ab, aB, and ab have frequencies denoted as $p_{11}, p_{10}, p_{01},$ and p_{00}, respectively. The two SNPs yield nine joint genotypes, $AABB$ (coded as 1),$AABb$ (coded as 2), ..., $aabb$ (coded as 9), which are actually observed. Each subject must bear one of these genotypes, and the parents in each family will be one of $9 \times 9 = 81$ possible genotype by genotype combinations. If each parent for a combination is homozygous for both SNPs, their offspring will have one genotype. As long as one parent is heterozygous for one SNP, the offspring will have two or more genotypes. However, only when both SNPs are heterozygous for at least one parent, the genotype frequencies of offspring will be determined by the recombination fraction between the markers (r). Tables show the structure and frequencies of mother by father genotype combinations under random mating and their offspring genotype frequencies in the second and third generation, respectively. For a double heterozygote $AaBb$, its observed genotype may be derived from two possible diplotypes, $AB|ab$ (with the relative proportion of $\phi = \dfrac{p_{11}p_{00}}{p_{11}p_{00} + p_{10}p_{01}}$) or $Ab|aB$) (with the relative proportion of $1 - \phi = \dfrac{p_{10}p_{01}}{p_{11}p_{00} + p_{10}p_{01}}$). Each of these two diplotypes produce four haplotypes AB, Ab, aB, and ab, whose frequencies are expressed as

		Haplotype			
Diplotype	Proportion	AB	Ab	aB	ab
$AB\|ab$	ϕ	$\frac{1}{2}(1-r)$	$\frac{1}{2}r$	$\frac{1}{2}r$	$\frac{1}{2}(1-r)$
$Ab\|aB$	$1-\phi$	$\frac{1}{2}r$	$\frac{1}{2}(1-r)$	$\frac{1}{2}(1-r)$	$\frac{1}{2}r$

A similar likelihood (5) cane be formulated for haplotype models. A complicated EM algorithm is derived to estimate haplotype frequencies using the parental information. Let N_{ij} denote the observation of mating type between genotype i for one parent and genotype j for the second parent. In the E step, calculate the proportion of a diplotype for a heterozygous genotype for a particular mating design by

$$\psi_1 = \frac{(1-\phi)r}{\omega_1}$$

$$\psi_2 = \frac{\phi r}{\omega_2}$$

$$\psi_3 = \frac{(1-\phi)r + \phi r}{\omega_1 + \omega_2}$$

$$\psi_4 = \frac{(1-\phi)^2 r^2 + \phi(1-\phi)r(1-r)}{\omega_1^2}$$

$$\psi_5 = \frac{\phi^2 r^2 + \phi(1-\phi)r(1-r)}{\omega_2^2}$$

$$\psi_6 = \frac{\phi(1-\phi)r^2 + [\phi^2 + (1-\phi)^2]r(1-r)}{2\psi_1\omega_2}$$

$$\psi_7 = \frac{\phi^2 r^2 + 2\phi(1-\phi)r(1-r) + (1-\phi)^2 r^2}{2(\omega_1^2 + \phi_2^2)},$$

where $\omega_1 = (1-\phi)r + \phi(1-r), \omega_2 = \phi r + (1-\phi)(1-r)$.

In the M step, estimate the haplotype frequencies and recombination fraction by

$$p_{11} = \{4N_{11} + 3(N_{12} + N_{21} + N_{14} + N_{41}) + 2(N_{22} + N_{44} + N_{13} + N_{31}$$
$$+ N_{16} + N_{61} + N_{17} + N_{71} + N_{18} + N_{81} + N_{19} + N_{91} + N_{24} + N_{42})$$
$$+ N_{23} + N_{32} + N_{26} + N_{62} + N_{27} + N_{72} + N_{28} + N_{82} + N_{29} + N_{92}$$
$$+ N_{34} + N_{43} + N_{46} + N_{64} + N_{47} + N_{74} + N_{48} + N_{84} + N_{49} + N_{94}$$
$$+ \phi[3(N_{15} + N_{51}) + 2(N_{25} + N_{52} + N_{45} + N_{54}) + N_{35} + N_{53} + N_{56}$$
$$+ N_{65} + N_{57} + N_{75} + N_{58} + N_{85} + N_{59} + N_{95}]$$
$$+ (1 - \phi)[2(N_{15} + N_{51}) + N_{25} + N_{52} + N_{45} + N_{54}] + 2\phi N_{55}\}$$
$$/(4 \sum_{i=1, j=1}^{9,9} N_{ij}),$$

$$p_{10} = \{4N_{33} + 3(N_{23} + N_{32} + N_{36} + N_{63}) + 2(N_{22} + N_{66} + N_{13} + N_{31}$$
$$+ N_{26} + N_{62} + N_{34} + N_{43} + N_{37} + N_{73} + N_{38} + N_{83} + N_{39} + N_{93})$$
$$+ N_{12} + N_{21} + N_{16} + N_{61} + N_{24} + N_{42} + N_{27} + N_{72} + N_{28} + N_{82}$$
$$+ N_{29} + N_{92} + N_{46} + N_{64} + N_{67} + N_{76} + N_{68} + N_{86} + N_{69} + N_{96}$$
$$+ \phi[2(N_{35} + N_{53}) + N_{25} + N_{52} + N_{56} + N_{65}]$$
$$+ (1 - \phi)[3(N_{35} + N_{53}) + 2(N_{25} + N_{52} + N_{56} + N_{65}) + N_{15} + N_{51}$$
$$+ N_{45} + N_{54} + N_{57} + N_{75} + N_{58} + N_{85} + N_{59} + N_{95}] + 2(1 - \phi)N_{55}\}$$
$$/(4 \sum_{i=1, j=1}^{9,9} N_{ij}),$$

$$p_{01} = \{4N_{77} + 3(N_{47} + N_{74} + N_{78} + N_{87}) + 2(N_{44} + N_{88} + N_{17} + N_{71}$$
$$+ N_{27} + N_{72} + N_{37} + N_{73} + N_{67} + N_{76} + N_{48} + N_{84} + N_{79} + N_{97})$$
$$+ N_{41} + N_{18} + N_{81} + N_{24} + N_{42} + N_{28} + N_{82} + N_{34} + N_{43} + N_{38}$$
$$+ N_{83} + N_{46} + N_{64} + N_{49} + N_{94} + N_{68} + N_{86} + N_{89} + N_{98}$$
$$+ \phi[2(N_{57} + N_{75}) + N_{45} + N_{54} + N_{58} + N_{85}]$$
$$+ (1 - \phi)[3(N_{57} + N_{75}) + 2(N_{45} + N_{54} + N_{58} + N_{85}) + N_{15} + N_{51}$$
$$+ N_{25} + N_{52} + N_{35} + N_{53} + N_{56} + N_{65} + N_{59} + N_{95}] + 2(1 - \phi)N_{55}\}$$
$$/(4 \sum_{i=1, j=1}^{9,9} N_{ij}),$$

$$p_{00} = \{4N_{99} + 3(N_{69} + N_{96} + N_{89} + N_{98}) + 2(N_{66} + N_{88} + N_{19} + N_{91}$$
$$+ N_{29} + N_{92} + N_{39} + N_{93} + N_{49} + N_{94} + N_{68} + N_{86} + N_{79} + N_{97})$$
$$+ N_{16} + N_{61} + N_{18} + N_{81} + N_{26} + N_{62} + N_{28} + N_{82} + N_{36} + N_{63}$$
$$+ N_{38} + N_{83} + N_{46} + N_{64} + N_{48} + N_{84} + N_{67} + N_{76} + N_{78} + N_{87}$$
$$+ \phi[3(N_{59} + N_{95}) + 2(N_{56} + N_{65} + N_{58} + N_{85}) + N_{15} + N_{51} + N_{25}$$
$$+ N_{52} + N_{35} + N_{53} + N_{45} + N_{54} + N_{57} + N_{75}]$$
$$+ (1 - \phi)[2(N_{59} + N_{95}) + N_{56} + N_{65} + N_{58} + N_{85}] + 2\phi N_{55}\}$$
$$/(4 \sum_{i=1, j=1}^{9,9} N_{ij}).$$

$$r = \{\psi_1(N_{51}^1 + N_{51}^5 + N_{52}^1 + N_{52}^6 + N_{53}^2 + N_{53}^6 + N_{54}^1 + N_{54}^8 + N_{56}^2$$
$$+ N_{56}^9 + N_{57}^4 + N_{57}^8 + N_{58}^4 + N_{58}^9 + N_{59}^5 + N_{59}^9 + N_{15}^1 + N_{15}^5 + N_{25}^1$$
$$+ N_{25}^6 + N_{35}^2 + N_{35}^6 + N_{45}^1 + N_{45}^8 + N_{65}^2 + N_{65}^9 + N_{75}^4 + N_{75}^8 + N_{85}^4$$
$$+ N_{85}^9 + N_{95}^5 + N_{95}^9) + \psi_2(N_{51}^2 + N_{51}^4 + N_{52}^3 + N_{52}^4 + N_{53}^3 + N_{53}^5$$
$$+ N_{54}^2 + N_{54}^7 + N_{56}^3 + N_{56}^8 + N_{57}^5 + N_{57}^7 + N_{58}^6 + N_{58}^7 + N_{59}^6 + N_{59}^8$$
$$+ N_{15}^2 + N_{15}^4 + N_{25}^3 + N_{25}^4 + N_{35}^3 + N_{35}^5 + N_{45}^2 + N_{45}^7 + N_{65}^3 + N_{65}^8$$
$$+ N_{75}^5 + N_{75}^7 + \psi_3(N_{52}^2 + N_{52}^5 + N_{54}^4 + N_{54}^5 + N_{56}^5 + N_{56}^6 + N_{58}^5$$
$$+ N_{58}^8 + N_{25}^2 + N_{25}^5 + N_{45}^4 + N_{45}^5 + N_{85}^6 + N_{85}^7 + N_{95}^6 + N_{95}^8) + N_{65}^6$$
$$+ N_{85}^5 + N_{85}^8) + \psi_4(N_{55}^1 + N_{55}^9) + N_{65}^5 + \psi_5(N_{55}^3 + N_{55}^7)$$
$$+ \psi_6(N_{55}^2 + N_{55}^4 + N_{55}^4 + N_{55}^6 + N_{55}^8) + \psi_7 N_{55}^5\}$$
$$/(\sum_{j=1}^{9} N_{5j} + \sum_{i=1}^{9} N_{i5} - N_{55})$$

In the M step, the equations for estimating additive, dominant, imprinting effects expressed in paternal and offspring generations are also derived. The E and M steps are iterated until the estimates converge to a stable value. These stable values are the maximum likelihood estimates (MLEs) of parameters. The estimated haplotype frequencies and recombination fraction are embedded into a mixture model for estimating genotypic values and variances for different generations.

AUTHOR CONTRIBUTIONS

Conceived and designed the experiments: RW. Performed the experiments: JW DL. Analyzed the data: YL YG WH JW MNC. Contributed reagents/materials/ analysis tools: YL WH JW. Wrote the paper: RW YL.

REFERENCES

1.   Reik W, Walter J (2001) Genomic imprinting: parental influence on the genome. Nat Rev Genet 2: 21–32.

2.   Wilkins JF, Haig D (2003) What good is genomic imprinting: The function of parent-specific gene expression. Nat Rev Genet 4: 359–368.

3.   Itier JM, Tremp G, Léonard JF, Multon MC, et al. (1998) Imprinted gene in postnatal growth role. Nature 393: 125–126.

4.   Li LL, Keverne EB, Aparicio SA, Ishino F, et al. (1999) Regulation of maternal behaviour and offpring growth by paternally expressed Peg3. Science 284: 330–333.

5.   Isles AR, Wilkinson LS (2000) Imprinted genes, cognition and behaviour. Trend Cogn Sci 4: 309–318.

6.   Constancia M, Kelsey G, Reik W (2004) Resourceful imprinting. Nature 432: 53–57.

7.   Wood AJ, Oakey RJ (2006) Genomic imprinting in mammals: Emerging themes and established theories. PLoS Genet 2(11): e147.

8.   Wilkinson LS, Davies W, Isles AR (2007) Genomic imprinting effects on brain development and function. Nat Rev Neurosci 4: 1–19.

9.   Wang CG, Wang Z, Luo JT, Li Q, et al. (2010) A model for transgenerational imprinting variation in complex traits. PLoS ONE 5(7): e11396.

10. Frost JM, Moore GE (2010) The importance of imprinting in the human placenta. PLoS Genet 6(7): e1001015.

11. Sha K (2008) A mechanistic view of genomic imprinting. Ann Rev Genom Hum Genet 9: 197–216.

12. De Koning DJ, Rattniek AP, Harlizius B, Arendonk JAM, et al. (2000) Genome-wide scan for body composition in pigs reveals important role of imprinting. Proc Natl Acad Sci U S A 97: 7947–7950.

13. Liu T, Todhunter RJ, Wu S, Hou W, et al. (2007) A random model for mapping imprinted quantitative trait loci in a structured pedigree: An implication for mapping canine hip dysplasia. Genomics 90: 276–284.

14. Cheverud JM, Hager R, Roseman C, Fawcett G, et al. (2008) Genomic imprinting effects on adult body composition in mice. Proc Natl Acad Sci U S A 105: 4253–4258.

15. Wolf JB, Cheverud JM, Roseman C, Hager R (2008) Genome-wide analysis reveals a complex pattern of genomic imprinting in mice. PLoS Genet 4: e1000091.

16. Li YC, Coelho CM, Liu T, Wu S, et al. (2007) A statistical strategy to estimate maternal-zygotic interactions and parent-of-origin effects of QTLs for seed development. PLoS ONE 3: e3131.

17. Morgan HD, Santos F, Green K, Dean W, et al. (2005) Epigenetic reprogramming in mammals. Hum Mol Genet 14: R47–R58.

18. Sasaki H, Matsui Y (2008) Epigenetic events in mammalian germ-cell development: reprogramming and beyond. Nat Rev Genet 9: 129–140.

19. Tal O, Kisdi E, Jablonka E (2010) Epigenetic contribution to covariance between relatives. Genetics 184: 1037–1050.

20. McGrath J, Solter D (1984) Inability of mouse blastomere nuclei transferred to enucleated zygotes to support development in vitro. Science 226: 1317–1319.

21. Surani MA, Barton SC, Norris ML (1984) Development of reconstituted mouse eggs suggests imprinting of the genome during gametogenesis. Nature 308: 548–550.

22. Morgan HD, Sutherland HG, Martin DI, Whitelaw E (1999) Epigenetic inheritance at the agouti locus in the mouse. Nat Genet 23: 314–318.

23. Cropley JE, Suter CM, Beckman KB, Martin DI (2006) Germ-line epigenetic modification of the murine Avy allele by nutritional supplementation. Proc Natl Acad Sci U S A 103: 17308–17312.

24. Skinner MK (2008) What is an epigenetic transgenerational phenotype? F3 or F2. Reprod Toxic 25: 2–6.

25. Dolinoy DC, Weidman JR, Waterland RA, Jirtle RL (2006) Maternal genistein alters coat color and protects Avy mouse offspring from obesity by modifying the fetal epigenome. Environ Health Perspect 114: 567–572.

26. Whitelaw NC, Whitelaw E (2008) Transgenerational epigenetic inheritance in health and disease. Curr Opin Genet Dev 18: 273–279.

27. Youngson NA, Whitelaw E (2008) Transgenerational epigenetic effects. Ann Rev Genom Hum Genet 9: 233–257.

28. Pembrey ME, Bygren LO, Kaati G, Edvinsson S, et al. (2006) Sex-specific, male-line transgenerational responses in humans. Europ J Hum Genet 14: 159–166.

29. Wu RL, Ma CX, Casella G (2007) Statistical Genetics of Quantitative Traits: Linkage, Map, and QTLs. New York: Springer.

30. Li Q, Wu RL (2009) A multilocus model for constructing a linkage disequilibrium map in human populations. Stat Appl Genet Mol Biol. 8, Iss. 1, Article 18.

31. Chan EY (2005) Advances in sequencing technology. Mutant Res 573: 13–40.

32. Beckmann JS, Estivill X, Antonarakis SE (2007) Copy number variants and genetic traits: closer to the resolution of phenotypic to genotypic variability. Nat Rev Genet 8: 639–646.

33. Weinberg CR, Wilcox AJ, Lie RT (1998) A log-linear approach to case-parent triad data: Assessing effects of disease genes that act directly or through maternal effects, and may be subject to parental imprinting. Am J Hum Genet 62: 969–978.

34. Cordell HJ, Barratt BJ, Clayton DG (2004) Case/pseudocontrol analysis in genetic association studies: a unified framework for detection of genotype and haplotype associations, gene-gene and gene-environment interactions and parent-of-origin effects. Genet Epid 26: 167–185.

35. Hager R, Cheverud JM, Wolf JB (2008) Maternal effects as the cause of parent-of-origin dependent effects that mimic genomic imprinting. Genetics 178: 755–1762.

36. Jirtle RL, Skinner MK (2007) Environmental epigenomics and disease susceptibility. Nat Rev Genet 8: 253–262.

37. Churchill GA, Doerge RW (1994) Empirical threshold values for quantitative triat mapping. Genetics 138: 963–971.

38. Dawson E, Abecasis GR, Bumpstead S, Chen Y, et al. (2002) A first-generation linkage disequilibrium map of human chromosome. Nature 418: 544–548.

39. Gabriel SB, Schaffer SF, Nguyen H, Moore JM, et al. (2002) The structure of haplotype blocks in the human genome. Science 296: 2225–2229.

40. Patil N, Berno AJ, Hinds DA, Barrett WA, et al. (2001) Blocks of limited haplotype diversity revealed by high-resolution scanning of human chromosome 21. Science 294: 1719–1723.

41. Zhang K, Deng M, Chen T, Waterman MS, Sun F (2002) A dynamic programming algorithm for haplotype block partitioning. Proc Natl Acad Sci U S A 99: 7335–7339.

42. Wu RL, Lin M (2008) Statistical and Computational Pharmacogenomics. London: Chapman & Hall/CRC.

43. Cheng Y, Berg A, Wu S, Li Y, Wu RL (2009) Computing genetic imprinting expressed by haplotypes. Method Mol Biol 573: 189–212.

44. Wang CG, Cheng Y, Liu T, Li Q, et al. (2008) A computational model for sex-specific genetic architecture of complex traits in humans. Mol Pain 4: 13.

45. Wu S, Yang J, Wang CG, Wu RL (2007) A general quantitative genetic model for haplotyping a complex trait in humans. Curr Genom 8: 343–350.

Chapter 5

LEVERAGING NON-TARGETED METABOLITE PROFILING VIA STATISTICAL GENOMICS

Miaoqing Shen[1,2], Corey D. Broeckling[3], Elly Yiyi Chu[2], Gregory Ziegler[4,5], Ivan R. Baxter[4] , Jessica E. Prenni[3], Owen A. Hoekenga[2]

[1] Boyce Thompson Institute for Plant Research, Ithaca, New York, United States of America

[2] United States Department of Agriculture, Agricultural Research Service, RW Holley Center for Agriculture and Health, Ithaca, New York, United States of America

[3] Colorado State University, Proteomics and Metabolomics Facility, Fort Collins, Colorado, United States of America

[4] United States Department of Agriculture, Agricultural Research Service, Plant Genetics Research Unit, St. Louis, Missouri, United States of America

[5] Donald Danforth Plant Science Center, St. Louis, Missouri, United States of America

ABSTRACT

One of the challenges of systems biology is to integrate multiple sources of data in order to build a cohesive view of the system of study. Here we describe the mass spectrometry based profiling of maize kernels, a model system for genomic studies and a cornerstone of the agroeconomy. Using a network analysis, we can include 97.5% of the 8,710 features detected from 210 varieties into a single framework. More conservatively, 47.1% of compounds detected can be organized into a network with 48 distinct modules. Eigenvalues were calculated for each module and then used as inputs for genome-wide association studies. Nineteen modules returned significant results, illustrating the genetic control of biochemical networks within the maize kernel. Our approach leverages the correlations between the genome and metabolome to mutually enhance their annotation and thus enable biological interpretation. This method is applicable to any organism with sufficient bioinformatic resources.

INTRODUCTION

Assembling increasingly large datasets due to the enhanced efficiency of various phenotyping technologies (e.g. metabolomic and proteomic profiling or nucleic acid sequencing) permits increasingly comprehensive views of biological processes. However, the problem of analysis and visualization in systems biology has led some commenters to question how best to "drink from a fire hose" [1]. Statistical methodologies that are highly inclusive can help solve the dual problem of analysis and visualization [2], [3]. Here, we describe the use of weighted correlation network analysis (WGCNA) as a method to integrate mass spectrometry-based simultaneous of the maize kernel, isolated from a diverse panel of inbred maize varieties previously utilized for genome wide association studies of multiple traits [4], [5], [6], [7], [8]. We assert that the use of such a study panel allows us to leverage the genetic and genomic resources already available to enhance our annotation and analysis of mass spectrometry results. Likewise, this approach also improves the annotation of a genome by providing specific metabolites and chemistries to describe the roles of predicted proteins. This approach relies heavily on software written in the R programming language, which should enable wide adoption by the scientific community due to the lack of associated cost [9].

Our choice of study system was deliberate. Maize has incredible genetic and phenotypic diversity, providing an ideal resource for systems biology studies [10]. Variation in plant yield, composition, and morphological traits has been reported in multiple collections of diverse inbred varieties and related biparental mapping populations [4], [5], [6], [7], [8], [11]. Much has also been learned about the structural and genetic variation within the maize genome [12], [13]. The quality of maize grain is a key factor for breeders and other stakeholders, but the development of biomarkers to assist breeding and transgenic crop improvement remains challenging [14],[15], [16], [17]. Metabolomics, relying on the use of mass signals as markers, provides a rapid approach to characterize related varieties and enable the description of existing and novel quality traits [18], [19]. The aim of metabolomics is to provide a comprehensive and quantitative analysis of a vast number of components in a specific biological sample, and identify as many metabolites as possible [20], [21], [22].

Metabolomic analyses of plants can be especially challenging, as plants contain great chemical diversity especially in secondary metabolites [18], [23]. These secondary metabolites help keep plants' systems working properly, play roles in the response to genetic or environmental changes, and have powerful physiological effects in humans or animals [20]. Although mass spectrometry-based metabolomics enables the measurement of hundreds or thousands of compounds from a single complex sample, the plant metabolome is still poorly defined and the identification process for specific compounds remains challenging [24], [25], [26], [27], [28],[29]. However, metabolite profiling is not mutually exclusive of statistical genetic and genomics-based approaches [30]; the combination of systems biology strategies is mutually supportive and beneficial.

Genome-wide association studies (GWAS) consider nucleotide variation patterns, relative to population structure, to identify correlations between particular genomic regions and phenotypes. Often, susceptibility to particular diseases or metabolic syndromes is analyzed using GWAS, as this statistical genomics approach is useful in both humans and model systems [7], [31], [32]. GWAS have been applied to metabolomic datasets, to identify SNPs that may be causally linked to particular biochemical processes or pathways [14], [33], [34],[35], [36]. However, in each of these cases the unit of analysis has been individual metabolites, which does little to improve the efficiency of calculation or maximize the benefit of measuring hundreds or thousands of analytes.

Here, we describe non-targeted metabolite profiling of whole maize kernels. The study of grain quality was approached from the perspective of maize as a foodstuff, thus methanolic extracts were isolated from cooked kernels. Our choice of study panel gave us access to more than 1 million SNPs to support GWAS [12]. We applied WGCNA to organize our data into modules that contained multiple markers that also enabled the identification of networks under genetic regulation. This condensation step allowed us to reduce the complexity of the dataset, addressing the multiple testing problem that is endemic to systems biology, and increase computational efficiency. GWAS were applied to weighted averages for each module (hereafter, module eigenvalues), to identify SNPs associated with collections of biochemical markers. We suggest that the WGCNA procedure does not excessively smooth the data, as SNPs correlated with module eigenvalues were significantly correlated with specific compounds assigned to those modules. Finally, module eigenvalues were used in linear regression models to analyze traits that were recalcitrant to GWAS.

MATERIALS AND METHODS

Materials

HPLC-grade acetonitrile, methanol, and formic acid were purchased from Fisher (Pittsburgh, PA); UPLC HSS C18 column (1.8 μm, 2.1 mm×100 mm), sample vials, UPLC column test mix and leucine enkephalin were purchased from Waters (Milford, MA); and all other reagents were available through Sigma (St. Louis, MO), or as indicated.

Sample Preparation

A maize inbred diversity panel was grown in 2010 on the Musgrave Research Farm of Cornell University (Poplar Ridge, NY)(Flint-Garcia et al., 2005). Duplicated trials were grown using a randomized field design with regular check rows of the B73 accession; 210 of the 282 accessions produced sufficient grain for subsequent analysis, largely due to flowering time issues. Whole maize kernels (n=50) were covered with an equal volume of 18 megaohm water and autoclaved for 30 minutes to fully cook the grain. Samples were then freeze-dried and ground to a fine flour using a consumer-grade grain mill (KoMo Medium Mill, Pleasant Hill Grain, Hampton, NE). Ground, cooked samples were frozen at $-20°C$ until extracted with a 1:1 mixture of water and methanol. After 10 min sonication, extracts were centrifuged for 10 min at 4000 rpm. The supernatant was filtered through 0.45 µm filter. Two independent biological replicates were obtained and analyzed, although only one is discussed here.

UPLC and Mass Spectrometer

Sample injections were performed with an ACQUITY UPLC system (Waters), equipped with a Waters Acquity UPLC HSS C18 column. The samples were injected by means of a 7.5 µL partial loop injection with 3 technical replicates by randomizing all injections. Mobile phase A consisted of 0.1% formic acid in water and mobile phase B contained 0.1% formic acid in acetonitrile. The following gradient was used: 4.5 min 2.4% B, 0.5 min 40% B, 3.5 min 64% B, and 3.5 min 97.6% B. Flow rate was 0.4 ml/min and column temperature was maintained at 40°C. The eluent from the column was delivered to a Xevo G2 TOF (Waters). The mass spectrometer operated in a positive mode using a samples cone voltage of 20 V and a capillary voltage of 2.5 kV with the temperature of source and desolvation at 120°C and 350°C and the flow rate of nitrogen desolvation gas at 850 L/h. Data were acquired in a centroid mode from 50 to1,200 m/z with scan time of 0.2 sec. MS data were collected at a collision energy of 6 V with alternative collection of MSE mode using a ramped collision energy of 20–40 V. Leucine enkephalin was used as the lock mass compound (m/z 556.2771 in positive) and infused at 10 µl/min with a concentration of 1 ng/µl. The lock mass was acquired in all injections of samples to ensure accuracy and reproducibility. The instrument was calibrated using sodium formate at a concentration of 5 mM with mass accuracy within 1 ppm.

Data Transfer and Statistics Approaches

A variety of statistical procedures were employed to analyze data using R (version 2.13.1) or JMP (version 9, SAS Institute, Cary NC). MarkerLynx (v4.1, Waters) was used to integrate and align MS data points and to convert them into exact mass and retention time signals. Principal component analysis (PCA) was performed using Pareto-scaled data on all detected features for initial charactering the separation of maize variables and checking repeatability for technical replicates. The MarkerLynx generated feature list was somewhat larger than that obtained using XCMS, but WGCNA produced highly similar outcomes from both datasets. To use XCMS to identify and annotate features, raw data files were converted to NetCDF format using the Waters DataBridge software [37]. Peak detection and alignment was performed on both the low and high collision energy channels (MS and MS^E) using XCMS software (version 1.22.1 [38]). Reconstruction of indiscriminant MS/MS spectra (idMS/MS) was performed as described [37], with exception that rather than utilizing CAMERA groupings, the grouping was centered around the retention time of the feature of interest, with a 2 second window on either side. Reconstructed spectra presented in this paper are supplied as File S1, an msp format spectral file suitable for viewing using NIST MS search program. A correlational filter was then used to find features that demonstrated similar quantitative patterns to the feature of interest. Reconstructed MS and idMS/MS spectra were exported as an '.msp' formatted spectral library using a custom R script. The library was batch searched against the MassBank database [39], and manually searched against the NIST (http://www.nist.gov/srd/nist1a.cfm) and Metlin databases [40]. Identification confidence scores were assigned as described [37]. For multiply charged peptides, the spectra were manually converted to.mgf format, and the precursor ion was manually interpreted based on the MS spectrum. idMS/MS spectra were searched against the NCBInr protein database using a taxonomy filter for maize (version 07/12/12) (43,920 sequence entries) using the Mascot database search engine (version 2.3). Search parameters were set as follows: monoisotopic mass, parent ion mass tolerance of 0.05 Da, fragment ion mass tolerance of 0.1 Da, no enzyme specificity, and variable modification of oxidation of Met.

The XCMS also fills empty cells based on retention time and mass specific signal, thus reducing the frequency of zero values in the dataset. The XCMS generated data matrix, with an intensity value for each feature and each sample, was used as input for the clustering analysis by WGCNA. Weighted correlation network analysis (WGCNA) was produced with the R package, creating unsigned networks where both positive and negative correlations could be clustered into a single module [2], [3]. WGCNA used autoscaled data in order to reduce the dominance of dynamic, high-concentration metabolites (Table S1). Module eigenvalues for each module were calculated for each of the 210 maize accessions, providing a condensed dataset of derived variables for subsequent genetic analysis. Cytoscape, the open source bioinformatics software, was used to illustrate metabolite networks [41].

Genome-wide Association Study (GWAS)

GAPIT, the genome association and prediction integrated tool (http://www.maizegenetics.net/gapit) was used to perform GWAS and genome prediction. Module eigenvalues from WGCNA were used as maize phenotypic traits in GWAS. The genotypic data are publicly available from panzea.org (http://www.panzea.org/lit/data_sets.html#genos). Previously, the maize diversity panel had been characterized using next-generation sequence analysis such that more than 1 million SNPs across the maize genome are available to characterize the genetic diversity [12]. A kinship matrix was calculated [42] and population structure was modeled as a fixed effect [43]. Significant SNPs (FDR corrected $p < 0.001$ and $p < 0.05$) were identified for each module eigenvalue with a requirement that the SNP be present with $\geq 5\%$ allele frequency.

RESULTS

Non-Targeted Metabolite Profiling

Our long-term goal is to characterize phenotypic variation in maize grain quality and to identify the genetic and environmental factors that influence the metabolomic composition of this important staple food and model plant. This effort will provide information to better describe the existing food supply and also project what new grain quality traits may be achievable in the future using conventional plant breeding. Towards this goal, we chose to use mass spectrometry based non-targeted metabolite profiling of maize meal prepared from cooked, whole kernels (Figure 1). While it may be counterintuitive to treat samples in this way, our dataset represents a genetically diverse sample of a foodstuff that could be consumed by either humans or animals. This choice helps to define the range of normal and acceptable variation within a highly diverse crop plant [17]. More than 8,710 metabolomic features were detected from the whole kernel methanolic extracts (Table S1). Principal component analysis (PCA) gave an initial characterization of the profiling results. PCA explains about 22% of the variance with 2 PC's (Figure S1). The performance of PCA for this dataset is typical as the composition varies very widely across genetically distant accessions.

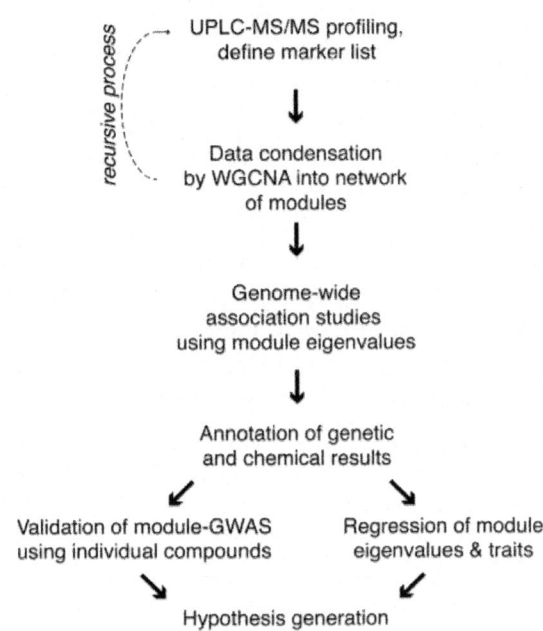

Figure 1: Genomics-assisted chemistry & chemistry-assisted genomics.

This flow chart describes the process by which statistical genetics and genomics can enable metabolite profiling to have greater power and impact.

Data Condensation by Weighted Correlation Network Analysis (WGCNA)

One of the endemic problems of systems biology is the multiple testing problem, wherein the number of variables measured dwarfs observations. One potential solution to this problem is to condense the dataset into a smaller number of distinct groups (hereafter, modules), normalizing the issue of observations and variables. WGCNA is an approach to display model network relationships, identifying co-regulated groups of features (hereafter, nodes) such as patterns of gene expression [2]. It can also be used to visualize metabolite networks and increase the comprehensiveness of non-targeted metabolomics [3]. WGCNA describes the relationships between all of the input variables, summarizing the correlation and connectivity of all nodes. The network can be more or less elaborate, depending on the rules set for inclusion into the network. A principal component is calculated for each module for each variety, summarizing the contribution of all nodes included into a particular module, which is referred to by a randomly assigned color. This principal component (hereafter, module eigenvalue) can be used for correlation tests or ANOVA.

For our dataset, 97.5% of the detected molecular features (nodes) could be included in a network with 56 defined modules (Table S2). The network was then pruned to require that the minimum connectivity between nodes exceed 4 standard deviations (SD) above the mean connectivity observed between all nodes. At this threshold, the network contained 48 modules and 4,102 nodes (47% of nodes, 3.1% of the theoretical connections; Figure 2). The network was redefined under even stricter terms, using a 6SD threshold (Table S2). As the modules were defined by the strength of the correlations among members, modules varied in size and membership according to the inclusion threshold. For example, the turquoise module in the initial description had 2,105 nodes and ~3.73 million edges (Table S2). At the 4SD threshold, the turquoise module reduced to 1,597 nodes with more than 0.62 million edges, while at the 6SD shrinking further to 635 nodes with 40,217 connections. Nodes within the turquoise module were also connected with members of the black module, which likewise contained connections to both the turquoise and purple modules. Other modules were much less elaborate; orange contained 81 nodes in its initial description, 63 nodes at 1 SD, dropping to 9 nodes and 56 connections at 4SD, and disappearing completely at 6SD (Table S2). At the 4SD threshold some modules broke into distinct clusters as connections that helped to define the original module, using the original definitions, dropped below the significance threshold (Figure 2). This facet of the WGCNA procedure represents both a strength and a weakness for the approach. Information can be applied to poorly connected members of a particular module using guilt by association on tightly connected central elements. However, as the module eigenvalues are estimated when the network was initially described, the poorly connected nodes may transmit an excessive degree of variance to these values and perhaps confound downstream applications.

This node and edge projection describes the grain metabolome observed in the methanolic extract from 210 inbred line varieties of maize. This network requires a minimum degree of connectivity between any two nodes (i.e. biochemical markers detected by mass spectrometry) that exceeds four standard deviations above the mean connectivity observed between detected markers. According to this threshold, 4,102 nodes are organized into 48 modules each represented by particular color. However, some modules have separated into multiple, distinct clusters as internal connectivity may fall beneath the 4 standard deviation cutoff, such that there are 101 objects in this projection.

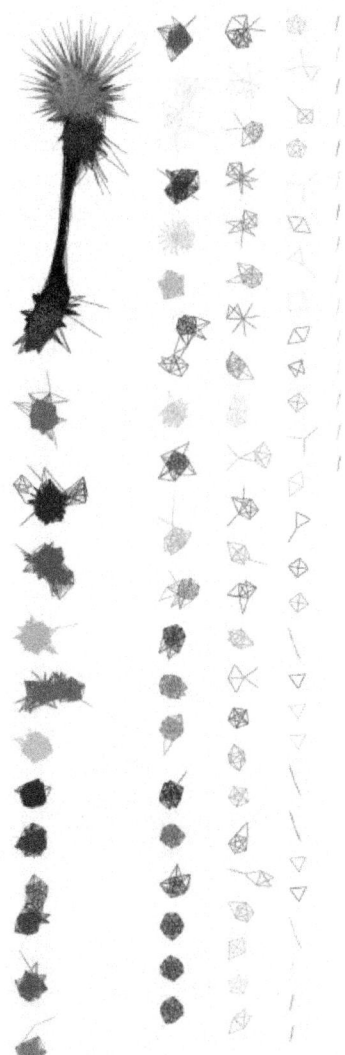

Figure 2: Visualization of maize grain metabolome.

Module Eigenvalues Drive Genome Wide Association Studies

We expected that using WGCNA for the analysis of mass spectrometry based non-targeted metabolite profiling data would accomplish two goals: (1) define the co-regulated networks of metabolites and peptides that contribute to maize kernel quality and composition and (2) reduce the number of variables for downstream analyses. One such analysis is a genome wide association study

(GWAS), to correlate particular genomic regions with phenotypes of interest. This approach has already been applied to maize but not on derived variables such as module eigenvalues, so far as we are aware. And while computational resources are improving, conducting GWAS with a SNP dataset as large as that available for the Buckler Diversity Panel using optimized procedures is still a time intensive procedure (0.5 hr/trait or >150 d for the original data) [12], [44]. Module eigenvalues for all 56 modules were analyzed, 19 of which found significant associations (FDR corrected p-value <0.05; Table S2). Modules that were detected under the most stringent membership conditions (>6SD) were more likely to produce significant GWAS outcomes than those present only under lesser requirements (14 of 27 versus 5 of 21; Table S2). However, modules with fewer connections at 4SD were more likely to identify significantly correlated SNPs with GWAS (χ^2=4.56, p=0.0328). While 4,830 SNPs were identified by GWAS, nearly two-thirds were associated with only two modules (plum2 and salmon). A variety of patterns were observed in the results, ranging from few to many SNPs and wide to narrow distribution across the genome (Figure 3).

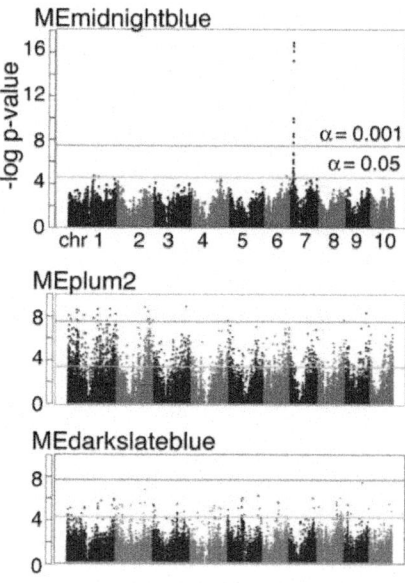

Figure 3: Genome-wide association studies on three module eigenvalues (ME).

Nineteen modules returned significantly correlated SNP markers according to GAPIT. Three are shown here. Significance thresholds were empirically calculated for each trait using GAPIT; FDR-corrected p-values at both a conservative (p<0.001; green line) and generous (p<0.05; aqua line) are displayed. MEmidnightblue identified one region of chromosome 7 with high confidence, with a second region of chromosome 1 with lower confidence. MEplum2 identified multiple genomic intervals with high confidence. MEdarkslateblue identified no significant regions at the conservative threshold, but several regions at the lower threshold.

The strongest associations between SNPs and module eigenvalues were found with the midnightblue module. Nearly all of the significant SNPs were identified at both conservative and relaxed FDR corrected p-value thresholds and were located in a single region of chromosome 7. Most SNPs were identified with variants of the α-zein 19C2 seed storage protein, a result that is supported by analysis of the mass spectrometry data, which are consistent with a C-terminal peptide derived from α-zein 19C2 protein (Figure S2; Table 1; File S1). The plum2 module gave results as one might optimistically hope for, with a variety of genomic regions identified under conservative significance thresholds. SNPs associated with plum2 included those within a putative mitogen-activated protein kinase, which suggested that signal transduction pathway components were detected by GWAS, and also a cytochrome P450 with significant similarity to flavonoid 3-monooxygenases (Table 1). The darkslateblue module returned far fewer SNPs above the α=0.001 significance threshold, but many above the 0.05 level. Much like plum2, these SNPs identified a mixture of genes with potential functions while others had no obvious connection to the regulation of maize kernel composition.

Table 1: Sample results from genome-wide association studies.

SNP	−log (p-value)	Module/Marker	Gene Annotation
S7_18857356	17.45	Midnightblue*	α-zein precursor 19C2[CDS]
S7_18857356	8.74	1056.528_226.57*	α-zein precursor 19C2[CDS]
S2_160151277	8.05	darkslateblue	α-amylase/protease inhibitor[CDS]
S9_24144378	7.85	darkslateblue	Protein phosphatase 2A regulatory subunit[CDS]
S2_184267091	9.59	plum2	Flavonoid 3-monooxygenase[6.6 kb-s]
S1_264986642	9.06	plum2	Mitogen-activated protein kinase[CDS]
S5_168853373	6.75	orange[#]	bZIP transcription factor[4.4 kb-s]
S5_168853373	9.28	138.092_70.421[#]	bZIP transcription factor[4.4 kb-s]

SNP indicates the chromosome and position (bp) within the maize genome (version 5b.60). The significance of the SNP association is indicated by the negative of the log for the false discovery rate corrected p-value. Module/marker indicates which module eigenvalue the SNP was associated. Single constituents of the midnightblue (*) and orange (#) modules were also analyzed by GWAS. Gene annotation describes the closest gene model relative to the SNP evaluated.
doi:10.1371/journal.pone.0057667.t001

A potential pitfall for the WGCNA procedure as a data condensation tool was the potential to excessively smooth the data, creating a false picture of the genetic regulation of the metabolome. In this instance, collapsing multiple metabolite markers into a single signal might obscure the effect of a particular locus for importance of a module constituent. To address this concern, we examined the orange module in greater detail. At 4SD, orange has 9 nodes, one of which we identified as tyramine (Figure S3). The abundance of tyramine alone was used as a trait for GWAS; this result was compared with the GWAS on the orange module eigenvalue (Figure 4). The orange module returned 27 significant SNPs, 7 of which were also identified as significant for tyramine (Table 1). While GWAS on tyramine returned more significant SNPs than for the orange module, our result does indicate that SNPs associated with a single compound can be identified from GWAS on the module eigenvalue.

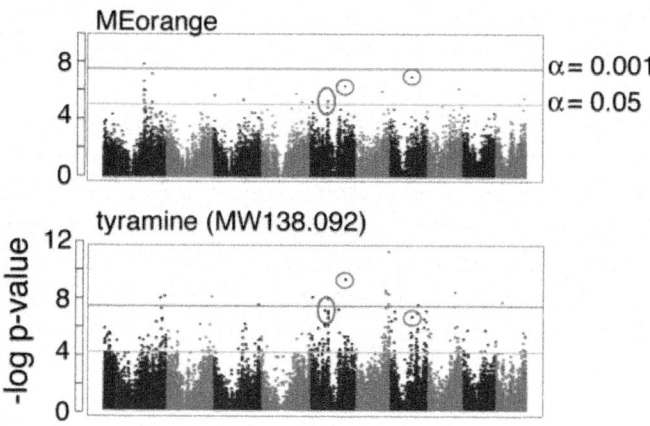

Figure 4: Module eigenvalues do not obscure the importance of single compounds.

MEorange was estimated from 81 molecular features, one of which was identified to be tyramine. GWAS on MEorange identified 27 significant SNPs at the FDR-corrected p<0.05 threshold. GWAS on tyramine alone identified 7 SNPs in common (red circles) with MEorange.

Leveraging Genomic Information to Assist Annotation of Mass Spectrometry Data

Compound identification in mass spectrometry based non-targeted metabolite profiling experiments represents a major challenge of this approach [45]. The utility of indiscriminant MS/MS (idMS/MS) was recently demonstrated to improve the rate and confidence of metabolite identification in non-targeted metabolite profiling experiments [37]. In the current study, the idMS/MS process was applied to selected features from modules with significant GWAS results. As described above, GWAS with the midnightblue module identified a region of maize chromosome 7 consistent with the α–zein 19C2 storage protein (UniProtKB P06677). Sixty-seven features were identified in this module with retention times between 226.35 and 227.06 sec, suggesting that they all represent the same compound. The reconstructed MS and idMS/MS spectra were highly suggestive of a peptide structure due the observation of multiple charge states. The molecular weight of the potential peptide was inferred from the two multiply charged isotope clusters in the MS spectra, and the corresponding idMS/MS spectrum was searched against the maize genome using the Mascot database search engine. This search returned a single peptide as the likeliest match (PAASYQQHIIGGALF), which represents the C-terminus for both the 19C1 and 19C2 variants of the α–zein storage protein. Taken together, these results suggest that *cis*-acting variation at the α-zein locus on chromosome 7 influences the quantitative expression of this protein and this variation is apparent in the cooked maize meal product. We achieved this peptide identification in spite of the fact that our MS data were collected with small molecules in mind and without the benefit of predictable proteolytic cleavage that most proteomic search engines rely upon. Further, we accomplished this from a single separation/MS experiment without the need for a second targeted MS/MS experiment, demonstrating the utility of the idMS/MS workflow. GWAS on this peptide alone returned significant SNPs common with the midnightblue module eigenvalue. However, midnightblue produced a far more significant p-value than the single feature, in contrast to the previous example (Table 1).

WGCNA alone, without the benefit of GWAS, can also assist in the annotation of mass spectra. The orange module contained 9 features with 4 different retention times at the 4SD network threshold, suggesting that there are only a few compounds contained in this module that behave similarly across this maize diversity panel. The first two eluting features demonstrated a strong similarity to tyramine, as described above. The third retention time group contained 3 isotopes of a molecular ion of 284.13, with a fragmentation patterns consistent with *p*-coumaric acid. A dehydration conjugation between tyramine and p-coumaric acid is consistent this molecular ion. The fourth retention time group contained 3 isotopes of the molecular ion 314.14; the idMS/MS patterns suggested tyramine, which lead us to hypothesize an additional tyramine conjugate. The molecular ion for 314.14 is consistent with a ferulic acid-tyramine dehydration conjugate. The final retention time group consisted of a single feature, where the idMS/MS spectra did not show strong matches in public databases. As a whole, the spectral annotations of the orange module members suggest that this cluster is focused on variation in tyramine and at least two of its phenylpropanoid conjugates. The module identification enabled annotation of the mass spectra, as we were able to restrict our decision space based on the data. Likewise, understanding the underlying chemistry should enable our analysis of the GWAS identified SNPs, to clarify how these genes would contribute to the synthesis and modification of tyramine.

Correlation of Module Eigenvalues with Potentially Related Phenotypes

One of the original applications of WGCNA estimated eigenvalues was correlation analysis with related phenotypes, such as using patterns of gene expression to predict disease risk [46]. WGCNA was used to condense large datasets into more computationally manageable ones, with the added advantage of creating new testable hypotheses regarding cause and effect between genetic networks and phenotypic outcomes. Here, we used the WGCNA estimated module eigenvalues for stepwise regression of kernel weight, a commonly studied quality trait. GWAS was applied to kernel weight, but failed to identify any significant SNPs at the α=0.05 significance threshold. The GWAS procedure, as implemented by GAPIT, estimated the heritability of kernel weight at 0.38, which suggested that genetic regulation of kernel weight was highly complex and resulted from the interaction of many genes of very small individual effect.

Table 2: Regression of kernel weight using module eigenvalues

Source	DF	SS	F-ratio	p-value	t Ratio
MEsalmon4[gwas]	1	30242.822	22.7874	<0.0001	−4.77
MEroyalblue[gwas]	1	21356.628	16.0918	<0.0001	−4.01
MEdarkslateblue[gwas]	1	14387.453	10.8407	0.0012	3.29
MEsalmon[gwas]	1	7384.471	5.564	0.0194	2.36
MEpink[4SD]	1	36952.627	27.8431	<0.0001	5.28
MEblue[4SD]	1	20934.891	15.774	0.0001	3.97
MEdarkgrey[4SD]	1	11517.166	8.678	0.0037	2.95
MEgreen[1SD]	1	53412.158	40.245	<0.0001	−6.34
MEred[1SD]	1	31513.736	23.745	<0.0001	4.87
MEsaddlebrown[1SD]	1	20056.221	15.112	0.0001	3.89
MEwhite[1SD]	1	11504.719	8.6686	0.0037	−2.94
Error	173	229601.4			
Model	184	539795.71	21.2477	<0.0001	Adj $r^2=0.548$

Superscripts indicate whether the module eigenvalues gave significant correlations with GWAS and/or was included in the network using the 4SD threshold or the initial description of the network (1SD threshold).
doi:10.1371/journal.pone.0057667.t002

Stepwise regression for kernel weight that included all module eigenvalues reduced to a model using 11 modules and explained more than half of the observed variance (Table 2). These modules included those with significant GWAS associated SNPs, modules that were well defined according to the connectivity rules, and still others that were more diffuse. While the regression model may have overestimated the fraction of variance due to genetic factors (0.548 vs. 0.38), the two analyses were consistent in their overall findings. The eigenvalue regression model summarized the contributions of 11 modules, which conservatively contained at least 600 features (Table S2). Both statistical methods indicated that kernel weight was determined by the interplay of many genetic factors, however the module regression model did quantify the relative input from defined entities and provide a logical framework from which to build additional hypotheses.

DISCUSSION

One of the promises of systems biology is that through the integration of analytical technologies, a more comprehensive and complete view of biological processes can be achieved. Here, we utilized mass spectrometry based non-targeted metabolite profiling to characterize the maize kernel metabolome. We chose to profile cooked maize ground in a consumer-grade grain mill to better understand the variation present in food product that might reasonably be encountered by a consumer, rather than to estimate the maximal genetic potential found in these fractions. We used WGCNA to identify the patterns that help determine composition and quality, and to resolve the multiple testing problem and rebalance the number of observations to variables under analysis. This data condensation step allowed us leverage investments made in maize genetics and genomics to assist the annotation of our mass spectra, through the application of simple (i.e. correlation and regression) and complex (GWAS) statistical procedures. While we chose to profile maize kernels, the statistical and bioinformatic process outlined here is applicable to any biological system with sufficient genetic and genomic investment and should enhance the impact of systems biology approaches in plant, animal and microbial model organisms.

A second promise of systems biology is that of translational genomics, to apply our increasingly deep view of biological processes in more applied contexts and to produce positive outcomes for society. One such application is genomic selection or whole genome prediction, where all available genetic markers are used to predict phenotypes [15]. In a recent example, a SNP microarray was used to genotype a panel of diverse maize varieties that had also been evaluated using metabolite profiling and standard agronomic evaluation [15]. Both genetic and metabolomic markers gave high accuracy predictions of agronomically important traits such as biomass accumulation and flowering time. Additionally, the use of metabolomics allowed a light to shine into the "black box" of genomic selection, where the goal is merely to apply abundant and anonymous genetic markers to predict the phenotype of interest. Through simultaneous genetic and metabolomic profiling, it is possible to correlate micro-scale phenotypes (i.e. glucose) with macro-scale phenotypes (i.e. biomass) while also generating hypotheses to address causality. The methodology we describe here is consistent with that reported by Riedelsheimer and colleagues [14], [15], although the scale of genetic and phenotypic data available to us is considerably larger.

As our understanding of biology is an accretive process, there will always be more data to include in future analyses. One of the limitations of this study is the size of the diversity panel characterized by mass spectrometry. Statistical power increases as a function of the size of the study panel, such that if we had surveyed a larger fraction of the 282 varieties we should have been able to resolve genes with smaller individual function. One of the advantages of our workflow is that it encourages reexamination (Figure 1). As we gain additional phenotypes with the study panel, we can recalculate the network and start the correlation and GWAS over again. As we identify particular metabolite or protein markers, we can apply "guilt by association" to improve the annotation of other members of the same modules and the annotation of the genome itself. SNP detection technologies are also rapidly improving in scale and price, such that repeating the GWAS in a year's time will likely identify new regions of the maize genome that were not adequately covered in the present set of SNPs [12]. Finally, even with incomplete knowledge of the maize genome and inadequate statistical power, we were able to create a logical framework to explain an otherwise recalcitrant trait. We know that the basis of kernel weight is complicated [47], however, we can build testable hypotheses out of the module regression model that can be more fully explored in either larger diversity panels, to repeat GWAS with more power, or to choose biparental mapping populations, to test the effect of particular SNPs with the power advantages that simple mapping populations offer [48].

ACKNOWLEDGMENTS

The authors would like to thank Alex Lipka and Zhiwu Zhang for their input regarding GWAS. The authors would also like to thank Ted Thannhauser and Kevin Howe for technical assistance with liquid chromatography and mass spectrometry, Meghan den Bakker, Jon Hart, and the staff of the Musgrave Research Farm for assistance with the field related portion of the workflow, Paul Armstrong for building the weighing robot, and Matthew DiLeo for his original implementation of WGCNA to support metabolomics datasets. The U.S. Department of Agriculture (USDA) prohibits discrimination in all its programs and activities on the basis of race, color, national origin, age, disability, and where applicable, sex, marital status, familial status, parental status, religion, sexual orientation, genetic information, political beliefs, reprisal, or because all or part of an individual's income is derived from any public assistance program. (Not all prohibited bases apply to all programs.) Persons with disabilities who require alternative means for communication of program information (Braille, large print, audiotape, etc.) should contact USDA's TARGET Center at (202) 720–2600 (voice and TDD). To file a complaint of discrimination, write to USDA, Director, Office of Civil Rights, 1400 Independence Avenue, S.W., Washington, D.C. 20250-9410, or call (800) 795–3272 (voice) or (202) 720–6382 (TDD). USDA is an equal opportunity provider and employer.

AUTHOR CONTRIBUTIONS

Conceived and designed the experiments: OAH. Performed the experiments: MS GZ. Analyzed the data: MS CB EYC GZ IRB JEP OAH. Contributed reagents/materials/analysis tools: OAH CB IRB. Wrote the paper: MS CB JEP OAH

REFERENCES

1. Hunter DJ, Kraft P (2007) Drinking from the Fire Hose: Statistical Issues in Genomewide Association Studies. New England Journal of Medicine 357: 437–439. doi: 10.1056/nejmp078120

2. Langfelder P, Horvath S (2008) WGCNA: an R package for weighted correlation network analysis. BMC Bioinformatics 9: 559. doi: 10.1186/1471-2105-9-559

3. DiLeo MV, Strahan GD, den Bakker M, Hoekenga OA (2011) Weighted correlation network analysis (WGCNA) applied to the tomato fruit metabolome. PLoS ONE 6: e26683. doi: 10.1371/journal.pone.0026683

4. Hansey CN, Johnson JM, Sekhon RS, Kaeppler SM, de Leon N (2011) Genetic diversity of a maize association population with restricted phenology. Crop Science 51: 704–715. doi: 10.2135/cropsci2010.03.0178

5. Yan J, Shah T, Warburton ML, Buckler ES, McMullen MD, et al. (2009) Genetic characterization and linkage disequilibrium estimation of a global maize collection using SNP markers. PLoS ONE 4: e8451. doi: 10.1371/journal.pone.0008451

6.   Buckler ES, Holland JB, Bradbury PJ, Acharya CB, Brown PJ, et al. (2009) The genetic architecture of maize flowering time. Science 325: 714–718. doi: 10.1126/science.1174276

7.   Poland JA, Bradbury PJ, Buckler ES, Nelson RJ (2011) Genome-wide nested association mapping of quantitative resistance to northern leaf blight in maize. Proceedings of the National Academy of Sciences of the United States of America 108: 6893–6898. doi: 10.1073/pnas.1010894108

8.   Kump KL, Bradbury PJ, Wisser RJ, Buckler ES, Belcher AR, et al. (2011) Genome-wide association study of quantitative resistance to southern leaf blight in the maize nested association mapping population. Nature Genetics 43: 163–168. doi: 10.1038/ng.747

9.   R Development Team (2009) R: A Language and Environment for Statistical Computing. Vienna, Austria: R Foundation for Statistical Computing.

10. Buckler ES, Gaut BS, McMullen MD (2006) Molecular and functional diversity of maize. Current Opinion In Plant Biology 9: 172–176. doi: 10.1016/j.pbi.2006.01.013

11. Cook JP, McMullen MD, Holland JB, Tian F, Bradbury P, et al. (2012) Genetic architecture of maize kernel composition in the nested association mapping and inbred association panels. Plant Physiology 158: 824–834. doi: 10.1104/pp.111.185033

12. Chia JM, Song C, Bradbury PJ, Costich D, de Leon N, et al. (2012) Maize HapMap2 identifies extant variation from a genome in flux. Nature Genetics 44: 803–807. doi: 10.1038/ng.2313

13. Schnable PS, Ware D, Fulton RS, Stein JC, Wei F, et al. (2009) The B73 maize genome: complexity, diversity, and dynamics. Science 326: 1112–1115.

14. Riedelsheimer C, Lisec J, Czedik-Eysenberg A, Sulpice R, Flis A, et al. (2012) Genome-wide association mapping of leaf metabolic profiles for dissecting complex traits in maize. Proceedings of the National Academy of Sciences of the United States of America 109: 8872–8877. doi: 10.1073/pnas.1120813109

15. Riedelsheimer C, Czedik-Eysenberg A, Grieder C, Lisec J, Technow F, et al. (2012) Genomic and metabolic prediction of complex heterotic traits in hybrid maize. Nature Genetics 44: 217–220. doi: 10.1038/ng.1033

16. Cellini F, Chesson A, Colquhoun I, Constable A, Davies HV, et al. (2004) Unintended effects and their detection in genetically modified crops. Food and Chemical Toxicology 42: 1089–1125. doi: 10.1016/j.fct.2004.02.003

17. Hoekenga OA (2008) Using metabolomics to estimate unintended effects in transgenic crop plants: problems, promises, and opportunities. Journal of Biomolecular Techniques 19: 159–166.

18. Weckwerth W (2003) Metabolomics in systems biology. Annual Review Of Plant Biology 54: 669–689. doi: 10.1146/annurev.arplant.54.031902

19. Fernie AR, Schauer N (2009) Metabolomics-assisted breeding: a viable option for crop improvement? Trends in Genetics: TIG 25: 39–48. doi: 10.1016/j.tig.2008.10.010

20. Last RL, Jones AD, Shachar-Hill Y (2007) Towards the plant metabolome and beyond. Nature reviews Molecular Cell Biology 8: 167–174. doi: 10.1038/nrm2098

21. Dettmer K, Aronov PA, Hammock BD (2007) Mass spectrometry-based metabolomics. Mass spectrometry reviews 26: 51–78. doi: 10.1002/mas.20108

22. Yanes O, Tautenhahn R, Patti GJ, Siuzdak G (2011) Expanding coverage of the metabolome for global metabolite profiling. Analytical Chemistry 83: 2152–2161. doi: 10.1021/ac102981k

23. Dixon RA, Strack D (2003) Phytochemistry meets genome analysis, and beyond. Phytochemistry 62: 815–816. doi: 10.1016/s0031-9422(02)00712-4

24. Tolstikov VV, Fiehn O (2002) Analysis of highly polar compounds of plant origin: combination of hydrophilic interaction chromatography and electrospray ion trap mass spectrometry. Analytical Biochemistry 301: 298–307. doi: 10.1006/abio.2001.5513

25. Huhman DV, Sumner LW (2002) Metabolic profiling of saponins in Medicago sativa and Medicago truncatula using HPLC coupled to an electrospray ion-trap mass spectrometer. Phytochemistry 59: 347–360. doi: 10.1016/s0031-9422(01)00432-0

26. Lei Z, Huhman DV, Sumner LW (2011) Mass spectrometry strategies in metabolomics. The Journal of Biological Chemistry 286: 25435–25442. doi: 10.1074/jbc.r111.238691

27. Bolleddula J, Fitch W, Vareed SK, Nair MG (2012) Identification of metabolites in Withania sominfera fruits by liquid chromatography and high-resolution mass spectrometry. Rapid communications in mass spectrometry 26: 1277–1290. doi: 10.1002/rcm.6221

28. Fiehn O, Wohlgemuth G, Scholz M, Kind T, Lee do Y, et al. (2008) Quality control for plant metabolomics: reporting MSI-compliant studies. Plant Journal 53: 691–704. doi: 10.1111/j.1365-313x.2007.03387.x

29. Oliver MJ, Guo L, Alexander DC, Ryals JA, Wone BW, et al. (2011) A sister group contrast using untargeted global metabolomic analysis delineates the biochemical regulation underlying desiccation tolerance in Sporobolus stapfianus. Plant Cell 23: 1231–1248. doi: 10.1105/tpc.110.082800

30. Zhang N, Gibon Y, Gur A, Chen C, Lepak N, et al. (2010) Fine quantitative trait loci mapping of carbon and nitrogen metabolism enzyme activities and seedling biomass in the maize IBM mapping population. Plant Physiology 154: 1753–1765. doi: 10.1104/pp.110.165787

31. Fong C, Ko DC, Wasnick M, Radey M, Miller SI, et al. (2010) GWAS analyzer: integrating genotype, phenotype and public annotation data for genome-wide association study analysis. Bioinformatics 26: 560–564. doi: 10.1093/bioinformatics/btp714

32. Zeggini E, Scott LJ, Saxena R, Voight BF, Marchini JL, et al. (2008) Meta-analysis of genome-wide association data and large-scale replication identifies additional susceptibility loci for type 2 diabetes. Nature Genetics 40: 638–645. doi: 10.1038/ng.120

33. Gieger C, Geistlinger L, Altmaier E, Hrabe de Angelis M, Kronenberg F, et al. (2008) Genetics meets metabolomics: a genome-wide association study of metabolite profiles in human serum. PLoS Genetics 4: e1000282. doi: 10.1371/journal.pgen.1000282

34. McCarthy MI, Abecasis GR, Cardon LR, Goldstein DB, Little J, et al. (2008) Genome-wide association studies for complex traits: consensus, uncertainty and challenges. Nature Reviews Genetics 9: 356–369. doi: 10.1038/nrg2344

35. Inouye M, Ripatti S, Kettunen J, Lyytikainen LP, Oksala N, et al. (2012) Novel Loci for metabolic networks and multi-tissue expression studies reveal genes for atherosclerosis. PLoS Genetics 8: e1002907. doi: 10.1371/journal.pgen.1002907

36. Kettunen J, Tukiainen T, Sarin AP, Ortega-Alonso A, Tikkanen E, et al. (2012) Genome-wide association study identifies multiple loci influencing human serum metabolite levels. Nature Genetics 44: 269–276. doi: 10.1038/ng.1073

37. Broeckling CD, Heuberger AL, Prince JA, Ingelsson E, Prenni JE (2013) Assigning precursor-product ion relationships in indiscriminant MS/MS data from non-targeted metabolite profiling studies. Metabolomics 9: 33–43. doi: 10.1007/s11306-012-0426-4

38. Smith CA, Want EJ, O'Maille G, Abagyan R, Siuzdak G (2006) XCMS: processing mass spectrometry data for metabolite profiling using nonlinear peak alignment, matching, and identification. Analytical Chemistry 78: 779–787. doi: 10.1021/ac051437y

39. Horai H, Arita M, Kanaya S, Nihei Y, Ikeda T, et al. (2010) MassBank: a public repository for sharing mass spectral data for life sciences. Journal of mass spectrometry 45: 703–714. doi: 10.1002/jms.1777

40. Sana TR, Roark JC, Li X, Waddell K, Fischer SM (2008) Molecular formula and METLIN Personal Metabolite Database matching applied to the identification of compounds generated by LC/TOF-MS. Journal of biomolecular techniques 19: 258–266.

41. Kohl M, Wiese S, Warscheid B (2011) Cytoscape: software for visualization and analysis of biological networks. Methods in Molecular Biology 696: 291–303. doi: 10.1007/978-1-60761-987-1_18

42. Loiselle BA, Sork VL, Nason J, Graham C (1995) Spatial genetic-structure of a tropical understory shrub, *Psychotria officinalis* (Rubiaceae). American Journal of Botany 82: 1420–1425. doi: 10.2307/2445869

43. Pritchard JK, Stephens M, Rosenberg NA, Donnelly P (2000) Association mapping in structured populations. American Journal Of Human Genetics 67: 170–181. doi: 10.1086/302959

44. Lipka AE, Tian F, Wang Q, Peiffer J, Li M, et al.. (2012) GAPIT: Genome Association and Prediction Integrated Tool. Bioinformatics.

45. **45.**Patterson SD (2003) Data analysis–the Achilles heel of proteomics. Nature Biotechnology 21: 221–222. doi: 10.1038/nbt0303-221

46. Plaisier CL, Horvath S, Huertas-Vazquez A, Cruz-Bautista I, Herrera MF, et al. (2009) A systems genetics approach implicates USF1, FADS3, and other causal candidate genes for familial combined hyperlipidemia. PLoS Genetics 5: e1000642. doi: 10.1371/journal.pgen.1000642

47. Sabelli PA, Larkins BA (2009) The development of endosperm in grasses. Plant Physiology 149: 14–26. doi: 10.1104/pp.108.129437

48. Flint-Garcia SA, Thuillet AC, Yu J, Pressoir G, Romero SM, et al. (2005) Maize association population: a high-resolution platform for quantitative trait locus dissection. Plant Journal 44: 1054–1064. doi: 10.1111/j.1365-313x.2005.02591.x

Chapter 6

GENETIC GEOSTATISTICAL FRAMEWORK FOR SPATIAL ANALYSIS OF FINE-SCALE GENETIC HETEROGENEITY IN MODERN POPULATIONS: RESULTS FROM THE KORA STUDY

A. N. Diaz-Lacava[1,2,3] M. Walier[1] D. Holler[1] M. Steffens[1] C. Gieger[4,5] C. Furlanello[6] C. Lamina[7] H. E. Wichmann[8,9,10] and T. Becker[1,11]

[1]Institute for Medical Biometry, Informatics, and Epidemiology, University of Bonn, 53127 Bonn, Germany

[2]Cologne Center for Genomics, University of Cologne, 50931 Cologne, Germany

[3]DNA Analysis Unit, Official College of Pharmacists and Biochemists, C1184ABA Buenos Aires, Argentina

[4]Research Unit of Molecular Epidemiology, Helmholtz Zentrum München, German Research Center for Environmental Health, 85764 Neuherberg, Germany

[5]Institute of Epidemiology II, Helmholtz Zentrum München, German Research Center for Environmental Health, 85764 Neuherberg, Germany

[6]FBK, 38122 Trento, Italy

[7]Division of Genetic Epidemiology, Department of Medical Genetics, Molecular and Clinical Pharmacology, Medical University of Innsbruck, 6020 Innsbruck, Austria

[8]Institute of Medical Informatics, Biometry and Epidemiology, Chair of Epidemiology, Ludwig-Maximilians-University, 81377 Munich, Germany

[9]Institute of Epidemiology I, Helmholtz Zentrum München, German Research Center for Environmental Health, 85764 Neuherberg, Germany

[10]Institute of Medical Statistics and Epidemiology, Technical University Munich, 81675 Munich, Germany

[11]German Center for Neurodegenerative Diseases (DZNE), 53127 Bonn, Germany

ABSTRACT

Aiming to investigate fine-scale patterns of genetic heterogeneity in modern humans from a geographic perspective, a genetic geostatistical approach framed within a geographic information system is presented. A sample collected for prospective studies in a small area of southern Germany was analyzed. None indication of genetic heterogeneity was detected in previous analysis. Socio-demographic and genotypic data of German citizens were analyzed (212 SNPs; $n = 728$). Genetic heterogeneity was evaluated with observed heterozygosity (H_O). Best-fitting spatial autoregressive models were identified, using socio-demographic variables as covariates. Spatial analysis included surface interpolation and geostatistics of observed and predicted patterns. Prediction accuracy was quantified. Spatial autocorrelation was detected for both socio-demographic and genetic variables. Augsburg City and eastern suburban areas showed higher H_O values. The selected model gave best predictions in suburban areas. Fine-scale patterns of genetic heterogeneity were observed. In accordance to literature, more urbanized areas showed higher levels of admixture. This approach showed efficacy for detecting and analyzing subtle patterns of genetic heterogeneity within small areas. It is scalable in number of loci, even up to whole-genome analysis. It may be suggested that this approach may be applicable to investigate the underlying genetic history that is, at least partially, embedded in geographic data.

INTRODUCTION

Accurate assessment of genetic heterogeneity is relevant to manifold fields, ranging from clinical research, pharmacogenetics, and statistical genetics, over forensic sciences up to evolution (for a review, cf. [1]). In planning genetic epidemiological studies or the collection of control cohorts for prospective studies it is crucial to prevent confounding effects due to undetected or disregarded population structure [2, 3]. Population-based association studies of unrelated individuals, involving case-control and cohort studies, are prone to population structure, which may lead to false positive results or to failure to reveal genuine associations [4, 5]. In family-based linkage analysis unknown population stratification may lower statistical power [6].

Uncovering the genetic basis of complex traits remains an immense and urgent challenge in genetic epidemiological research. Great efforts are set to establish well-designed cohorts and large control samples, intended to serve as basis for genetic epidemiological studies. Besides restricting recruitment to individuals of uniform ancestry, a common strategy applied to efficiently gain a representative sample of the inspected population and to control for potential unknown population substructure is to collect samples in smaller geographical areas, usually in medium to large urban centers (e.g., [7–9]).

Even well-characterized or supposedly homogeneous regions may still account for subtle genetic structure with potential geographical components [3, 10]. Sloan et al. [3] shortly reviewed studies related to geographic genetic structure of human populations and pointed out clear lack of research focusing on genetic heterogeneity of smaller geographic regions or those focused on more urban, highly admixed populations.

Most available well-standardized methods in geographical genetics [11] were developed for other research areas and may not be suited for assessing subtle genetic heterogeneity of modern populations inhabiting geographically restricted areas. Modern humans account per se for the lowest species genetic diversity among primates [12]. A typical western population inhabiting a geographically restricted area sets additional difficulties. Such populations are typically outbred and account for a large degree of admixture, product of older and recent regional, interregional, and even international migration. It is to expect that genetic evolutionary forces, such as selection, mutation, drift, or barriers to gene flow, would play a relatively insignificant role in modeling fine-scale variation of genetic heterogeneity. At this geographical scale, it is more likely that neighborhood preferences and modern mating behavior would have a central role in modeling recent admixture, consequently, having strong influence on the observed pattern of genetic variation of modern small areas (i.e., [13]). In other words, within modern western circumscribed areas, socio-demographic factors would probably explain a large proportion of the observed pattern of genetic heterogeneity.

With the aim of unveiling modest amounts of population substructure in a small, admixed area we (a) searched for subtle patters of genetic heterogeneity and (b) explored potential predictors of the observed patterns. To this end, we combined statistical genetics with spatial statistics (geostatistics) within the framework of a Geographic Information System (GIS). A GIS provides a computational environment designed for spatial analysis of geographic data, therefore the most suitable framework to detect, to model, and to analyze the geographic variation of genetic diversity.

We analyzed a well-characterized cohort collected for prospective studies in a small area of southern Germany. The sampling area included the middle-size city of Augsburg, the surrounding suburban area, and the neighboring countryside. As previously reported by Steffens et al. [14], the KORA S4 sample shows a minimal but measurable increase of the inbreeding factor ($8.4E - 5$% heterozygotes deficit) measured in terms of F_{IS} values [15], that is, withingroup deviation from expected heterozygosity but no indication of population substructure. Despite extensive search of potential population stratification with the software package STRUCTURE [16], in this cohort no signals could be detected [14]. The STRUCTURE program implements a model-based clustering method. It estimates the proportion of individuals' genome that may originate from differential populations, the probability that an individual belongs to a certain population as well as allele frequency differences in terms of Wright's F_{ST} statistics [17]. Regardless of intensive computations under several models, STRUCTURE results did not provide any indication of a potential pattern of genetic heterogeneity in the KORA S4 survey [14].

MATERIAL AND METHODS

Subjects and Genotypes

Our analysis is based on a subset of the KORA cohort (Kooperative Gesundheitsforschung in der Region Augsburg; in English: Cooperative Health Research in the Region of Augsburg; [7, 9]). The KORA survey is an ongoing study, which takes place in a circumscribed region of southern Germany: Augsburg City and the two neighboring districts. The KORA cohort was recruited for prospective studies. In 4 surveys (S1–S4), a total of 18,000 participants were randomly selected from the adult population of resident German citizens (25–74 years) [7, 9]. Phenotypic, socio-economic information, and residence locality were gathered. The KORA cohort is a sample of the extant German population in the region.

The analysis was conducted on a random set of the KORA S4 survey (n = 4261), recruited in the period between the years 1999 and 2001 [7, 9]. The data set consisted of 728 unrelated healthy German citizens, which included subjects born within and outside of Germany as well. The graphical method GRR (Graphical Representation of Relationships; [18]) was used to exclude the presence of biological relationship of individuals based on genetic data (see Supplement 7). In this paper we distinguished the portion of Germans citizens born outside of Germany as "immigrants" and those born in Germany as "natives." The immigrant group (n = 179) included subjects born in twenty worldwide distributed countries, half of these countries represented only once. Four countries, Czech Republic, Romania, Poland, and Ukraine, corresponded to the land of birth of 82 percent of all immigrants. The group of subjects born in one of these four countries was classified as "major immigrant group." Based on the information "land of birth" we differentiated between data sets: (a) ALL: the complete data set of 728 subjects (resident German citizens),

(b) GER: the total set of 549 natives, and (c) MAIN IMG: the subset of 146 immigrants, born in either Czech Republic, Romania, Poland, or Ukraine

The KORA S4 sample was genotyped for 212 single nucleotide polymorphisms (SNPs) (Supplementary Table S6 in supplementary material available online at http://dx.doi.org/10.1155/2015/693193) [14]. These SNPs can be differentiated in two sets. The first set includes 68 coding SNPs located in exons of functional genes. These SNPs either cause an amino acid exchange or an effective promotor alteration in respect to the resulting protein. Assuming evolutionary times these SNPs may be subject to selective forces. The second set comprises 144 neutral SNPs. These loci were chosen at random throughout the genome in putative "genomic deserts," pursuing to achieve uniform distribution across the genome (setting a minimum of 500 Kbp intermarker distance). For this selection only SNPs presenting a minor allele frequency between 10 and 50% in Caucasians were considered. The markers included in the final intergenic set were uniformly spaced and located >100 Kbp apart from any known genes and >1 Mbp apart from centromeres and telomeres. This procedure followed the set of rules proposed by Devlin and Roeder for genomic control markers [19]. Accordingly, these intergenic SNPs are assumed to be neutral to selection forces in the absence of any specific information. In this sense, these loci are expected to reflect the effects of demographic processes involving migration (gene flow) and even drift, if evolutionary times are considered. Steffens et al. [14] undertook an extensive quality assessment to this data set. The averaged call rate over all samples was 97.3%; intragenic SNPs achieved an average call rate of 96.2% and intergenic SNPs, an average call rate of 97.9% [14]. Details of the genetic properties of the full set of 212 SNPs are listed in Supplementary Table S6.

Study Area

The study area comprised three administrative regions: the municipality of Augsburg City and its two neighboring districts, Aichach-Friedberg District and Augsburg District (Figure 1). It covered an area of approximately 2,970 km^2. This is a surface comparable with the Grand Duchy of Luxembourg (Figure 1). The area is located in the Swabia administrative region of Bavaria, southern Germany, between the coordinates 10.491°E/48.091°N and 11.310°E/48.642°N. The population had approximately 630,000 members in 2004. The mean population density is 212 inhabitants/km^2, a figure that is comparable with the German average.

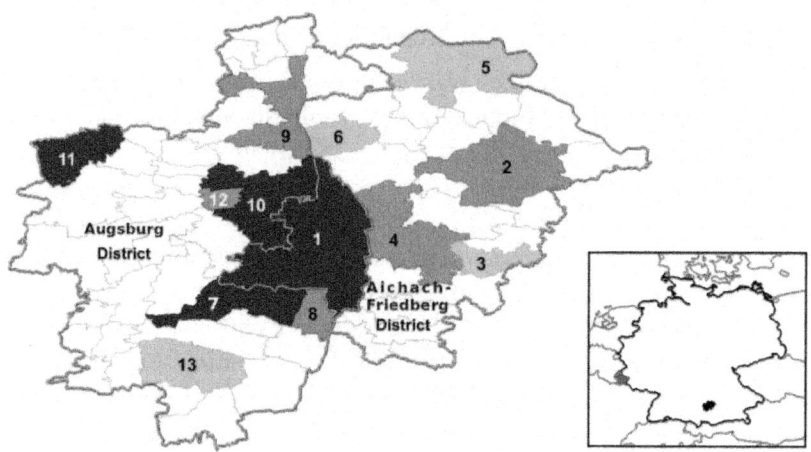

■ Sample count <20
■ Sample count <50
■ Sample count >50

Figure 1: Study area. Land units: (1) Augsburg; (2) Aichach; (3) Eurasburg; (4) Fried-berg; (5) Pottmes; (6) Rehling; (7) Bobin- ¨ gen; (8) Konigsbrunn; (9) Langweid (Me-itingen); (10) Neus ¨ aß¨ (Gersthofen; Stadtbergen); (11) Altenmunster; (12) Aystetten; (13) ¨ Schwabmunchen. Location of the study area within Germany is ¨ indicated in the inset in black; German boundaries are displayed with a black line, neighboring-country boundaries are displayed with a gray line; Luxembourg, a country covering an extension similar to the study area, located on the western boundary of Germany, is displayed in gray.

Augsburg City is a typical middle-size German urban area. The Aichach-Friedberg and Augsburg Districts include a suburban area neighboring Augsburg City and a periurban area, a patchy pattern of smaller cities and villages widespread across a rural landscape.

Regionalization Methods

The spatial analysis required diverse types of regionalization of the study area. Three regionalization methods were applied: (a) a subdivision of the total study area into minimal representative spatial units of analysis; (b) a subdivision of the total study area into contiguous sampled units using a polygon-based method; and (c) modification of the first regionalization in order to achieve a set of contiguous spatial analytical units while retaining the original geometry defined in (a).

Land Units

Genetic landscapes, in this work referred to matricial representations of genetic variation in the geographic space, were created with geostatistic methods of surface interpolation (see Section 2.7.1). For this objective, it is convenient to define a minimal spatial unit of analysis which is representative of data coverage and it covers a spatial surface much smaller that the phenomenon of interest.

The basic spatial unit of analysis of the genetic landscapes was the postal area. The postal area corresponds to the smallest district or region defined by the German postal system (the German postal system divides Germany in ca 28.700 postal areas). We considered this an appropriate analytical area because the German postal system divides the territory into spatial units with a similar number of inhabitants, independent of the extension of the spatial unit. Similar population size among land units allows adequate comparisons from socio-demographic perspective. Postal areas include as well a population size large enough in order to guarantee subjects' anonymity. A finer geographical reference of subjects, that is, postal address, was not available and it would not be in agreement with local official restrictions in respect to personal anonymity. We considered the subdivision of the study area into postal areas adequate to identify and to analyze fine-scale patterns of genetic variation. The study area included a total of 64 postal areas. The spatial extension of the postal areas ranged from $1.8 \, km^2$ to $93 \, km^2$, with an average of $26 \, km^2$.

The sampled area covered about 20% of the total study area, that is, approximately $600 \, km^2$ (Figure 1). It included Augsburg City and 15 settlements located in AichachFriedberg District and Augsburg District. Each sampled settlement corresponded to one postal area, except for Augsburg City. Augsburg City itself contains 14 postal areas. In summary, out of a total of 64 postal areas, data was available in Augsburg City (including 14 postal areas) and in another 15 postal areas. Augsburg City samples were pooled together for frequency computations, since no information about postal area of residence was available for residents in this city. A subdivision of Augsburg City into postal areas was only considered in the step of spatial interpolation to improve interpolation results (see Section 2.7.1). Postal areas with a very low number of samples were aggregated to neighboring sampled areas in order to exclude bias due to low number of samples per land unit. Explicitly, the quarters Stadtbergen ($n=7$) and Gersthofen ($n = 6$) were aggregated to Neusaß" ($n = 47$); Meitingen ($n = 12$) was aggregated to Langweid ($n = 22$) (Figure 1). In the final geostatistical analysis the sampled area included 13 analytical land units. In this work analytical land units (areal representing sampled data) defined on the basis of the geographical coverage of postal areas are further referred to as land unit (LU). LUs were labeled with the sampling-location name; aggregated LUs were labeled with the name of the location accounting for the largest number of samples. Augsburg City included the maximum number of samples ($n = 359$). The remaining 15 sampled postal areas (aggregated into 12 LUs) included a total of 369 samples. Letting aside Augsburg City, sample size per LU ranged between $n=9$ samples (Eurasburg) and $n = 60$ samples (Neusaß). The mean sample size per LU was 30.8 and ¨ the standard deviation was 17.4 (Table 1).

Table 1: Description of LUs and sampling locations, total count of samples, natives, immigrants, MAIN_IMG (subset of German citizens born in either Czech Republic, Romania, Poland, or Ukraine), and values of MAIN_IMP (percentage of individuals out of the total count of immigrants per land unit corresponding to the major immigrant group). Sampling locations with a low number of samples, which were aggregated to a contiguous sampled unit, are indicated in square brackets.

LU-ID	LU/sampling location	All (n)	Natives (n)	Immigrants (n)	MAIN_IMG (n)	MAIN_IMP (%)
1	Augsburg	359	258	101	79	78
2	Aichach	23	18	5	3	60
3	Eurasburg	9	9	0	0	0
4	Friedberg	25	21	4	2	50
5	Pöttmes	12	12	0	0	0
6	Rehling	13	12	1	1	100
7	Bobingen	51	36	15	13	87
8	Königsbrunn	42	24	18	17	94
9	Langweid [Meitingen]	34	22	12	9	75
10	Neusäß [Gersthofen] [Stadtbergen]	60	52	8	8	100
11	Altenmünster	53	48	5	5	100
12	Aystetten	31	23	8	7	88
13	Schwabmünchen	16	14	2	2	100

Polygon-Based Regionalization

The implementation of the spatial autocorrelation tests performed in this study (see Section 2.7.2) required to count with a set of adjacent spatial analytical units. This means that only spatial units with at least one contiguous neighbor could be included in the analysis.

The total study area was divided into 13 Thiessen polygons (designation given to Voronoi diagrams used to analyze spatially distributed data) [20]. Each polygon corresponded to one LU defined in the first regionalization (see Section 2.3.1). We chose this simple type of regionalization since many natural patterns may be closely approximated to this type of areal structure. Thiessen polygons were delimited based on the centroids (polygon geometrical center) of the 13 sampled land units (Figure 2(a)). The Voronoi tessellation was created with the method v.voronoi of the open-source software package GRASS 6.4 (Geographic Resources Analysis Support System, http://grass.osgeo.org/).

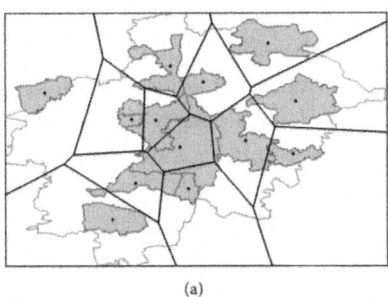 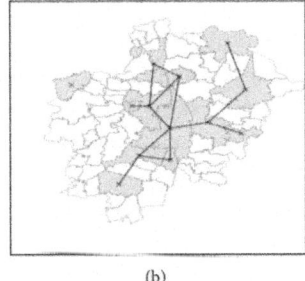

(a)   (b)

Figure 2: (a) Regionalization of the study area in 13 Thiessen polygons; each polygon represents a LU. LU centroids were used to delimit the Thiessen polygons; (b) net of contiguous LUs; three LUs were spatially connected to the next sampled LU performing a geometrical correction of LUs' boundaries: Schwabmunchen was connected to Bobingen, Aichach to Friedberg, and P ¨ ottmes to Aichach; vectors show pairs of LU ¨ assigned a spatial weight equal to unity in the matrix of spatial weights; pairs of LUs not connected with vectors received a value equal to zero in this matrix.

Net of Contiguous Sampled Units

The implementation of the algorithms used in this study to search for best predictors of spatial variation fitting the data (see Section 2.7.3) required as well contiguous analytical units. For such more complex analysis, the coverage of each LU was retained. In this case the first step was to verify the presence of direct neighbors for all LUs. Four LUs did not account for contiguous neighbors: Aichach, Pöttmes, Schwabmünchen, and Altenmünster (Figure 1). Of these, the first four LUs were not further than 5 km away from the next closest LU border. This geographical distance was considered negligible in the context of connectivity and human interaction between modern settlements. In order to get maximal information of the available data we modified slightly the geometry of these four LUs and of their closest neighbors in order to meet the contiguity condition for at least 12 LUs. Schwabmünchen was connected to Bobingen, Aichach to Friedberg, and Pöttmes to Aichach (Figure 2(b)). With this step, 12 LUs conformed a continuous geographical space. The most peripheral sampled land unit, Altenmünster, without a close sampled contiguous neighbors (distance to the closest LU > 10 km), was not included in the computations (Figure 2(b)).

Matrix of Spatial Weights

Both implementations of spatial dependence analysis performed in this study required the definition of a matrix of spatial weights representing the interaction between LUs (see Sections 2.7.2 and 2.7.3).

On the basis of the previously defined regionalization, either the Thiessen polygons (Figure 2(a)) or the geometrically modified LUs, for each analytical unit the geographical central point (the centroid), were specified. These two tests were performed in this study based on a binary representation of the spatial weight matrix, which assigned a weight of unity for neighbors, and zero otherwise. A binary encoding was chosen since not enough information was available to set assumptions about the assumed spatial process. The function poly2nb was used to construct the neighbor list with default parameters and the function nb2listw to construct the weight matrix, setting the function parameter style = B for a binary system. Pairs of LUs assigned a spatial weight equal to unity are indicated in Figure 2 with a vector net. All other pairwise combinations of LUs received a spatial weight equal to zero.

Socio-Demographic Parameters

Socio-demographic information collected during recruitment included age, education years, degree of professional training and education, and place of birth. Age ranged between 25 and 74 years. Education years ranged between 8 and 17 years old. Detailed descriptions of demographic features are provided in Supplementary Table S1.

As described in the Introduction the KORA S4 survey mirrors the case of plenty of study designs in human genetic research, in which control cohorts are used for population studies. In the context of these studies it could be crucial to assume genetic homogeneity of controls. One strategy is to collect samples in small areas and to restrict recruitment to individuals of same ancestry (see Section 1). In the concrete case of the KORA S4 it could be verified that the presence of immigrants introduced a small but significant effect on the total amount of genetic variation (see Supplement 4). In this paper we focus on the case of a population that may be considered a priori to be genetically homogeneous and may account for subtle genetic substructure and if this is the case, which factors may be regarded as best predictors. To perform our study in accordance with these objectives and assumptions we worked with the two data sets. We first considered the total sample (ALL). We used this set with the purpose of inspecting the effect of immigrants on the total genetic variation among other factors. On the other side we excluded immigrants and analyzed the subset of natives (GER). This analysis is intended to specify best predictors of sublet genetic substructure and to estimate their effect in an admixed modern small population. The total set of immigrants did not include enough individuals of similar ancestry to perform further separated geostatistical analysis. Therefore no further group with homogeneous ancestry could be identified.

Measures related to age, education years, and education level were computed only on the native GER data set. Variables related to immigrant representation in the total sample were computed for the total data set (ALL). All measures were computed per land unit.

(I)   Variables related to age, education years, and education level, computed only for the GER data set: AGE25_39: percentage of subjects in the age category of 25 to 39 years; AGE40_54: percentage of subjects in the age category of 40 to 54 years; AGE55_74: percentage of subjects in the age category of 55 to 74 years; AGE_MEAN: mean age; EY8_11: percentage of subjects achieving a maximum of 11 school years; this variable indicates the fraction of the sample which did not achieve the education level required to access to academic studies; EY_MEAN: mean years of school attendance; EDU_MEAN: mean education level, scored according to the degree of professional training and education, ranging from 0 = no school degree up to 9 = graduate degree (M.S. equivalent or higher).

(II)   In order to inspect the effect of immigrant representation on total population from a model building perspective, variables related to birth land were included; these variables were computed for the ALL data set.

The representation of the total immigrant fraction in relation to the total sample was modeled with the variable: GER_P: percentage of natives over all subjects.

(III)   on the same line and for purpose of ascertaining a potential effect related to the presence of major fraction of immigrants incoming from a reduced number of countries, which could be acting as a differentiated population within the migrant group, a further variable was included: MAIN_IMP: percentage of the major group of immigrants (subjects born in Czech Republic, Romania, Poland, or Ukraine) over all immigrants (German citizens born outside of Germany).

Measure of Genetic Diversity per Land Unit

We attempted to achieve a reduction in form of genetic landscapes of the complex georeferenced data available for this cohort (genotype per SNP and sample and geocoordinates of LUs). These genetic landscapes should allow visualization of the estimated distribution of genetic diversity across geographic space and further assessment of associations between spatial patterns of genetic diversity and other georeferenced data, for example, average socio-demographic characteristics per LU.

We chose to create maps based on indices calculated per LU. The reason for this is that we considered these types of genetic landscapes easier to interpret than those based on relative measures, for instance, the genetic differences between LUs. For this step it was necessary to select a genetic measure of diversity referred to each single LU, in opposition to relative measures such as Wright's F_{ST} [17] or alike, The average heterozygosity is a usual measure of the genetic variability of a group [21]. We chose this simple measure of genetic diversity, the observed heterozygosity (H_O) [22], to summarize sample genetic attributes of each LU. H_O, the observed frequency of heterozygotes averaged over loci, was estimated using

$$H_O = \frac{\sum_j^l h_j}{l},$$

(1)

where l is the number of loci and h_j indicates the proportion of heterozygote individuals per locus j_{th} [22]. Observed heterozygosity was computed separately for each land unit with the total data set ((ALL)) and with the subset of natives (H_O(GER)).

Multivariate Analysis of Spatial Population Structure

In an attempt to frame the challenge embedded in this sample, further genetic measures were computed with well-standardized tools for detecting population structure. First exploratory analysis with geostatistical methods indicated a potential differentiation of the periurban areas from Augsburg city and its periphery (see Supplement 2). Potentially, the fine-scale patterns of genetic diversity observed in these first exploratory evaluations could be explained by various simple models of spatial variation. For instance, the observed pattern (Supplementary Figure S2a-b) may be the result of a simple process of isolation by distance. As a result, genetic diversity would follow a pattern of gradual variation (e.g., gradients of allele or genotype frequencies). It must be noted that in such case, the observed pattern would correspond to a small fraction of the geographical landscape where the process occurs. This is so because both geographic extension of the study area and evolutionary times of the study population (here it refers to the number of generations necessary to fix the effects of any gene-flow process [23, 24]) are jointly, most probably, not large enough to have generated a local process of fixation of gradual variation of genetic features. Spatial correlation methods (e.g., spatial autocorrelation) and Mantel tests would be the first methods of choice to detect spatial correlations of genetic distance with geographic distance.

The observed pattern (Supplementary Figure S2a-b) could as well be the product of undetected population clustering. In this case, individuals of similar genetic features tend to reside in distinct areas than individuals less similar. Clustering would also require that individuals of distinct groups present reduced interaction with individuals of other groups. At larger geographical scales this could be observed when cultural, linguistic, or political limits set a barrier to gene flow. It is important to note that this scenario is less probable. For instance, it is improbable that a modern western population inhabiting such a small area would be composed of several groups with reduced exchange (low migration rates among the subareas and low predisposition to mate with individuals of other groups). This situation is even less probable if considered that spatial patterns of genetic variation were even detected within the group of natives. The result is supported by previous analysis undertaken with this sample: an exhaustive evaluation of population clustering was conducted by Steffens et al. [14] with the well-known software STRUCTURE [16]. Despite the large number of runs with varying models and parameters there was no indication of any population substructure. Results indicated that the model assuming a number of populations equal to unity (k=1) showed the highest posterior probability for the KORA S4 data.

The following software packages were used in this step: (a) GENELAND [20, 25]; (b) EIGENSTRAT EIGENSOFT [26, 27]; (c) PLINK, version v0.99s (http://pngu.mgh.harvard.edu/purcell/plink; [28]), with additional multidimensional scaling using the R software package, version 2.12.1 (R Foundation for Statistical Computing, 2010); (d) SPAGeDI [29]. The widespread used software STRUCTURE [16] was not considered, since in a previous study [14] no evidence for genetic substructure was found with this tool. Methods (a) to (c), as well as STRUCTURE, share the possibility to search for groups of genetically similar individuals. SPAGeDI (d) is a tool for detecting dependency between genetic and geographic distances among individuals or populations. GENELAND (b) and SPAGeDI (d) are individual-based methods and require including in the computation the geographic reference of each individual. As mentioned above (see Section 2.3.1), available data and anonymity restrictions did not allow a more precise georeference of subjects than sampling location. For these reasons, all individuals sampled in one location were georeferenced to the same geographical coordinates. This data aggregation consequently involves loss of power when applying these methods. Therefore, in our case and as it most probably would occur in this type of human genetic studies, the full capabilities of software making use of individual geographical coordinates could not be exploited.

With each tool (a–d), several exploratory runs were performed. In each case runs were started with default parameters and recommended model assumptions. Following, multiple runs with varied parameter values and model assumptions were conducted. For computations demanding a priori definition of an assumed number of subpopulations, runs were repeated for incremental number of subpopulations not larger than ten.

Geostatistical Analysis

Geostatistical analysis was conducted using the open-source software package GRASS 6.4 and spatial packages contributed to R software package, version. 2.12.1 (R Foundation for Statistical Computing, 2010) within the GRASS environment.

Generation of Genetic Landscapes

In this framework, genetic landscapes were defined as matrix representations of genetic variation in the geographic space. Spatial matrices were created by the transformation of sampling-point data to an elevation surface by spatial interpolation. An elevation surface is a 3D layer of continuous data (grid or raster layer) with elevation information at each point of the area. GRASS defines this type of spatial object as 2.5 dimensions (2.5D). As a simplification, the usual denomination for this type of spatial object: "3 dimensions (3D)" is adopted. The elevation parameter characterizes the estimated statistic. We decided to perform interpolation based on spline function. Interpolation based on splines proved to be a better choice for phenomena which combine a random component as well as processes which minimize energy, as it could be considered socio-demographic processes [30]. We chose the function "regularized spline with tension" implemented in the GRASS-method v. surf.rst [31]. This method computes the continuous 3D layer (raster data) simulating a thin flexible plate passing through or close to the measured data points; it is the most general and accurate method available in GRASS [30].

In order to run v.surf.rst, point-data layers are required. For each LU, we first specified its geometrical center (centroid) with a GRASS basic module. Statistic values were linked to the centroids. We obtained one point-data layer for each measured statistic. In case of Augsburg City, which contains 14 postal areas and covers a disproportionately large area, centroids of all postal areas were used. Computed values for Augsburg City data were assigned to all its 14 centroids. With this step, we smoothed spatial interpolation results in the area of Augsburg City and surroundings, while avoiding interpolation artifacts. We modeled genetic landscapes based on tuned values of v.surf.rst parameters. In order to be able to adequately execute the v.surf.rst procedure the H_0 values computed using the raw data ($H_{0\text{raw}}$) were transformed into percentage as follows:

$H_{O\text{raw}} \cdot 100 = H_O,$ (2)

where $H_{O\text{raw}} \in [0, 1]$, $H_O \in [0, 100]$.

Interpolation surfaces based on H_0 values were created for the following data sets: (ALL), and H_0(GER). Interestingly, since the KORA S4 genotypes conform a control population pool for genetic studies [7, 9], (ALL)landscape may be examined as a representative estimation of the spatial variation of genetic diversity of the extant population and H_0(GER) landscape of the native fraction in the region of Augsburg.

Spatial Autocorrelation

The presence of simple association between the variability of an attribute and the geographical space was tested by means of spatial autocorrelation. In this case, the null hypothesis is that the feature of interest is spatially distributed at random among other attributes within the study area. This analysis was based on the Moran's I tests. Spatial correlation measured with the test statistic Moran's I is inferential, which implies that results must be interpreted in dependence of the null hypothesis. For this analysis we used a Global Moran's I statistic, which means that we tested for spatial autocorrelation in the study area as a whole, assuming that the spatial process is the same everywhere.

Spatial autocorrelation of each of the genetic and sociodemographic variables defined in this study was tested with the R package spdep [32].

Moran's I tests were performed using the function implementations moran. test and moran.mc. Accounting for normality deviation of the data, moran. testwas run under the specification of randomization assumption in computing the variance of the statistic. This test specification allows relaxing the simpler normality assumption by introducing a correction term based on the kurtosis of the inspected variable.

The second implementation, moran.mc, is a permutation-based test. With this implementation spatial autocorrelation is evaluated independently of normality and randomization assumptions. The function moran.mc uses a Monte Carlo test, based on a permutation bootstrap. Observed values are randomly assigned to areal entities, and the value of the observed Moran's I is computed nsim times [33]. We set nsim = 10 000. These tests were run using a binary matrix of spatial weights (see Section 2.3.4). Both implementations, moran.test and moran.mc, were used to test for spatial autocorrelation in measures of genetic variation: (H_0(ALL), H_0(GER)), as well as on the socio-demographic variables: GER P, MAIN IMP, AGE25 39, AGE40 54, EA55 74, AGE MEAN, EY8 11, EY MEAN, and EDU MEAN.

Search of Best Predictors

Socio-demographic measures were inspected as predictors of the observed pattern of H_0(GER) under the assumption that socio-demography would provide useful indication of spatial arrangement of recent migration processes, specially regional and national migration, which we assumed that it must have had a strong influence on fine-scale genetic variation. The contribution of socio-demographic factors to explain the observed spatial pattern was analyzed under the assumption of spatial dependence. Best-fit spatial autoregressive models (SAR) predicting heterozygosity (H_m) were selected. A stepwise forward search was conducted using the function spautolm of the package R spdep [32].The function spautolm computes a regression on the values from the other areas to estimate the spatial dependence of the residuals of the specified linear predictor. The spatial dependence is estimated with a maximum likelihood test, computing a spatial autocorrelation parameter, λ. The p value of the likelihood ratio test compares the model with no spatial autocorrelation ($\lambda=0$) to the one which allows for it [33]. A binary matrix of spatial weights was used for this analysis (see Section 2.3.4).

Model selection was started with following parameters: GER P, AGE25 39, AGE40 54, EA55 74, AGE MEAN, EY8 11, EY MEAN, and EDU MEAN. In order to test if the spatial distribution of the major immigrant group (MAIN IMG) improves model prediction, the influence of the parameter MAIN IMP on the selected model was tested. A set of models best fitting the data were selected according to the p values of the covariates (p value < 0.05).

Evaluation of Model Accuracy

Finally, the goodness of fit of the selected SAR models was analyzed. In this step, the pixelwise divergence between predicted (H_m) and observed values (H_0) was quantified. Interpolation surfaces were created based on the predicted values (H_m) by each selected model. The pixelwise divergence in absolute values of these interpolation surfaces from the (GER) landscape was used to compare prediction accuracy among the selected SAR models. In order to facilitate comparison among models, a standardized difference was computed. The standardization was performed based on the maximal range of pixel values (max rg_{GER}) measured in the (GER) landscape. The parameter max rgGER was computed as follows:

$$\mathrm{max\_rg}_{GER} = z\mathrm{max\_int}_{GER} - z\mathrm{min\_int}_{GER},$$

(3)

where zmax int_{GER} is the maximal value measured in the H_0(GER) landscape and zmin intGER the minimal value. For each SAR model, a new elevation surface (raster layer) storing the respective pixelwise difference was created. Following pixelwise computation was performed with the GRASS basic module r.mapcalc:

$$\left(\frac{\text{abs}\left[H_O(\text{GER}) - H(m_n)\right]}{\text{max\_rg}_{\text{GER}}} \right) 100,$$

(4)

where abs implies absolute value, $H_0(\text{GER})$ refers to the pixel values of the $H_0(\text{GER})$ landscape, $H(m_n)$ refers to the pixel values of the interpolated surface created on the bases of the predicted H values of the mn SAR model, and the parameter max $\text{rg}_{\text{GER}} = 1.64$ (zmin int = 43.36; zmax int = 45.00).

For each one of these elevation surfaces, spatial global statistics were computed. For this purpose, elevation surfaces were imported into the spatial R environment provided by the packages sp, rgdal, spdep, and spgrass6 (R Foundation for Statistical Computing, 2010, http://www.r-project .org/ foundation/). Mean, standard deviation (sd), median, minimum (min), and maximal values (max) of the elevation surfaces were computed with the R function summary(). These statistics were applied as global quantitative measures of prediction goodness of each selected SAR model and were used to select the model best fitting the data.

The model with the lowest global difference between observed and predicted H values was selected as the one best fitting the data. Based on this model, maps representing the spatial variation of predicted H values and the estimated divergence between observed and predicted H values were created. The former map represents the estimated variation in heterozygosity according to predictions obtained by the SAR model best fitting the data. The latter maps allows a visual estimation of the agreement between observed and predicted heterozygosity in each land unit as well as the estimated spatial variation of divergence in the total study area.

RESULTS

Descriptive statistics of all measures, including mean, standard deviation, median, minimum value, and maximum value, are presented in Table 2.

Table 2: Descriptive statistics of genetic diversity and socio-demographic measures per LU (mean, standard deviation, median, minimum value, and maximum value)

Variable	Mean	SD	Median	Min	Max
H_O (ALL)	43.76	0.54	43.70	42.96	45.02
H_O (GER)	43.84	0.49	43.68	43.36	45.02
GER_P	81.5	13.2	84.0	57.0	100.0
MAIN_IMP	71.7	35.5	87.0	0.0	100.0
AGE25_39[1]	26.9	11.3	22.7	12.5	44.4
AGE40_54[1]	21.3	8.5	21.4	9.5	33.3
AGE55_74[1]	51.8	13.1	52.4	25.0	75.0
AGE_MEAN[1]	52.0	3.6	53.2	44.8	55.5
EY8_11[1]	66.0	10.6	65.2	50.0	83.3
EY_MEAN[1]	11.3	0.7	11.7	10.1	12.1
EDU_MEAN[1]	4.3	0.6	4.6	3.2	4.8

[1] Age and education-related variables refer only to the native group (GER).

Spatial Variation of Socio-Demographic Factors

Agerelated parameters showed a heterogeneous spatial distribution (Figure 3). Younger individuals (25 to 39 years old) comprised more than 30 percent of the GER sample in the eastern sector and reached a proportion of 45 percent in Pottmes ¨ (Figure 3(a)). Subjects corresponding to the intermediate age category (40 to 54 years old) showed a lower proportion (less than 30 percent) in LUs contiguous to Augsburg City in the South and in the East (Figure 3(b)). The upper age category (55 to 74 years old) accounted for more than 50 percent in eight of the 13 LUs, presenting the higher proportions (>70%) southern from Augsburg City (Figure 3(c)). The lowest mean age values were recorded in Pottmes, Rehling, and Neus ¨ aß¨ (Figure 3(d)). The percentage of individuals of the intermediate and the upper age category (AGE40 54, AGE55 74) showed a deviation of the expected Moran's *I* value with both tests (moran.test, moran. mc) on a significance level of $\alpha = 0.05$. The percentage of individuals of the lower age category (AGE25 39) showed a significant deviation of the expected Moran's *I* value with moran.test, but with the MC permutation bootstrap test did not reach a significance at $\alpha = 0.05$ (Table 3).The mean age (AGE MEAN) did not show any indication of spatial dependence with either of both tests (Table 3).

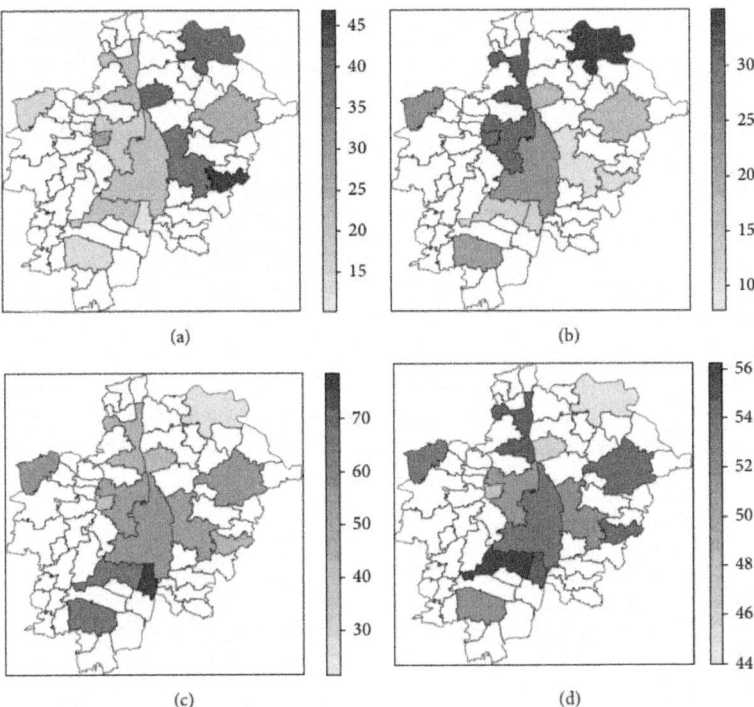

Figure 3: Values of age-related parameters; values refer only to German native subjects. (a) Percentage of subjects in the age category: 25–39 years (AGE25 39); (b) percentage of subjects in the age category: 40–54 years (AGE40 54); (c) percentage of subjects in the age category: 55–74 years (AGE55 74); (d) mean age per land unit (AGE MEAN).

With regard to the spatial distribution of education level in the study area, the largest values of education years, that is, lower values of EY8 11, were observed in Augsburg City and in neighboring LUs in the East (Friedberg) and in the West (Aystetten, Neusaß), as well as in the southern ¨ LU of Schwabmunchen (¨ Figure 4(a)). Both mean variables, means of education years and education level (EY MEAN, EDU MEAN), showed the largest values in the center and in the South of the study area, while the peripheral LUs Aichach and Altenmunster showed the lowest values (Figures ¨ 4(b) and 4(c)). Whereas education level (EDU MEAN) showed a significant deviation from the expected Moran's I value with both tests (moran.test, moran.mc) at $\alpha =$ 0.05, the education years (EY MEAN) presented just an indication of potential spatial dependency at this significance level (Table 3).

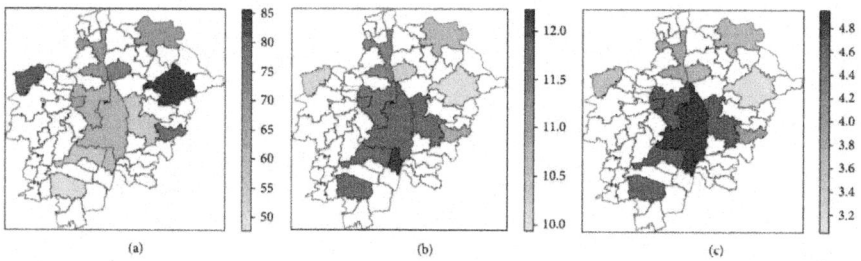

Figure 4: Values of education-related parameters; values refer only to German native subjects. (a) Percentage of subjects achieving a maximum of 11 school years, where the maximum in the sample is 17 school years (EY8 11); (b) mean years of school attendance per LU (EY MEAN); (c) mean score of the education level per LU, ranging from 0 = no school degree up to 9 = graduate degree (M.S. equivalent or higher).

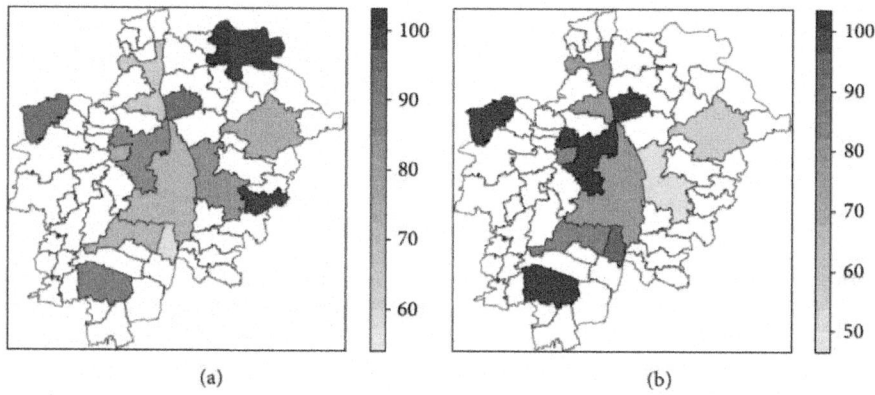

Figure 5: Spatial distribution of samples according to land of birth. (a) Spatial distribution of GER P; (b) spatial distribution of MAIN IMP.

The percentage of natives (GER P) showed a complex pattern (Figure 5(a)). Considering LUs with an intermediate number of samples, the percentage of natives decreased with an increase of the absolute number of samples per land unit with the exception of Altenmunster and Neus ¨ aß" (Table 1). In Augsburg City, with a quite larger number of samples in comparison with all other LUs (Table 1), natives composed ca. 70 percent of the total samples (Figure 5(a)). The lowest percentages were measured around Augsburg City, in Konigsbrunn, followed by Langweid. Bobingen, ¨ contiguous to Augsburg City on the South, and Aystetten, relatively peripheral to Augsburg City, presented the next lower frequencies of natives (Figure 5(a)). Samples included only natives in the eastern peripheral LUs, Pottmes and Eurasburg ¨ (Figure 5(a)), both accounting for the lowest sample counts as well (Table 1). The distribution of the proportion of natives (GER P) did not show any indication of a potential spatial dependency (Table 3).

Table 3: Estimated Moran's *I* values and *p* values of two Moran's *I* tests for spatial autocorrelation computed for all defined variables

Measure	Moran's I	p value[1]	p value (MC)[2]
H_O (ALL)	0.1132	*0.074*	*0.090*
H_O (GER)	0.2095	**0.015**	**0.024**
AGE25_39	0.1823	**0.039**	0.054
AGE40_54	0.3220	**0.004**	**0.010**
AGE55_74	0.2326	**0.013**	**0.020**
AGE_MEAN	−0.1092	0.571	0.550
EY8_11	0.0624	0.164	0.159
EY_MEAN	0.1425	*0.066*	*0.075*
EDU_MEAN	0.2509	**0.013**	**0.029**
GER_P	−0.1945	0.775	0.762
MAIN_IMP	0.1194	*0.072*	*0.091*

[1]Computed using the R package **spdep** moran.test based on a randomisation assumption.
[2]Computed using the R package **spdep** moran.mc, consisting of a Monte Carlo test, based on a permutation bootstrap test; *p* values were obtained on 10 000 runs.

The major group of immigrants (MAIN IMG) showed a higher ratio in the western LUs (Figure 5(b)). More than half of the units showed values of MAIN IMP larger than 80 percent. In Augsburg City, where a considerably larger total number of immigrants were sampled, the major immigrant group composed 78 percent of the total immigrant data set (Table 1). The results of both Moran's *I* tests of spatial autocorrelation (moran.test, moran.mc) pointed to a potential simple spatial dependency of the parameter MAIN IMP, which however did not reach a significance level of $\alpha = 0.05$ (Table 3).

Table 4: Best spatial autoregressive models fitting the data. The spatial autocorrelation left in the residuals (λ) and the p value of the likelihood ratio test, comparing the residuals of the fitted model with the one with no spatial autocorrelation (i.e., λ=0), are indicated for each model. In order to compare these four models, landscapes were created as the pixelwise difference between the observed and the predicted genetic landscape for each n model (H_O (GER) $-H$ (m_n)). Differences were computed in percentage to the maximal range of values of the H_O (GER) landscape. Mean, standard deviation (SD), and maximal values (Max) of the differences are indicated

	Model	λ	p value	Mean	SD	Max
m_1	40.768 + GER_P (0.014) + AGE40_54 (0.031) + EY8_11 (0.021)	0.203	0.449	17.0	14.9	70.6
m_2	48.903 + AGE25_39 (0.028) + EY8_11 (−0.036) + BILD_MN (−0.802)	−0.309	*0.018*	10.0	7.6	36.6
m_3	38.897 + GER_P (0.014) + EA55_74 (−0.013) + EY_MN (0.881) + BILD_MN (−1.293)	0.130	0.590	13.2	8.4	38.4
m_4	52.457 + EA55_74 (−0.031) + AGE_MN (0.090) + EY_MN (−0.811) + EY8_11 (−0.038)	−0.159	0.451	11.8	8.2	39.9

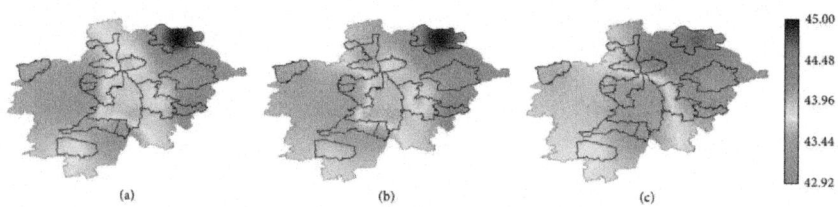

Figure 6: Landscapes estimating the genetic variation in the study area (a) H_O(ALL): observed heterozygosity of the total sample (officially registered German citizens); (b) H_O(GER) observed heterozygosity of the native data set (individuals born in Germany); (c) predicted heterozygosity according to the best-fit spatial autoregressive model (Table 4).

Geographic Variation of Genetic Diversity

The search for indications of spatial patterns of genetic heterogeneity with well-established procedures did not provide any positive results. Although a high number of different explorative runs with different parameters were performed, tests based either on EIGENSTRAT EIGENSOFT, PLINK, GENELAND, or SPAGeDI did not provide any indication of a potential geographic variation of genetic heterogeneity in the study area. A brief summary of a representative extract of these computations is presented in Supplement 5.

Geostatistical analysis based on the statistic observed heterozygosity (H_O) [22] provided indication of spatial patterning. Table 3 presents results of the spatial autocorrelation analysis performed with test statistic Moran's I. On the one side, the variable (ALL) showed an indication of association between genetic variation and geographic coordinates. On the other, in the native sample, tests of global spatial autocorrelation showed a significant deviation (on a significance level of $\alpha = 0.05$) of a random spatial distribution of (GER) values (Table 3). This result was obtained with both function implementations moran.test and moran.mc. Significant results obtained in the native sample may indicate that the additional genetic variability contributed by the immigrant fraction of the sample could introduce noise, which diluted a subtle patterning of the genetic attributes of the native sample.

(ALL) landscape, which estimates the variation of the genetic heterogeneity of the extant German population in the study area, presented a marked depression in the East (Figure 6(a)). The highest H_O values were measured in the eastern area. Intermediate H_O values covered the centralnorthern sectors, including some areas of Aichach-Friedberg District, Langweid, Augsburg City, and Schwabmunchen. ¨ The lowest H_O values were observed in the western area. The minimum values were found in Neusaß and K ¨ onigsbrunn.

Figure 6: Landscapes estimating the genetic variation in the study area (a) : observed heterozygosity of the total sample (officially registered German citizens); (b) observed heterozygosity of the native data set (individuals born in Germany); (c) predicted heterozygosity according to the best-fit spatial autoregressive model (Table4).

The (GER) landscape, estimating the spatial variation of genetic heterogeneity of the native population, showed similar values to the H_0(ALL) landscape in the western and in the eastern periphery. In the central belt, running across the study area in north-south direction, this landscape showed higher values than the (ALL) landscape (Figure 6(b)).

Four spatial autoregressive models were selected according to the p value (p value < 0.05) of the covariates (Table 4). The four models included as covariates variables related to age and education; two models (m_1, m_3) included as well the proportion of natives per LU (GER P), which is at the same time an indication of the proportion of immigrants per LU. The inclusion of the variable MAIN IMP, which involves a differentiation of subgroups of immigrants, did not improve any of the selected models. Out of these four selected models m_2 showed a significant p value when the likelihood ratio of λ was tested, indicating left spatial correlation in the residuals (p value = 0.018). The $H(m_2)$ landscape (Figure 6(c)) showed as well the lowest mean and standard deviation of pixelwise difference to the H_0(GER) landscape (Table 4), which we used as indicators of model goodness. The pixelwise difference between (GER) and $H(m_2)$ surfaces showed a good agreement over a large area as well. Areas where both surfaces showed very similar values are indicated in Figure 7 with a white or light grey and correspond to a pixelwise H difference close to zero. The maximal differences were measured in Augsburg City and in Rehling, a small residential area in the country side. Best fitting was obtained in the peripheral ring surrounding Augsburg City (Figure 7).

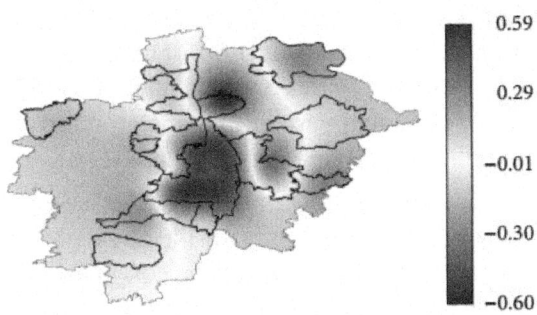

Figure 7: Difference between the interpolated surfaces of the observed and the predicted heterozygosity of the native subset ((GER) − $H(m_2)$)

DISCUSSION AND CONCLUSIONS

Fine-scale variation of genetic heterogeneity within a small region was detected and analyzed applying a geographic perspective. Population genetics and geostatistics were combined with the open-source geographic information system GRASS. The capabilities of this approach were tested on a subset of the KORA S4 survey [7,9], collected in southern Germany for prospective studies. Tests on this data set with the well-known software STRUCTURE previously reported by others [14] did not provide evidence for population substructure. We assumed that within small urbanized areas of modern western countries, as is the case in Germany, genetic composition may be strongly affected by migratory movements of the last half century, which may be still estimated by means of socio-demographic measures.

Genotypes (212 autosomal SNPs) and socio-demographic information (age, education, place of residence, and birth land) of 728 healthy German citizens were analyzed.

Socio-demographic and genetic measures showed heterogeneous distribution across the study area. The estimated values of the observed heterozygosity showed to some extent a cline of decreasing values from east to west. In a first step to analyze spatial processes controlling the observed patterns, Moran's tests for spatial autocorrelation were performed for all available parameters. Indications of global spatial dependencies were observed in socio-demographic variables related to age categories and education level. While significant deviation of the expected Moran's was obtained with the observed heterozygosity of the native subset (GER), the observed heterozygosity of the total data set showed just an indication of a potential deviation of the expected Moran's . In other words, these results could be interpreted as an indication that heterozygosity values of this small area may be regulated by a global spatial process, but such subtle process could only be observed when the subset of native Germans is considered. This is consistent with the elemental assumption that because immigrants may differ to some extent in their genetic background, and this would be reflected in their observed heterozygosity, they would introduce additional variability which hinders detection of spatial variability pattern of the most frequent group, the German natives.

Spatial dependencies of socio-demographic variables at this local level could be axiomatically interpreted as a result of neighborhood preferences of recent local and regional migratory processes. It is to expect that social preferences affecting recent migration, such as regional and international migratory movements of the last half century, may be reflected by socio-demographic parameters. Individuals would not choose new residence randomly and this may be reflected in similar socio-demographic attributes of contiguous neighborhoods. This expectation, which goes along with common sense, would also be in agreement with the fact that the level of admixture would associate to some extension with socio-demographic attributes of the location. This would be the result of new and recurrent resettlement of some areas or that inhabitants of areas presenting certain socio-demographic features would show higher predisposition to admixture. Indeed, the fact that patterns of socio-demographic structure and genetic admixture account for spatial autocorrelation could indicate the occurrence of further unobserved phenomena influencing both, as it could be "social preferences" or "official urban planning." Nevertheless, it went beyond the scope of this work to search for such unobserved processes.

Spatial autoregressive models (SAR) fitting the data were selected by a forward search. The four best-fit SAR models contained as covariates socio-demographic measures related to age, school-attendance years, and education level. Two of the four selected models included as well the percentage of natives per unit. This parameter directly accounts for degree of demographic admixture in sense of proportion of immigrants as measured by the parameter "land of birth." The prediction strength of the selected model was estimated with quantitative comparisons between the genetic landscapes created on basis of observed and predicted measures of heterozygosity. We found a good agreement between the predicted and the observed patterns, which supports the assumption of a certain relationship between genetic admixture degree and socio-demographic structure. Predicted and observed values showed the highest agreement in the surrounding belt of Augsburg City. The largest deviations were measured in Augsburg City itself and in two small residential areas in the eastern countryside. The differentiation of Augsburg City to the countryside (see Supplement 3) may be interpreted as the expected divergence of middle-size industrial settlements and countryside [34]. This effect could still be detected even after excluding immigrants, subjects born outside of Germany. This expected differentiation between urban and countryside areas in relation to genetic heterozygosity may be considered as a confirmation that the observed patterns are not artifacts.

The differentiated area in the countryside, Rehling, corresponded to a land unit with relatively low number of samples. The lower predictive capability of the best-fit model could be attributed to sampling bias. A replication would be necessary to verify this conjecture. A further possibility could be that this settlement offers an attractive residential area for individuals working in any of the larger urban centers located eastern from Augsburg, as it could be Ingolstadt or Munich. Both urban centers offer highly profitable working alternatives, in sense of carrier opportunities and higher income, and act as an attraction pole for domestic and foreign migration. As well, both cities are also among the ones with the most expensive living costs. Rehling showed also the largest proportion of younger adults. This may reinforce the idea that this location may be attractive for newly settled workers willing to commute between their working place and a less expensive residential area, relatively close to a middle-size urban center as Augsburg. If this would be the case, the inclusion of distance and accessibility to attractive urban centers could considerably improve model prediction. Further studies could test this possibility.

The exclusion of immigrants increased the global mean of observed heterozygosity. This effect was especially strong in areas accounting for the largest proportion of immigrants. About 15 percent of the samples were immigrants, born worldwide. Remarkably, almost 80 percent of them corresponded to individuals born either in Czech Republic, Romania, Poland, or Ukraine. Some areas situated in the periphery of Augsburg City showed a stronger component of immigrants, mostly or totally represented by this major immigrant group. Although this group involved four birth lands, it may be speculated that the concentration of these individuals is not casual. It could be expected that, within each provenance group, individuals could be originals of nearby regions or belonging to large related families. In total, a concentration of small groups, each one showing a higher degree of homozygosity, would stand out over a much more admixed group, as it is expected for the native German population.

Our additional analysis with vastly cited multivariate methods, GENELAND, EIGENSTRAT EIGENSOFT, PLINK, and SPAGeDI, did not also submit indication of population stratification in the study area. This outcome is consistent with our expectations. These tools proved to be successful in studies of groups with considerably larger genetic differences, significantly more polymorphic loci, or much larger number of loci than in the present work. The KORA study was carefully designed for prospective studies aiming to reduce any type of genetic structuring. Following this objective, only German citizens were included in the sample. The sampling area was kept very circumscribed as well. Consequently, the genetic differentiation of this subset of the KORA cohort is expected to be considerably lower than in humans studies conducted at broader geographical scales or in further studies acknowledging larger evolutionary histories and dimensions (i.e., nonhuman species or samples with large population differences). A further aspect to be considered is the population informativeness of the available SNPs. These loci [14] were not specifically selected by their informativeness for distinguishing among major regional groups (i.e., [35, 36]). For this reason, the number of available SNPs was probably too low for detecting fine-scale population differences with these standard tools.

Using full capabilities of tools such as SPAGeDI or GENELAND, both offering individual-based analysis tools, was not possible either since the search of patterns of variation of genetic heterogeneity of the KORA sample cannot be carried out on an individualized geographical basis.

Data of human genetic studies would most probably not include an individualized geographical reference. Official restrictions concerning personal anonymity forbid the use of data which could individualize a subject, such as postal address or any other precise geographical reference. Therefore, for human studies, precision of geographic references must be kept low with the consequence that data might be spatially aggregated. In opposite to studies analyzing other species with a continuous geographical distribution, the identification of spatial structures of humans inhabiting small areas may not be carried out on an individual basis. Such an approach would jeopardize personal anonymity and would go against most official restrictions of human studies. Accordingly, it is not surprising that well-established tools to detect human population stratification on a broader level or those which make use of geographic references on an individual basis (mostly developed for studies of other species) may not detect fine-scale patterns of genetic variation in small areas.

Most genetic studies on modern human groups address recruitment of control samples on extant populations or make use of available control cohorts. These samples may present some degree of heterogeneity even if recruitment was restricted to small areas, by citizenship, or in combination with homogeneous place of birth. The degree of spatial heterogeneity of small geographic areas, frequently assumed to be neglectful in the context of genetic studies of modern human groups, should be evaluated on a case-by-case basis. Based on our outcomes, it could be stated that genetic heterogeneity could not be automatically assumed to be negligible. These results support the elemental assumption that within multiethnic, urban, and suburban groups, as found in medium-sized German cities and surroundings, the socio-economic parameter "birth land" allows a first reduction of genetic heterogeneity.

As it was presented in this study (see Supplement 2), even after removing immigrants from the KORA S4 data set, the degree of genetic differentiation in natives still overlapped with the spatial frequency distribution of immigrants. If future studies verify our exploratory results obtained with visual examination of genetic landscapes based on analogous Reynolds' F_{ST} genetic distance,R potentially, the proportion of immigrants may be used as subrogate of degree of natives' admixture, which could actually reflect native's behavior in regard to choice of area of residence and tolerance or predisposition to admix.

Our results about a differentiation among urban, suburban, and periurban population are a suggestion of a true effect in the sense of subtle population differentiation (cf. [34]). As mentioned above in this section, it is to assume that such differentiation would not result from isolation between neighboring areas in the sense of evolutionary processes, but from differences in regard to the degree of migration, residential preferences and willingness to intermix.

However, it must be stressed that our analysis is preliminary and it is predominantly aimed at a methodological evaluation. In particular, the coverage of the area is patchy and far from complete. In order to make use of these results for further genetic studies, first, the postulated fine-scale variation of genetic heterogeneity should be confirmed with a larger data set. Second, the magnitude of the detected bias for the corresponding analysis should be evaluated. A comprehensive analysis on an augmented data set is in preparation.

Knowledge of fine-scale patterns of genetic variation could provide information about areas where expected genetic heterogeneity could introduce undesired bias. Areas with an observed higher genetic heterogeneity than tolerable could be avoided. In case that spatial heterogeneity would be assessed after recruitment, examining the spatial pattern of genetic heterogeneity could serve as a basis to decide about a stratified analysis (e.g., grouping samples according to residence or any other relevant spatial reference) or to correct for population stratification (cf. [26, 27]). Our vision is to further develop our approach in order to be capable of testing as well as detecting and correcting, if it is applicable, for spatial patterns of genetic heterogeneity within the study sample (cf. [26, 27]). In contrast to the method implemented in EIGENSOFT EIGENSTRAT [26,27], which infers strata based on genetic data alone, such approach would make use of information on subject area membership to define the strata. This usage of additional a priori information potentially leads to improve strata definition (cf. [37]).

Taking these findings of the KORA S4 sample altogether we can state that fine-scale spatial genetic variation may be assumed in the study area. Our results indicate that patterns of genetic heterogeneity can be present in small regions within Germany. In conclusion, it may be stated that the presented genetic geostatistical approach has the potential of being a powerful tool for detecting, modeling, and analyzing spatial patterns of genetic heterogeneity even within populations inhabiting small regions.

ACKNOWLEDGMENTS

The authors wish to thank Professor Dr. Thomas F. Wienker for helpful comments and criticisms. The study was supported by the German BMBF/GEM and national genotyping platform of the NGFN.

REFERENCES

1. M. J. Bamshad, S. Wooding, W. S. Watkins, C. T. Ostler, M. A. Batzer, and L. B. Jorde, "Human population genetic structure and inference of group membership," American Journal of Human Genetics, vol. 72, no. 3, pp. 578–589, 2003.

2. L. R. Cardon and L. J. Palmer, "Population stratification and spurious allelic association," The Lancet, vol. 361, no. 9357, pp. 598–604, 2003.

3. C. D. Sloan, E. J. Duell, X. Shi et al., "Ecogeographic genetic epidemiology," Genetic Epidemiology, vol. 33, no. 4, pp. 281–289, 2009.

4. M. L. Freedman, D. Reich, K. L. Penney et al., "Assessing the impact of population stratification on genetic association studies," Nature Genetics, vol. 36, no. 4, pp. 388–393, 2004.

5. J. Marchini, L. R. Cardon, M. S. Phillips, and P. Donnelly, "The effects of human population structure on large genetic association studies," Nature Genetics, vol. 36, no. 5, pp. 512–517, 2004.

6. M. D. Shriver, G. C. Kennedy, E. J. Parra et al., "The genomic distribution of population substructure in four populations using 8,525 autosomal SNPs," Human genomics, vol. 1, no. 4, pp. 274–286, 2004.

7. R. Holle, M. Happich, H. Löwel, and H. E. Wichmann, "KORA—a research platform for population based health research," Gesundheitswesen, vol. 67, supplement 1, pp. S19–S25, 2005.

8. Lindenberg, J. Brinkmeyer, N. Dahmen et al., "The German multi-centre study on smoking-related behavior-description of a population-based case-control study," Addiction Biology, vol. 16, no. 4, pp. 638–653, 2011.

9. H.-E. Wichmann, C. Gieger, and T. Illig, "KORA-gen—resource for population genetics, controls and a broad spectrum of disease phenotypes," Gesundheitswesen, vol. 67, supplement 1, pp. S26–S30, 2005.

10. Helgason, B. Yngvadòttir, B. Hrafnkelsson, J. Gulcher, and K. Stefánsson, "An Icelandic example of the impact of population structure on association studies," Nature Genetics, vol. 37, no. 1, pp. 90–95, 2005.

11. K. Epperson, Geographical Genetics, Princeton University Press, Princeton, NJ, USA, 2003.

12. H. Kaessmann, V. Wiebe, G. Weiss, and S. Pääbo, "Great ape DNA sequences reveal a reduced diversity and an expansion in humans," Nature Genetics, vol. 27, no. 2, pp. 155–156, 2001.

13. K. N. Laland, J. Odling-Smee, and S. Myles, "How culture shaped the human genome: bringing genetics and the human sciences together," Nature Reviews Genetics, vol. 11, no. 2, pp. 137–148, 2010.

14. M. Steffens, C. Lamina, T. Illig et al., "SNP-based analysis of genetic substructure in the German population," Human Heredity, vol. 62, no. 1, pp. 20–29, 2006.

15. S. Wright, Evolution and the Genetics of Populations, Vol. 2: Theory of Gene Frequencies, University of Chicago Press, Chicago, Ill, USA, 1969.

16. J. K. Pritchard, M. Stephens, and P. Donnelly, "Inference of population structure using multilocus genotype data," Genetics, vol. 155, no. 2, pp. 945–959, 2000.

17. S. Wright, "The genetical structure of populations," Annals of Eugenics, vol. 15, pp. 323–354, 1951.

18. G. R. Abecasis, S. S. Cherny, W. O. C. Cookson, and L. R. Cardon, "GRR: graphical representation of relationship errors," Bioinformatics, vol. 17, no. 8, pp. 742–743, 2001.

19. B. Devlin and K. Roeder, "Genomic control for association studies," Biometrics, vol. 55, no. 4, pp. 997–1004, 1999.

20. G. Guillot, A. Estoup, F. Mortier, and J. F. Cosson, "A spatial statistical model for landscape genetics,"Genetics, vol. 170, no. 3, pp. 1261–1280, 2005.

21. M. Nei and A. K. Roychoudhury, "Sampling variances of heterozygosity and genetic distance," Genetics, vol. 76, no. 2, pp. 379–390, 1974.

22. M.Nei,"Analysisofgenediversityinsubdividedpopulations,"Proceedings of the National Academy of Sciences, vol. 70, no. 12, pp. 3321–3323, 1973.

23. S. Wright, "Isolation by distance," Genetics, vol. 28, pp. 114–138, 1943.

24. S. Wright, "Isolation by distance under diverse systems of mating," Genetics, vol. 31, pp. 39–59, 1946.

25. G. Guillot, F. Santos, and A. Estoup, "Analysing georeferenced population genetics data with Geneland: a new algorithm to deal with null alleles and a friendly graphical user interface," Bioinformatics, vol. 24, no. 11, pp. 1406–1407, 2008.

26. N. Patterson, A. L. Price, and D. Reich, "Population structure and eigenanalysis," PLoS Genetics, vol. 2, p. e190, 2006.

27. L. Price, N. J. Patterson, R. M. Plenge, M. E. Weinblatt, N. A. Shadick, and D. Reich, "Principal components analysis corrects for stratification

in genome-wide association studies," Nature Genetics, vol. 38, no. 8, pp. 904–909, 2006.

28. S. Purcell, B. Neale, K. Todd-Brown et al., "PLINK: a tool set for whole-genome association and population-based linkage analyses," American Journal of Human Genetics, vol. 81, no. 3, pp. 559–575, 2007.

29. O. J. Hardy and X. Vekemans, "SPAGeDi: a versatile computer program to analyse spatial genetic structure at the individual or population levels," Molecular Ecology Notes, vol. 2, no. 4, pp. 618–620, 2002.

30. M. Neteler and H. Mitasova, Open Source GIS: A GRASS GIS Approach, Kluwer Academic, Boston, Mass, USA, 2004.

31. H. Mitášová and L. Mitáš, "Interpolation by regularized spline with tension: I. Theory and implementation," Mathematical Geology, vol. 25, no. 6, pp. 641–655, 1993.

32. R. Bivand, spdep: Spatial Dependence: Weighting Schemes, Statistics and Models, R package version 0.5-21, 2010, http://cran.r-project.org/web/packages/spdep/index.html.

33. R. S. Bivand, E. J. Pebesma, and V. Gomez-Rubio, Applied Spatial Data Analysis with R, Springer, New York, NY, USA, 2008.

34. V. Vitart, A. D. Carothers, C. Hayward et al., "Increased level of linkage disequilibrium in rural compared with urban communities: a factor to consider in association-study design," American Journal of Human Genetics, vol. 76, no. 5, pp. 763–772, 2005.

35. N. Liu, L. Chen, S. Wang, C. Oh, and H. Zhao, "Comparison of single-nucleotide polymorphisms and microsatellites in inference of population structure," BMC Genetics, vol. 6, supplement 1, article S26, 2005.

36. N. A. Rosenberg, L. M. Li, R. Ward, and J. K. Pritchard, "Informativeness of genetic markers for inference of ancestry," American Journal of Human Genetics, vol. 73, no. 6, pp. 1402–1422, 2003.

37. S. Manel, F. Berthoud, E. Bellemain et al., "A new individual-based spatial approach for identifying genetic discontinuities in natural populations," Molecular Ecology, vol. 16, no. 10, pp. 2031–2043, 2007.

Chapter 7

EXPLORING STATISTICAL TOOLS IN MEASURING GENETIC DIVERSITY FOR CROP IMPROVEMENT

C. O. Aremu

Department of Crop Production and Soil Science Ladoke Akintola University of Technology, Ogbomoso, Oyo state Landmark University, Omu-Aran Nigeria

INTRODUCTION

Increase in global numerical population especially in developing nations has gradually led to food shortage and hence increase in poverty. Addressing and tackling the issue and causes of poverty in the developing nations is one major challenge to breeders (Fu and Somers 2009). The different theories of econometircs have identified the human and material resources traceable to poverty, but fail to identify the crop improvement techniques in addressing world food shortage (Baudoin and Mergeai 2001). Crop improvement techniques therefore remains a major concern to plant breeders (Akbar and Kamran, 2006; Aremu et al, 2007a). Several factors affect crop improvement for specific or general environment performance. Such factors include climate, weather, soil, edaphic and biological and more importantly crop genotype (Aremu, et al, 2007b). Crop genotypes are composed of different crop forms including inbred or pure lines hybrids, landraces, wildraces germplasm accessions, cultivars or varieties. These crop genotypes have wide and diverse origin and genetic background known as genetic diversity. Genetic diversity study is a major breakthrough in understanding intraspecie crop performance leading to crop improvement (Aremu, 2005). Knowledge of crop performance in genetic diverse population reveals the differences in the nature of genetic materials used. Genetic diversity studies therefore, is a step wise process through which existing variations in the nature of individual or group of individual crop genotypes are identified using specific statistical method or combination of methods (Christini et al. 2009; Warburton and Crossa 2000; Aremu, 2005; Weir 1996). It is expected that the identified variations would form a pattern of genetic relationship useable in grouping genotypes. Several researchers including breeders have employed different data source and type from diverse crops in their methods to study genetic diversity. Such data source include morphological and agronomic, pedigree, proximate or biochemical and molecular data (Aremu, et al., 2007a in cowpea; Liu et al., 2000 in cotton; Mostafa et al., 2011 in wheat; Adewale et al., 2010 in African Yam bean; Christine et al., 2009 in bentgrass. The choice of statistical method to be used is dependent on the achievable objectives laid out in the studies. This chapter reveals the underlying importance of genetic diversity and reviews useable statistical techniques for identifying and grouping genotypes for intraspecies crop improvement.

NEED FOR GERMPLASM RESOURCE IN GENETIC DIVERSITY PRESERVATION

Crop genotypes sourced as germplasm accessions, landraces, breeding lines, wild species, have rich and variable genetic integrity explorable for breeding programmes. The first step of any meaningful breeding programme is to identify crop plants that exhibit exploitable variation for the trait(s) of interest. However, these genetic diverse crops are under threat. Continuous hybridization and crossing systems have reduced the genetic variations in cropping programmes and leave a dearth in harvesting and utilization of novel crop types with exploitable traits. Also, the continuous threat or loss of genetic diversity as a result of replacement of landraces, wild species and other primitive term of crop species by exotic high- yielding varieties remains an insurmountable problem to plant breeders. Another major source of loss of genetic diversity is by changes and or increase in population size, resulting in land use acts promoting deforestation, wars, industrialization, urbanization and other factors. According to Brown (1989), preservation of genetic diversity is possible when genetic or germplasm resource is realized as the most precious asset in conserving genetic diversity. Germplasm therefore is an essential resource for successful plant breeding. Certain areas of the world exhibit high level of genetic variability for crops (Vavilov, 1950). Falconer and Mackay (1996); Eivazi et al; (2007); reported that such areas are considered as regions or center of genetic diversity. Therefore genetic diversity in crop may be associated with the origin of the crop. This is supported by Christine et al. (2009), who reported genetic diversity to be associated with origin. Potter and Doyle, (1992) reported Tropical Africa to be the centre of diversity for African yam bean. Van Bueningen and Busch (1997), reported genetic diversity of wheat to be centered in North America. Ariyo and Odulaja (1991), found correlation between genetic diversity and eco-geographic background in okro. Some grouping methods in genetic diversity studies identified origin and geographical diversity not important in measuring genetic diversity. Nair et al. (1998) discovered diversity in sugarcane not to be associated with origin. Aremu et al. (2007a), discovered that center of origin is not a measure of genetic diversity in cowpea. If crop origin is somewhat not important in the measure of genetic diversity a resource centre is therefore needed to preserve and maintain the wide genetic sources exploitable in breeding programmes. Genetic relationship and diversity are useful for developing germplasm conservation strategies and utilization of crop genetic resources. The use of genetic diversity resource centre cannot be under estimated as earlier discussed.

IMPORTANCE OF GENETIC DIVERSITY STUDIES

Study on genetic diversity is critical to success in plant breeding. It provides information about the quantum of genetic divergence and serves a platform for specific breeding objectives (Thompson et al, 1998). It identifies parental combinations exploitable to create segregating progenies with maximum genetic potential for further selection, as proven by Akoroda (1987), Weir, (1996), Liu et al.(2000); Dje et al.(2000), (Aremu et al, 2007b). Genetic diversity exposes the genetic variability in diverse populations and provides justification for introgression and ideotype breeding programmes to enhance crop performance. Mostafa al et. (2011), postulated that genetic diversity studies provides the understanding of genetic relationships among populations and hence directs assigning lines to specific heterogeneous groups useable in identification of parents and hence choice selection for hybridization. Choice of parent has been identified to be the first basic step in meaningful breeding programme (Akoroda 1987); (Aremu et al. 2007a); (Islam 2004), (Rahim et al, 2010). Furthermore, the choice of parent selection in diversity studies is valuable because it is a means of creating useful variations in subsequent progenies.); Dje et al. (2000), discovered that the higher the genetic distance between parents, the higher the heterosis in the developed progenies. Hence the heterotic progenies can be further hybridized and selections based on transgressive segregation. Akbar and Kamran, (2006). exploited this parental selection technique in wheat breeding program through hybridization. Mostafa et al. (2011), investigated genetic distance among 36 winter wheat genotypes cultivated in different regions of Iran using principal component analysis and discovered five major groups in the genotypes to distantly related. Comprehensive and significant emphasis are made by researchers especially plant breeders on the analysis of genetic diversity in a number of field crops white and yellow yam, (Akoroda, 1987); cowpea, (Adewale and Aremu, 2010); African yam bean, (Baudoin and Mergeai 2001); Flax, (Mohammadi et al. 2010); wheat, (Mostafa et al. 2011) and several other crops. The diversity studies on these crops at their respective primitive levels (Landrace, wildtype, accessions, lines etc) led to the development of their widely distributed cultivars and varieties with proven characteristics based on stability and adaptability of performance with consistent tolerance to adverse weather conditions and resistant to diseases around the world. Fu and Somers (2009) supported that the use of identified wheat parents resistant to environmental stress under different growing conditions has led to increased world wheat production. The early report of Mohammadi and Prasna (2003) revealed that appropriate parent selection for hybridization in maize using a definite diversity study technique, Bohn et al (1999), identified six groups of wheat land races in the Western Iran that can be grown in different geographical locations for improved yield. Martin et al., (2008) discovered 42 cultivars of bentgrass in the mancet city and that only diversity studies would identify reliable and definite cultivar(s) with varietal purity and ensure protection of breeder and consumer rights. Understanding the inter and intra specie genetic relationships as provided by diversity studies has proven to increases hybrid vigor and reduce or avoid re-selection within existing germsplasm. It is worthy of note that existing cultivar populations have narrow genetic bases, hence need for creating variability within and among cultivars using genetic diversity methods.

GENETIC DIVERSITY MEASUREMENT TOOLS

Genetic diverse populations arising from pure lines, accessions, landraces, wild or weed races are analyzed using a number of methods. Such method can be single or in combination of two or more methods. Franco et al. (2001) stressed the need for careful considerations to be made when measuring genetic diversity within and between crop populations in research. Such considerations include:

1. Use of multivariate data collected from morphological or agronomic traits. Such data may effectively display discrete, continuous, binomial ordinal etc. variables

2. Use of multiple data sets arising from morphological, biochemical and DNA-based collections. The use of such multiple data sets in diversity study helps to reveal the adequacy in terms of strength and constraints in the choice of each of the data sets. The use of multiple data pose some puzzles including can analysis and result interpretation be based on individual or combined data sets? And more worrisome is the puzzle on how to effectively combine the different data sets and still achieve meaningful result. To provide answers to these puzzles, Wrigley et al. (1982), studied phylogenetic relationships among triticeae species using individual and combined analysis of data sets consisting of morphological and DNA-based traits and discovered divergent results in the analysed individual and combined data. The discrepancies in the

results may be attributed to the discrete nature of DNA-based data and the continuous variable nature of the morphological data. No wonder Hillis 1987; Chippindale and Wein (1994) suggested the assignment of specific numbers to both quantitative and qualitiative traits in morphological, biochemical and molecular data set. In view of this, Pedersen and Seberg (1998) advised that both individual and combined data sets can be analyzed in many possible and meaningful ways to draw conclusions on genetic divergence. In 1999 and 2001, Taba et al. and Franco et al., respectively utilized the modified Location Model (MLM) which combines all variables into one multinomial variable called "W" to classify maize accessions from the genetic resource centres of Latin America. Better still, this MLM can combine molecular and morphological data to classify data better than when individual data set is employed. Individual data from morphological, biochemical or molecular data set can be analyzed using one or a combination of techniques. These techniques shall be discussed.

3. Expected objective to be achieved. This dictates choice of statistical tool in measuring genetic distance and the level of clustering of the intragenic factors in use. Such objective(s) include to determine the quantum of variation and grouping such genotype based on genetic distance, identify action following parental selection. In essence, breeding focus determines applicable method in explaining the nature of genetic divergence.

Variations are recorded in the measurement of genetic diversity in genotype relationships based on genetic distances and grouping populations from individual genotypes such as accessions, lines, wild races etc. The recorded variations are primarily because of the differences in the nature of genetic materials. Therefore, the basis or genetic variance theories which identifies genotype relationships based on genetic distance estimating genetic diversity depends largely on statistical genetic variance theories which identifies genotype relationships based on genetic distance / variance.

THE USE OF MORPHOLOGICAL DATA TO MEASURE GENETIC DISTANCE

Nei, (1973), first defined Genetic distance as the difference between two entities that can be described by allelic variation. This definition was later in 1987, modified to "extent of gene differences among populations that are measured using numerical values. Betterstill, in 1998, Beaumont et al., provided a more comprehensive definition of genetic distance as any quantitative measure of genetic difference at either sequence or allele frequency level calculated between genotype individuals or populations. The first early work of Anderson (1957), proposed the use of metrogliph and index-score to study the pattern of morphological variations in individual data set. In the early seventies (Singh and Chaudhary 1985) used this method to study morphological variation in green gram. This method uses a range of variations arising from trait such that extent of trait variation is determined by the length of rays on the glyph. The performance of a genotype is adjudged by the value of the index score of that genotype. The score value determine the length of ray which may be small, medium or long Akoroda (1987); Ariyo and Odulaja (1991) and Van Bueningen and Busch (1997), extensively explored the use of metroglyph and index-score to morphological variations in yellow yam, Okro and wild rye accessions respectively. Similar to metroglyph and the score index is Euclidian Distance (ED) measurement. According to Nei (1987), Euclidian distance measures similarity between two genotypes, populations or individuals using using statistical measures where two individuals i and j, having observations on morphological traits (p) denoted by x_1, x_2, x_3,......x_n and y_1, y_2,......y_n for i and j individuals respectively. Metroglyph and index-score methods measures genetic distance by use of morphological traits. Euclidian distance measurements utilize both morphological and molecular based marker data sets. Smith et al. (1991), applied the following statistic to measure ED.

$$\text{dij} = \varepsilon[(T_{1(i)} - T_{2(i)}^2)/\sigma^2 T_{(i)}]^{1/2}.$$

Where T_1 and T_2 are the values of the ith trait for 1 lines and 2 and $\sigma^2 T(i)$ is the variance for the ith trait over all the lines used. Much later, Weir (1996) developed a formula for calculating genetic distance to be.

$$d(I,j) = [(x_1 - y_1)^2 + (x_2 - y_2)^2 + ...(x_p - y_p)^2]^{1/2}$$

where i and j is the ED between two individuals lines having morphological traits (p)

x_1, x_2......x_p is the traits for i individuals and y_1, y_2......x_p is the traits for j individuals

from here, the individual character distances are summed and then divided by the total number of characters scored in both individuals. ED measurement allows the use of both qualitative and quantitative data several workers identified genotype distances using ED. Van Bueningen and Busch (1997) in wheat, smith et al, 1987 in sorghum and Ajmone – Marsan (1998) in maize.

THE USE OF MOLECULAR DATA IN MEASURING GE-NETIC DISTANCES

The advent and explorations in molecular genetics led to a better definition of Euclidean distance by Beaumont et al; (1998) to mean a quantitative measure of genetic difference calculated between individuals, populations or species at DNA sequence level or allele frequency level. Various genetic distance measurements are proposed for analyzing DNA-based data for the purpose of genetic diversity studies. Powel et al. (1996), identified different DNA-based marker techniques to include Random Amplified Polymorphic DNA (RAPD), Amplified Fragment Length Polymorphism (AFLP), Restriction Fragment Length Polymorphic (RFLPs) and the most recent Simple Sequence Repeats (SSR) and Microsatellite (MT) of single nucleotide polymorphism (SNPs). The above nucleotide differences can be used effectively to run individual or combined data sets of morphological, biochemical or DNA based data. For DNA based data, where the amplification products are equated to alleles, the allele frequencies can be calculated and the genetic distance between i and j individuals estimated as follows.

$$d(ij) = 1 < \left[\sum_{i}^{n} (X_{ai} - X_{aj}) \right]^{1/r}$$

Where X_{ai} is frequency of the allele a for individual I, and n is the number of alleles per loci; r is the constant based on the coefficient used. In its simple form, i.e $r = 1$, genetic distance can be calculated as:

$$d\int (ij) = 1/2 \left[\sum_{i}^{n} (X_{ai} - X_{aj}) \right]$$

Where $r = 2$, $d(i,j)$ is referred to as Rogers (1972) measure of distance (RD), where

$$RD_{ij} = 1/2[\Sigma (x_{ai} - x_{aj})^2]^{1/2}$$

Where allele frequencies are to be calculated for some of the molecular markers, the data must first generate a binary matrix for statistical analysis. Binary data has been long and widely used before the advent of molecular marker data to measure genetic distance by Rogers (1972); Nei and Chesser (1983) coefficient and known as GDMR and GDNL respectively. In the use of any given statistical formula to determine genetic diversity in molecular based data, one specific problem usually encountered is the failure of some genotypes to show amplification for some primer pairs. Robinson and Harris

(1999) noted that lack of amplification may be due to "null alleles". Most often, it is difficult to ascribe lack of amplification to "null allele". It is therefore the reposed confidence of the researcher, that a "null allele" status of a genotype will not be considered as missing data during computation of genetic similarity-distance matrix so as to avoid gross error during result interpretation. DNA based marker data have been successfully used to measure genetic distance in some crops (Pritchard et al. (2000) in pigeon pea; Beaumont et al. (1998) in wheat; Franco et al., (2001) in maize; Dje et al. (2000) in Sorghum.

GROUPING TECHNIQUES IN MEASURING GENETIC DIVERSITY

Genetic relationship among and with breeding materials can be identified and classified using multivariate grouping methods. The use of established multivariate statistical algorithms is important in classifying breeding materials from germplasm, accessions, lines, and other races into distinct and variable groups depending on genotype performance. Such groups can be resistant to diseases, earliness in maturity, reduced canopy drought resistant etc. The widely used techniques irrespective of the data source (morphological, biochemical and molecular marker data) are cluster analysis, Principal Component Analysis (PCA), Principal Coordinate Analysis (PCOA) Canonical Correlation and Multidimensional Scaling (MDS). Cluster analysis presents patterns of relationships between genotypes and hierarchical mutually exclusive grouping such that similar descriptions are mathematically gathered into same cluster (Hair et al. 1995); (Aremu 2005). Cluster analysis have five methods namely unweighted paired group method using centroids (UPGMA and UPGMC), Single Linkages (SLCA), Complete Linkage (CLCA) and Median Linkage (MLCA). UPGMA and UPAMC provide more accurate grouping information on breeding materials used in accordance with pedigrees and calculated results found most consistent with known heterotic groups than the other clusters (Aremu et al., (2007a). Principal components, canonical and multidimentional analyses are used to derive a 2-or 3- dimensonal scatter plot of individuals such that the geometrical distances among individual genotypes reflect the genetic distances among them. Wiley (1981), defined principal component as a reduced data form which clarify the relationship between breeding materials into interpretable fewer dimensions to form new variables. These new variables are visualized as different non correlating groups. Principal components analysis first determines Eigen values which explain the amount of total variation displayed on the component axes. It is expected that the first 3 axes will explain a large sum of the variations captured by the genotypes.

Cluster and principal component analysis can be jointly used to explain the variations in breeding materials in genetic diversity studies.

CONCLUSION

Genetic diversity studies is in no measure the first basic step in meaningful breeding programme and therefore require accurate and reliable means for estimation. Data sets sourced can morphological biochemical several workers successfully utilized various statistical tools in analysis diverse data sets and identified two major framework to really explain divergence in genotype performance. Genetic distance among and within individual data sets can be conveniently determined using specific tools while classificatory and cluster analysis require principal component and polymorphic sequence tools. Since each data set provide different molecular type of information, based marker data set is visualized to provide more reliable differentiate information on the genotypes. Analysis of data sets can be complex. Many software packages are available. There is still a need for a comprehensive and user-friendly software packages that would integrate different data set for analysis and generate reliable and useable information about genetic relationship. Equally important in genetic diversity studies is the need for a genetic resource centre. Studies should incorporate utilization of genetic diversity information in developing genetic resource centre accessible to breeders.

ACKNOWLEDGEMENT

Many thanks to Ibirinde Olalekan for the secretariat assistance. I also appreciate Olayinka Olabode the Head of Department of Agronomy LAUTECH for the technical contributions given to this chapter.

REFERENCES

1.   Adewale, B.D.,Kehinde, O.B., Aremu, C. O. Popoola, and Dumet, J. 2010. Seed metrics for genetic and shape determinations in African yam bean [Fabaceae] (Sphenostylis stenocarpa) African Journal of Plant Science Vol. 4(4): 107-115

2.   Ajmone-Marsan P., Carstiglioni, P., Fusari, F., Kuiper, M. and Motto, M. 1998. Genetic diversity and its relationship to hybrid performance in maize as revealed by RFLP and AFLP markers. Theor. Appl. Genet. 96:219-227.

3.   Akbar, A.A. and Kamran, M. (2006). Relationship among yield components and selection criteria for yield improvement of Safflower – Carthamustinctorious L. J. Appl. Sci. 6:2853-2855.

4. Akoroda, M.O. 1987. Principal component analysised metroglyph of variation among Nigerian yellow yam. Euphytica 32:565-573.

5. Aremu C.O. 2005. Diversity selection and genotypes Environment interaction in cowpea unpublished Ph.D Thesis. University of Agriculture, Abeokuta, Nigeria. P. 210.

6. Aremu, C.O., Adebayo M.A., and Adeniji O.T. 2008: Seasonal performance of cowpea in humid tropics using GGE biplot analysis world jour. of Biol.. Research. 1: (1) 8-13.

7. Aremu, C.O., Adebayo, M.A., Ariyo O.J., and Adewale B.D. 2007b. Classification of genetic diversity and choice of parents for hydridization in cowpea vigna unguiculata (L) walip for humid savanna ecology. African journal of biotechnology 6: (20) 2333-2339.

8. Aremu, C.O., Adebayo, M.A., Oyegunle, M. and Ariyo, J.O.2007a. TRhe relative discriminatory abilities measuring Genotype by environment interaction in soybean (Glycine max). Agricultural journ. 2 (2).: 210-215

9. Ariyo, O.J. and Odulaja, A. 1991. Numerical analysis of variation among accessions of Okra. Ann. Bot. 67:527-531.

10. Barrett, B.A. and Kidwell, K.K. 1998. AFLP-based genetic diversity assessment among wheatcultivars from the pacific Northwest crop Sci. 38:1261-1271.

11. Baudoin, J. and Mergeai G. 2001. Yam bean sphenostylis stendcarpa. In: R.H. Raemaekers (ed) crop production in tropical Africa Directorate general for international (DGIC) Brussels, Belgium. Pp. 372 – 377.

12. Bearmont, M.A., Ibrahim, K.M., Boursot, P. and Bruqord, M.W. 1998. Measuring genetic distance. P. 315-325. In A. karp et al. (ed) molecular tools for screening biodiversity. Chapman and Hall, London.

13. Bohn, M.,H.F. Hutz and A.E.Melchinegr. 1999. Genetic similarities among winter wheat cultivars determined on the basis of RFLPs, AFLPs ans SSRs and their used ion predicting progeny variance. Crop Sci. 39; 228-237.

14. Brown, A.H.D. 1989. The case for core collection. P. 135-156. In A.H.D England Brown et al. (ed). The use of plant genetic resources. Cambridge Univ. press Cambridge.

15. Brown-Guedira, G.L; Thompson J. Ajnelson, R.L. and Warburton M.L. 2000. Evaluation of genetic diversity of soybean introduction and North American ancestors using RAPD and SSR markers crop Sci. 40:815-823.

16. Bueninger, L.T. and Busch, Ritt. 1997. Genetic diversity among North American spring wheat cultivars. Analysis of the coefficient of parentage matrix. Crop Sc. 37:570-579.

17. Chippindale, P.T. and Weins J.I. 1994. Weighting portioning and combining characters in phylogenetic analyses. Syst. Biol. 43:273-287.

18. Christine, Joshua, H., William, A., Stacy, A. 2009. Genetic diversity of creeping bentgrass cultivars using SSR markers. Intern. Turfgrass Soc. Research Jour. 11.

19. Dje, Y, Hevretz, M., Letebure, C. and Vekemans, X. 2000. Assessment of genetic diversity within and among germplasm accessions in cultivated sorghum using microsatellite marked theor. Appl. Gent. 100:918-925.

20. Eivazi, A.R, Naghavi, M.R, Hajheidari, M. Mohammadi, S.A, Majidi, S.A, Salakdeh, I. and Mardi, M. 2007. assessing wheat genetic diversity using quality traits, amplified fragment length polymorphism simple sequence repeats and proteome analysis Ann. Appl. Biol. 152:81-91.

21. Falconer, D.S. and Mackay T.F. 1996. Introduction to Quantitative Genetics,Longman, Harlow.

22. Franco, J., Crossa, J., Ribaot, M., Betran, J., Warburton, M. land Khairallah, M. 2001. A method for combinary molecular markers and phenotypic attributes for classifying plant genotype Theor-Appl. Gent. 103: 944-952.

23. Fu, Y. and Somers D. 2009. Genomo-wide reduction of genetic diversity in wheat breeding Crop Sci. 49:161-168.

24. Hair, J.R., Anderson, R.E., Tatham, R.L., and Black, W.C. 1995. Multivariatev data analysis with Readings. 4th edition, Prentice- Hall, Englewood Cliffs, NJ.

25. Hillis, D.M. 1987. Molecular versus morphological approaches to systematics. Annu. Rev. Ecol. Systems. 18:23-42.

26. Islam. M.R. 2004. Genetic diversity in irrigated rice pak. Jorn of Biol. Sci. 2:226-226.

27. Joshi, B.K, Mudwari, A., Bhatta, M.R, Ferrara, G.O. 2004. Genetic diversity in Nepalese wheat cultivars based on agro-morphological traits and coefficient of parentage. Nep Agric Res. J. 5:7-17.

28. Liu, S., Cantrell, R.G., Mccarty, J.C. and Stewart, M.D. 2000. simple sequence repeat based assessment of genetic diversity in cotton race accessions. Crop Sci. 40:1459-1469.

29. Martin, E., Cravero, V., Esposito A, Lozez F, Milanebi, L. and Cointry, E. 2008. Identification of markers linked to agronomic traits in globe artichoke. Aust. J. Crop Sci. 1:43-46.

30. Mohammadi A.A, Saeidi, G. and Arzuni, G. 2010. Genetic analysis of some agronomic traits in flax. Aust. J. Crop sci. 4:343-352.

31. Mostafa K., Mohammad, H. and Mohammad, M. 2011 genetic diversity of wheat genotype baspdon cluster and principal component analyses for breeding strategies Australian Jour. of Crop Sc. 5 (1): 17-24.

32. Nei, M. 1973. Analysis of gene diversity in subdivided populations. Proc. Natil. Acad. Sci. (USA) 70. 3321-3323.

33. Nei, M. and Chesser, R.K. (1983). Estimation of fixation indices and gene diversities. Ann. Hum. Genet. 47: 253.-259

34. Pedersen, G. and Seberg, O. 1998. Moleculesvs Morphology. P. 359-365. In A. karp et al (ed) Molecular tools for screening Biodiversity.. Chapman and Hall, London.

35. Potter, D. and Doyle, J. J. 1992. Origin of African yam bean (Sphenostylisstenocarpa, Leguminosae): evidence from morphology, isozymes,chloroplast DNA and Linguistics. Eco. Bot. 46: 276-292.

36. Powell, W., Morgante, M., Andre, C., Hanafey, M., Vogel, J.Tinjey, S. and Rafalsky, A. 1996. The comparison of RFLP, RAPD, AFLP and SSR markers for germplasm analysis. Mol.breed. 2: 225-238.

37. Rahim, M.A., Mia, A.A, Mahmud, F., Zeba, N. and Afrin, K. 2010. Genetic variability, character association and genetic divergence in murgbean platn Omic 3:1-6.

38. Rogers, J.S. 1972. Measures of genetic similarity and genetic distance studies in genetics. VII. Univ. Tex. Publ. 2713: 145-153.

39. Singh R.K. and Chaudhary B.D. 1985. Biometrical methods in quantitative genetic analysis.Kalyani publishers, New Delhi. India. P 38.

40. Smith, J.S., Paszkiewics, S. Smith,O.S.and Schaoffer, J. 1987. Electorphoretic, chromatographic and genetic techniques diversity among corn hybrids. P. 187-20. In Proc. Am.Seed Trade Assoc.Washington DC.

41. Smith, J.S., Smith, O.S., Boven, S.L., Tenburg, R.A.and Wall, S.J. 1991. The description and ssessment of distances between inbred lines of maize III: A revised scheme for the testing of distinctiveness between inbred lines utilizing DNA RFLPs. Maydica, 36. 213-226.

42. Swoftord D.L, Olsen, G.J. Wadell, P.J. and Hillis D.M. 1996. Phylogenetic inference P. 407- 514. In D.M. Hillis et al (ed) Molecualr systematics. 2nd edition Sinaver Associates, Sunderland, M.A.

43. Taba, S., Diaz, J., Franco, J., Crossa, J.and Eberhart, S.A. 1999. A core subject of LAMP, from the Latin American maize project. CD-ROM, CIMMYT, Mexico, D.F., Mexico.

44. Thompson, J.A., Nelson, R.L. and Vodkin, L.O.. 1998. Identification of diverse soybean germplasm using RAPD markers. Crop Sci. 38: 1348-1355.

45. Van Bueningen, L.T. and Busch, R.H. 1997. Genetic diversity among North American spring wheat cultivars: I. Analysis of the coefficient of parentage matrix. Crop Sci. 37: 570-579.

46. Vavilov, N.I. 1950. The origin, variation, immunity and breeding of cultivated plants. Chronica botanica, B. chronica Botanica company, Waltham, Massachusetts. P 364.

47. Warburton, M. and Crossa, J. 2000. Data analysis in the CIMMYT. Applied Biotechnology Center for fingerprinting and Genetic Diversity Studies. CIMMYT, Mexico.

48. Weir, B.S. 1996. Intraspecific differentiation P. 385-403. in D.M. Hillis et al. (ed). Molecular systematics 2nd edition sunderlands M.A.

49. Wiley, E.O., 1981. Phylogenetics: The theory and practice of phylogenetics and systemic.John Wiley, New York.

50. Wrigley, C.W, Autran, J.C. and Bushuk, W. 1982. Identification of cereal varieties by gel electorphoresis of the grain proteins. Adv. Cereal. Sci. Techolog. 5:211-259.

Chapter 8

GENETICS OF RHEUMATIC DISEASE

Alex Clarke and Timothy J Vyse
Imperial College London, Faculty of Medicine, Section of Molecular Genetics and Rheumatology, Hammersmith Hospital

ABSTRACT

Many of the chronic inflammatory and degenerative disorders that present to clinical rheumatologists have a complex genetic aetiology. Over the past decade a dramatic improvement in technology and methodology has accelerated the pace of gene discovery in complex disorders in an exponential fashion. In this review, we focus on rheumatoid arthritis, systemic lupus erythematosus and ankylosing spondylitis and describe some of the recently described genes that underlie these conditions and the extent to which they overlap. The next decade will witness a full account of the main disease susceptibility genes in these diseases and progress in establishing the molecular basis by which genetic variation contributes to pathogenesis.

GENETICS OF RHEUMATIC DISEASE

The spectrum of rheumatic disease is wide and includes conditions with diverse pathology, although most have in common a heritable risk with a complex genetic basis. There has therefore been intense effort to understand the contribution of genotype to the expression of disease in terms of both basic pathogenesis and clinical characteristics. Recent technical advances in genotyping and statistical analysis and international collaborations assembling large cohorts of patients have led to a wealth of new data. In this review we describe insights gained into the pathogenesis of autoimmune rheumatic disease by the techniques of modern genetics, in particular evidence from genome-wide association (GWA) studies, which provide support for the existence of a common genetic risk basis to several diseases. To reflect the new data from GWA studies, our discussion will be confined to rheumatoid arthritis (RA), systemic lupus erythematosus (SLE), and ankylosing spondylitis (AS), which in some cases share a common autoimmune pathogenesis. Osteoarthritis and osteoporosis are also complex genetic traits but limitations of space are such that these two conditions will not be considered in this review.

The concept of a systematic, GWA study became practical with the cataloguing of libraries of common polymorphisms. Currently, over 20 million single nucleotide polymorphisms (SNPs) have been identified [1] and platforms are available to type up to 1 million of these in a single reaction. Although not all SNPs are currently genotyped, as the human genome is arranged into haplotype blocks in linkage disequilibrium, it is only necessary to type so-called tag SNPs, which identify these areas of limited variability [2], to achieve good representation of the total amount of genetic variation. Most typed SNPs are relatively common (minor allele frequency of > 5%) and if associated with disease are likely, therefore, to have only modest pathogenic effects (odds ratios (ORs) usually between 1.2 and 2), as otherwise they would become depleted in a population due to natural selection. It is necessary, therefore, to invoke the 'common-disease common-variant' (CD-CV) model [3], which assumes an accumulation of risk caused by the carriage of multiple deleterious alleles, to explain current experimental findings.

One of the revolutionary advantages of the GWA study is the freedom from a required gene-centric hypothesis, which provides an unprecedentedly effective technique for risk gene discovery. Many disease-associated genes identified by GWA studies were completely unsuspected to be relevant - for example, the autophagy system in Crohn's disease [4]. However, because in essence up to 1 million independent hypotheses are being tested in each genotyping reaction, sample sizes powered to detect even the stronger associations must be large, and criteria for significance stringent. The general consensus is that significance can be defined as a P-value smaller than 5×10^{-7}, which in a cohort such as the Wellcome Trust Case Control Consortium (WTCCC) of 2,000 cases, for example, approximates to a power of 43% rising to 80% to detect alleles with ORs of 1.3 and 1.5, respectively [5]. However, the genome is subject to variation at more than the SNP level, and individuals also differ in the copy number of sections of DNA of greater than several kilobases in size, so called copy number variation (CNV), which in fact accounts for more total nucleotide difference between individuals than SNPs [6,7]. CNV can affect gene expression levels [8] and has been linked to autoimmune disease [9, 10], including SLE [11]. Whilst the latest genotyping platforms include assessment of CNV, earlier products actively excluded SNPs within regions of the most variation as they were more likely to fail quality control steps. Association studies based on CNV are, therefore, in their relative infancy. Finally, the genome is subject to modification without a change in DNA sequence; epigenetic mechanisms can have profound effects on gene expression. These include DNA methylation and changes in chromatin structure [12].

It has become apparent that SLE, RA, and AS, which have divergent clinical features, may share a common genetic risk framework, and we aim in our review to illustrate this.

THE MHC REGION AND ANTIGEN PROCESSING

The major histocompatibility complex (MHC) region on chromosome 6 contributes to the risk of almost all autoimmune diseases, and its role in immunity in mice was recognized over 60 years ago. In humans, the MHC locus is also known as the *HLA* (human leukocyte antigen) region, reflecting the initial identification of MHC gene products on the surface of white blood cells. The classical MHC extends over around 4 megabases, and comprises three clusters: class I, II, and III. Class I and II regions include genes that encode the α- and β-chains of the MHC I and II complexes, and flank the class III region, which contains an assortment of immunologically relevant genes. Despite extensive study, the mechanisms that link the MHC to disease are largely unknown, although it is supposed that variation in the MHC peptide binding cleft facilitates presentation of self-antigen to autoreactive lymphocytes.

These difficulties in understanding the MHC are not without reason; it contains some of the most polymorphic loci described in the genome, and has a highly complicated genetic architecture, with some regions exhibiting extended linkage disequilibrium [13].

In RA, the MHC accounts for around a third of the genetic liability [14]. Alleles at *HLA-DRB1* contribute much of this risk - for example, *DRB1*0401* carries an OR of 3. GWA studies confirm the strong association with MHC variants; risk alleles confer an OR of around 2 to 3 in homozygotes [15], with very high statistical significance ($P < 10^{-100}$). Additional loci contributing to the risk of RA identified by high-density genotyping include HLA-DP in patients with anti-cyclic citrullinated peptide antibodies [16]. SLE not only has strongly associated alleles in the class II region, *HLA-DR2 (DRB1*1501)* and *DR3 (DRB1*0301)* [14], with ORs of 2 [17], but also risk variants in the class III cluster, which encodes genes such as *TNF* and the complement components *C2, C4A* and *C4B*. C4 is crucial in the classical and mannose-binding lectin pathways of complement activation, and complete deficiency of C4 or indeed other components of the classical pathway are rare, but strong, risk factors for SLE [18]. The *C4* gene is subject to CNV and is of two isotypes, *C4A* and *C4B*. It is an attractive hypothesis that CNV at *C4* affects expression and contributes to SLE risk. However, it remains to be established whether haplotypes carrying partial C4 deficiency exert their risk via an influence on complement or through other genetic variants that are in linkage disequilibrium. Other loci in the class III region have been implicated in

SLE, including the *SKIV2L* gene, SNPs in which carry an OR of 2 in a family-based analysis [19]. *SKIV2L* encodes superkiller viralicidic activity 2-like, the human homologue of which is a DEAD box protein that may have nucleic acid processing activity. The second MHC III signal for SLE we will consider was identified in the International Consortium on the Genetics of Systemic Lupus Erythematosus (SLEGEN) GWA study [17, 20]. The SNP rs3131379 in mutS homologue 5 (*MSH5*) has an OR of 1.82. There is evidence that MSH5 has a role in immunoglobulin class switch variation [21]. Again, further work is required to definitively implicate this gene rather than variants in linkage disequilibrium, which include HLA-DRB1*0301 and *C4A* deletions.

Clearly, *HLA-B27* is the overwhelming association in AS, with an OR of 200 to 300. In the MHC, other genetic risk variants have been identified, including *HLA-B60* (OR 3.6) [22] and various *HLA-DR* genes with relatively minor contributions [23]. The pathogenic mechanism for these risk alleles is unknown. Outside of the MHC, two significant genes have so far been identified in AS: *ARTS1* and *IL-23R* [24], the latter of which will be discussed below and has been associated with several different autoimmune diseases. ARTS1 has two identified functions. Its first is in the processing of peptide for presentation via MHC I. It is localised in the endoplasmic reticulum, and is upregulated by IFNγ. It acts as an amino-terminal aminopeptidase and in mice is essential for the display of the normal peptide repertoire. In its absence, many unstable and highly immunogenic MHC-peptide complexes are presented [25]. A hypothetical connection with HLA-B27 can thus be drawn. Its other function is to downregulate signalling by IL-1, IL-6, and TNFα through surface receptor cleavage [26–28]. The most associated SNP rs30187 risk allele has an OR of 1.4, and is of unknown functional significance.

INNATE-ADAPTIVE INTERFACE

Interferon Signalling: *IRF5*

It is clear that type 1 interferons (IFNα and IFNβ) are of great importance in the pathogenesis of SLE. Patients with active disease have high levels of IFNα, which has multiple immunomodulatory actions [29], including the induction of dendritic cell differentiation, the upregulation of innate immune receptors such as toll-like receptors (TLRs), the polarization of T cells towards a $T_H 1$ phenotype, and the activation of B cells. Type I interferons are produced by all cells in response to viral infection, but particularly by plasmacytoid dendritic cells in response to unmethylated CpG oligonucleotides binding to TLR-9, or RNA to TLR-7. Using a candidate gene approach targeting the IFN signalling pathway, the SNP rs2004640 in *IRF5*(interferon regulatory factor 5) was found to be significantly associated with SLE (OR 1.6) [30], a risk gene confirmed in several other studies [17, 31–35]. The functional consequences for IRF5 of the identified mutations are variable, but include the creation of a 5' donor splice site in an alternative exon 1, allowing the expression of several isoforms [35], a 30 base-pair in-frame insertion/deletion variant of exon 6, a change in the 3' untranslated region, and a CGGGG insertion-deletion (indel) polymorphism, the latter two affecting mRNA stability [32, 36]. Interestingly, these mutations may occur together in a haplotype, with varying degrees of associated risk. The exact role of IRF5 in IFN signalling has not been fully elucidated, but it is also critical for the gene induction programme activated by TLRs [37], providing further biological plausibility for its importance in the pathogenesis of SLE. Haplotypes of IRF5 are also implicated in RA, and may confer either protection (OR 0.76) or predisposition (OR 1.8) [38]. The same CGGGG indel allele described above also carries risk for multiple sclerosis and inflammatory bowel disease [36].

TNF-Associated Signalling Pathway: *TNFAIP3* and *TRAF1-C5*

TNF-associated signalling pathway genes play a prominent role in the risk for both SLE and RA, and associations with variants in *TNFAIP3*, and the *TRAF1-C5* locus have been identified [39, 40]. TNFα-induced protein-3 (TNFAIP3; also known as A20) is a ubiquitin editing enzyme that acts as a negative regulator of NFκB. A20 can disassemble Lys63-linked polyubiquitin chains from targets such as TRAF6 and RIP1. A second region of A20 catalyses Lys48-linked ubiquitination that targets the molecule for degradation by the proteasome [41]. A20 modifies key mediators in the downstream signalling of TLRs that use MyD88, TNF receptors, the IL-1 receptor family, and nucleotide-oligomerization domain protein 2 (NOD2) [42]. *Tnfaip3* knockout mice develop severe multi-organ inflammatory disease, and the phenotype is lethal [43]. The SNP rs10499194 in *TNFAIP3* carries an OR of 1.33 for RA, and rs5029939 an OR of 2.29 for SLE [44], the latter also conferring an increased risk of haematologic or renal complications [45].

On chromosome 9, the region containing *TRAF1* (TNF receptor associated factor 1) and *C5* (complement component 5) genes is associated with significant risk for RA (risk SNP OR of approximately 1.3) in most [15, 40, 46–48], but not all [5], studies. Due to linkage disequilibrium, the functional variant remains elusive. TRAF1 is principally expressed in lymphocytes, and inhibits NFκB signalling by TNF. This pathway is blocked in TRAF1 overexpression [49] whilst, conversely, *Traf1-/-* mice are sensitized to TNF and have exaggerated TNF-induced skin necrosis [50].

The complement system has long been known to be involved in the pathogenesis of RA. In the collagen-induced arthritis model of RA, C5 deficiency prevents disease *de novo* and ameliorates existing symptoms and signs [51, 52]. Interestingly, GG homozygotes at the *TRAF1-C5* SNP rs3761847 with RA have a significantly increased risk of death (hazard ratio 3.96, 95% confidence interval 1.24 to 12.6, $P = 0.02$) from malignancy or sepsis, potentially allowing identification of patients for appropriate screening [53].

Immunomodulatory Adhesion Molecule: *ITGAM*

Integrin-α-M (ITGAM), variants of which are strongly associated with SLE, forms a heterodimer with integrin-β-2 to produce $\alpha_M\beta_2$-integrin (also known as CD11b, Mac-1, or complement receptor-3), which mediates the adhesion of myeloid cells to the endothelium via ICAM-1 (Intercellular adhesion molecule-1) and recognizes the complement component iC3b. It not only has a role in cell trafficking and phagocytosis [54], but also has other immunomodulatory functions. Antigen-presenting cells produce tolerogenic IL-10 and transforming growth factor-β on iC3b binding to CD11b [55], and mice deficient in this receptor upregulate expression of IL-6, favouring a pro-inflammatory T_H17 response [56]. Despite its implication in defective immune complex clearance in SLE, experimental evidence for a role was lacking. GWA studies, however, demonstrate a strong and significant association [17, 33, 44], with an OR of 1.83 ($P = 7 \times 10^{-50}$) in meta-analysis [57]. The implicated SNP rs1143679 is non-synonymous, causing the substitution of histidine for arginine at amino acid 77, although this change does not affect the iC3b binding site [58]. Furthermore, although this SNP is disease associated in European and Hispanic patients, it is monomorphic in Japanese and Korean populations [59]; an explanation of its effect is therefore outstanding. It has been mentioned that CNV is important in C4 expression; the same is true for the Fcγ receptor IIIb (FCGR3B) [60], which relies on CD11b for function. Fcγ receptor IIIb is principally present on neutrophils and is important in the binding and clearance of immune complexes, therefore marking itself as a potential SLE risk gene. There is a significant association between low *FCGR3B* copy number and SLE. Patients with two or fewer copies of *FCGR3B* have an OR of 2.43 for SLE with nephritis, and 2.21 for SLE without nephritis [61].

LYMPHOCYTE DIFFERENTIATION

T cell Receptor Signalling: *PTPN22*

Outside the HLA region, the first reproducible genetic association for RA came with the implication of *PTPN22* from a candidate gene approach [62] based on linkage analysis identification of a susceptibility locus at 1p13 [63]. It has remained the strongest and most consistent association mapped by GWA studies in RA. A role in SLE has also been identified [17]. The OR for the risk allele is around 1.75 in RA, and 1.5 in SLE. However, it should be noted that this allele (encoding the R620W mutation) is monomorphic or not disease associated in Korean or Japanese patients [64, 65]. *PTPN22* encodes lymphoid tyrosine phosphatase (LYP), a protein tyrosine phosphatase that inhibits T cell receptor signalling, decreasing IL-2 production. The disease associated SNP is responsible for a change from arginine to tryptophan at position 620, which inhibits binding to the SH3 domain of carboxy-terminal Src kinase. This in turn appears to enhance dephosphorylation of tyrosine residues in the Src family kinases Lck, FynT, and ZAP-70 [66, 67]. The overall effect of the mutation is a reduction in T cell receptor signalling. The pathogenic effect of this is unclear, but may relate to impaired negative selection in the thymus, or lead to a reduction in regulatory T cells [68]. Conversely, the R623Q variant of *PTPN22*, which is a loss-of-function mutation affecting the phosphatase activity of LYP, is protective against SLE [69]. *PTPN22* does not appear to be a risk gene for AS [70].

Polarization Towards T_H1 and T_H17 Phenotypes: *STAT4* and *IL23R*

STAT4 encodes signal transducer and activation of transcription factor-4, responsible for signalling by IL-12, IL-23, and type 1 IFNs [71]. STAT4 polarizes T cells towards T_H1 and T_H17 phenotypes, which has the potential to promote autoimmunity [72]. In RA the OR for the risk allele of SNP rs7574865 is 1.32 in one case-control study [73], with a less strong disease association at rs11893432 in a meta-analysis of GWA studies (OR 1.14) [15]. There is convincing evidence that *STAT4* is a risk locus for SLE in multiple racial groups [33, 74], and it may be theorized that interference in type I IFN signalling may be the underlying pathogenic mechanism in this case. Distinctive disease pathways could, therefore, emerge from mutations in a single gene. The WTCCC AS study identified *IL23R* as a risk gene in AS [24]. IL-23 is instrumental in the development of T cells with the pro-inflammatory T_H17 phenotype [75], and *IL23R* has been linked to psoriasis, ulcerative colitis, and Crohn's disease in GWA studies [5, 76, 77]. An interesting connection between these conditions, all of which may share common clinical features, is thus made. In AS the risk SNP rs11209032 confers an OR of 1.3.

B cell Activation

B cells are a population long suspected to be important in autoimmune rheumatic disease, and the benefits of their depletion in RA and SLE has resurrected interest in their pathogenic role. The risk genes identified so far are involved in signalling from the B cell receptor (BCR). *BLK* encodes a Src family tyrosine kinase restricted to the B cell lineage and is poorly understood. Risk alleles in the region upstream of the transcription initiation site are associated with SLE (OR 1.39, $P= 1 \times 10^{-10}$) and reduce levels of *BLK* mRNA [33]. BANK1 (B cell scaffold protein with ankyrin repeats-1) undergoes tyrosine phosphorylation upon B cell activation by the BCR, leading to an increase in intracellular calcium through the inositol trisphosphate mechanism [78]. The non-synonymous SNP rs10516487 in *BANK1*, which substitutes histidine for arginine at amino acid 61, also has disease association (OR 1.38) [79]. The functional consequence of this may be higher affinity for the inositol trisphosphate receptor, as the substitution is located in the binding site.

Lyn, another Src tyrosine kinase, is important in determining signalling thresholds for myeloid and B cells. On BCR ligation, it phosphorylates tyrosine residues of Syk, an activating tyrosine kinase, CD19, and the immunoreceptor tyrosine-based activation motifs (ITAMs) of the Igα/Igβ subunits of the BCR. However, it also has a critical regulatory role, mediated by phosphorylation of the inhibitory motifs of CD22 and Fcγ RIIB, which in turn activate SH2-domain containing phosphatases, leading to dephosphorylation and deactivation of a number of signalling intermediaries [80]. *Lyn-/-* mice develop severe autoimmunity associated with glomerulonephritis [81]. An association between SNPs in *LYN* and SLE, identified initially in the SLEGEN GWA study [17], has been recently confirmed in a case-control study [82]. The most associated SNP, rs6983130, is near the primary transcription initiation site.

OX40L, a member of the TNF super-family encoded by *TNFSF4* (TNF superfamily 4), is associated with SLE. The cross-talk between B lymphocytes and dendritic cells expressing OX40L, and T cells that express its receptor, OX40, serves to enhance the adaptive immune response [83]. An upstream *TNFSF4* haplotype, associated with SLE, enhances gene expression *in vitro* [84, 85], although the mechanism responsible for the deleterious effects observed remains to be established.

Despite the importance of B cells in the pathogenesis of RA, none of the gene effects described above have been identified in the current generation of

GWA studies. However, variants at *CD40* in European patients do carry risk [15]. CD40 expressed on B cells, via interaction with its ligand CD154 on CD4+ T cells, promotes immunoglobulin class switching, and germinal centre formation. B cells, however, also have a regulatory role, likely to be mediated by IL-10, and disruption of this function may be another route to autoimmune disease [86].

Post-Translational Modification: *PADI4*

Peptidyl arginine deiminase-4 (PADI4) is a member of the enzyme family responsible for the post-translational citrullination of arginine residues in RA synovium, subsequently recognized by anti-cyclic citrullinated protein antibodies. In Japanese [87] and Korean patients [88], case-control association studies have identified functional haplotypes of *PADI4* conferring risk of RA. However, in Caucasian populations this association is inconsistent [89–91].

CONCLUSION

Even with the proliferation of new genetic associations discovered in the past few years by GWA studies, only around 10 to 15% of the inherited risk for SLE and RA can be currently explained. This may be accounted for, in part, by a number of factors, some related to limitations of recent study design. As mentioned above, even the largest current GWA cohorts have limited power to detect associations with ORs < 1.3, potentially losing multiple risk genes. By definition, most genotyped SNPs are common, and so rare but causal variants have a tendency to be missed. These rarer SNPs may be either those with a low minor allele frequency (< 5%), or occur *de novo*, of which 200 to 500 non-synonymous SNPs are expected per individual [92]. In many cases, it is far from certain if the associated SNP is functional, or in linkage disequilibrium with the true cause. Finally, the great majority of GWA studies have been conducted on European populations, thereby excluding carriers of many potential risk variants from analysis. However, it is unfortunately the case that current genotyping platforms often have poor coverage of tagging SNPs within populations that exhibit low levels of genomic linkage disequilibrium, such as those of African ancestry [93]. For example, the latest high-density genotyping chips from Affymetrix (6.0) and Illumina (1 M) may capture fewer than half the SNPs identified through re-sequencing in Yoruban Nigerians [94]. Given that clear differences exist in the risk of autoimmune disease according to ethnicity, and that not all disease risk alleles are in common, it is imperative that full account of this variation is made. Structural genetic differences have only recently begun to be assessed by modern genotyping platforms, and the contribution of, for example, CNV to inherited disease risk is largely unquantified. Even more difficult to appreciate is the influence of heritable epigenetic factors, and the exact relationship between genotype and phenotype. Nevertheless, although it will probably not be possible to explain all the observed genetic risk in the near future, we are rapidly moving towards the ability to quickly and cheaply fully sequence individual genomes [95], with all the advantages that brings [96]. In the meantime, understanding the functional basis of the disease risk variants so far identified presents an outstanding challenge. Integration of genotypic with RNA and protein expression data in a systems biologic approach represents one potentially valuable methodology [97]. Exploring and therapeutically utilizing the genetic differences between individuals is axiomatic to personalized medicine, and will undoubtedly lead to better outcomes in the management of autoimmune disease.

REFERENCES

1.   International HapMap Consortium, Frazer KA, Ballinger DG, Cox DR, Hinds DA, Stuve LL, Gibbs RA, Belmont JW, Boudreau A, Hardenbol P, Leal SM, Pasternak S, Wheeler DA, Willis TD, Yu F, Yang H, Zeng C, Gao Y, Hu H, Hu W, Li C, Lin W, Liu S, Pan H, Tang X, Wang J, Wang W, Yu J, Zhang B, Zhang Q, et al.: A second generation human haplotype map of over 3.1 million SNPs. Nature 2007, 449:851–861.

2.   Risch N, Merikangas K: The future of genetic studies of complex human diseases. Science 1996, 273:1516–1517.

3.   Rioux JD, Xavier RJ, Taylor KD, Silverberg MS, Goyette P, Huett A, Green T, Kuballa P, Barmada MM, Datta LW, Shugart YY, Griffiths AM, Targan SR, Ippoliti AF, Bernard EJ, Mei L, Nicolae DL, Regueiro M, Schumm LP, Steinhart AH, Rotter JI, Duerr RH, Cho JH, Daly MJ, Brant SR: Genome-wide association study identifies new susceptibility loci for Crohn disease and implicates autophagy in disease pathogenesis. Nat Genet 2007, 39:596–604.

4.   Wellcome Trust Case Control Consortium: Genome-wide association study of 14,000 cases of seven common diseases and 3,000 shared controls. Nature 2007, 447:661–678.

5.   Sebat J, Lakshmi B, Troge J, Alexander J, Young J, Lundin P, Månér S, Massa H, Walker M, Chi M, Navin N, Lucito R, Healy J, Hicks J, Ye K, Reiner A, Gilliam TC, Trask B, Patterson N, Zetterberg A, Wigler M: Large-scale copy number polymorphism in the human genome. Science 2004, 305:525–528.

6.   Redon R, Ishikawa S, Fitch KR, Feuk L, Perry GH, Andrews TD, Fiegler H, Shapero MH, Carson AR, Chen W, Cho EK, Dallaire S, Freeman

JL, González JR, Gratacòs M, Huang J, Kalaitzopoulos D, Komura D, MacDonald JR, Marshall CR, Mei R, Montgomery L, Nishimura K, Okamura K, Shen F, Somerville MJ, Tchinda J, Valsesia A, Woodwark C, Yang F, et al.: Global variation in copy number in the human genome. Nature 2006, 444:444–454.

7.   Stranger BE, Forrest MS, Dunning M, Ingle CE, Beazley C, Thorne N, Redon R, Bird CP, de Grassi A, Lee C, Tyler-Smith C, Carter N, Scherer SW, Tavaré S, Deloukas P, Hurles ME, Dermitzakis ET: Relative impact of nucleotide and copy number variation on gene expression phenotypes. Science 2007, 315:848–853.

8.   Aitman TJ, Dong R, Vyse TJ, Norsworthy PJ, Johnson MD, Smith J, Mangion J, Roberton-Lowe C, Marshall AJ, Petretto E, Hodges MD, Bhangal G, Patel SG, Sheehan-Rooney K, Duda M, Cook PR, Evans DJ, Domin J, Flint J, Boyle JJ, Pusey CD, Cook HT: Copy number polymorphism in Fcgr3 predisposes to glomerulonephritis in rats and humans. Nature 2006, 439:851–855.

9.   Hollox EJ, Huffmeier U, Zeeuwen PL, Palla R, Lascorz J, Rodijk-Olthuis D, Kerkhof PC, Traupe H, de Jongh G, den Heijer M, Reis A, Armour JA, Schalkwijk J: Psoriasis is associated with increased beta-defensin genomic copy number. Nat Genet 2008, 40:23–25.

10. Yang Y, Chung EK, Wu YL, Savelli SL, Nagaraja HN, Zhou B, Hebert M, Jones KN, Shu Y, Kitzmiller K, Blanchong CA, McBride KL, Higgins GC, Rennebohm RM, Rice RR, Hackshaw KV, Roubey RA, Grossman JM, Tsao BP, Birmingham DJ, Rovin BH, Hebert LA, Yu CY: Gene copy-number variation and associated polymorphisms of complement component C4 in human systemic lupus erythematosus (SLE): low copy number is a risk factor for and high copy number is a protective factor against SLE susceptibility in European Americans. Am J Hum Genet 2007, 80:1037–1054.

11. Felsenfeld G, Groudine M: Controlling the double helix. Nature 2003, 421:448–453.

12. Horton R, Wilming L, Rand V, Lovering RC, Bruford EA, Khodiyar VK, Lush MJ, Povey S, Talbot CC Jr, Wright MW, Wain HM, Trowsdale J, Ziegler A, Beck S: Gene map of the extended human MHC. Nat Rev Genet 2004, 5:889–899.

13. Fernando MM, Stevens CR, Walsh EC, De Jager PL, Goyette P, Plenge RM, Vyse TJ, Rioux JD: Defining the role of the MHC in autoimmunity: a review and pooled analysis. PLoS Genet 2008, 4:e1000024.

14. Raychaudhuri S, Remmers EF, Lee AT, Hackett R, Guiducci C, Burtt NP, Gianniny L, Korman BD, Padyukov L, Kurreeman FA, Chang M, Catanese JJ, Ding B, Wong S, Helmvan Mil AH, Neale BM, Coblyn J, Cui J, Tak PP, Wolbink GJ, Crusius JB, Horst-Bruinsma IE, Criswell LA, Amos CI, Seldin MF, Kastner DL, Ardlie KG, Alfredsson L, Costenbader KH, Altshuler D, et al.: Common variants at CD40 and other loci confer risk of rheumatoid arthritis. Nat Genet 2008, 40:1216–1223.

15. Ding B, Padyukov L, Lundström E, Seielstad M, Plenge RM, Oksenberg JR, Gregersen PK, Alfredsson L, Klareskog L: Different patterns of associations with anti-citrullinated protein antibody-positive and anti-citrullinated protein antibody-negative rheumatoid arthritis in the extended major histocompatibility complex region. Arthritis Rheum 2009, 60:30–38.

16. International Consortium for Systemic Lupus Erythematosus Genetics (SLEGEN), Harley JB, Alarcón-Riquelme ME, Criswell LA, Jacob CO, Kimberly RP, Moser KL, Tsao BP, Vyse TJ, Lange-feld CD, Nath SK, Guthridge JM, Cobb BL, Mirel DB, Marion MC, Williams AH, Divers J, Wang W, Frank SG, Namjou B, Gabriel SB, Lee AT, Gregersen PK, Behrens TW, Taylor KE, Fernando M, Zidovetzki R, Gaffney PM, Edberg JC, Rioux JD, et al.: Genome-wide association scan in women with systemic lupus erythematosus identifies susceptibility variants in ITGAM, PXK, KIAA1542 and other loci. Nat Genet 2008, 40:204–210.

17. Lewis MJ, Botto M: Complement deficiencies in humans and animals: links to autoimmunity. Autoimmunity 2006, 39:367–378.

18. Fernando MM, Stevens CR, Sabeti PC, Walsh EC, McWhinnie AJ, Shah A, Green T, Rioux JD, Vyse TJ: Identification of two independent risk factors for lupus within the MHC in United Kingdom families. PLoS Genet 2007, 3:e192.

19. Harley IT, Kaufman KM, Langefeld CD, Harley JB, Kelly JA: Genetic susceptibility to SLE: new insights from fine mapping and genome-wide association studies. Nat Rev Genet 2009, 10:285–290.

20. Sekine H, Ferreira RC, Pan-Hammarström Q, Graham RR, Ziemba B, de Vries SS, Liu J, Hippen K, Koeuth T, Ortmann W, Iwahori A, Elliott MK, Offer S, Skon C, Du L, Novitzke J, Lee AT, Zhao N, Tompkins JD, Altshuler D, Gregersen PK, Cunningham-Rundles C, Harris RS, Her C, Nelson DL, Hammarström L, Gilkeson GS, Behrens TW: Role for Msh5 in the regulation of Ig class switch recombination. Proc Natl Acad Sci USA 2007, 104:7193–7198.

21. Brown MA, Pile KD, Kennedy LG, Calin A, Darke C, Bell J, Wordsworth BP, Cornélis F: HLA class I associations of ankylosing spondylitis in the white population in the United Kingdom. Ann Rheum Dis 1996, 55:268–270.

22. Brown MA, Kennedy LG, Darke C, Gibson K, Pile KD, Shatford JL, Taylor A, Calin A, Wordsworth BP: The effect of HLA-DR genes on susceptibility to and severity of ankylosing spondylitis. Arthritis Rheum 1998, 41:460–465.

23. Wellcome Trust Case Control Consortium; Australo-Anglo-American Spondylitis Consortium (TASC), Burton PR, Clayton DG, Cardon LR, Craddock N, Deloukas P, Duncanson A, Kwiatkowski DP, McCarthy MI, Ouwehand WH, Samani NJ, Todd JA, Donnelly P, Barrett JC, Davison D, Easton D, Evans DM, Leung HT, Marchini JL, Morris AP, Spencer CC, Tobin MD, Attwood AP, Boorman JP, Cant B, Everson U, Hussey JM, Jolley JD, Knight AS, et al.: Association scan of 14,500 nonsynonymous SNPs in four diseases identifies autoimmunity variants. Nat Genet 2007, 39:1329–1337.

24. Hammer GE, Gonzalez F, James E, Nolla H, Shastri N: In the absence of aminopeptidase ERAAP, MHC class I molecules present many unstable and highly immunogenic peptides. Nat Immunol 2007, 8:101–108.

25. Cui X, Rouhani FN, Hawari F, Levine SJ: Shedding of the type II IL-1 decoy receptor requires a multifunctional aminopeptidase, aminopeptidase regulator of TNF receptor type 1 shedding. J Immunol 2003, 171:6814–6819.

26. Cui X, Rouhani FN, Hawari F, Levine SJ: An aminopeptidase, ARTS-1, is required for interleukin-6 receptor shedding. J Biol Chem 2003,278:28677–28685.

27. Cui X, Hawari F, Alsaaty S, Lawrence M, Combs CA, Geng W, Rouhani FN, Miskinis D, Levine SJ: Identification of ARTS-1 as a novel TNFR1-binding protein that promotes TNFR1 ectodomain shedding. J Clin Invest 2002, 110:515–526.

28. Ronnblom L, Eloranta ML, Alm GV: The type I interferon system in systemic lupus erythematosus. Arthritis Rheum 2006, 54:408–420.

29. Sigurdsson S, Nordmark G, Göring HH, Lindroos K, Wiman AC, Sturfelt G, Jönsen A, Rantapää-Dahlqvist S, Möller B, Kere J, Koskenmies S, Widén E, Eloranta ML, Julkunen H, Kristjansdottir H, Steinsson K, Alm G, Rönnblom L, Syvänen AC: Polymorphisms in the tyrosine kinase 2 and interferon regulatory factor 5 genes are associated with systemic lupus erythematosus. Am J Hum Genet 2005,76:528–537.

30. Sigurdsson S, Padyukov L, Kurreeman FA, Liljedahl U, Wiman AC, Alfredsson L, Toes R, Rönnelid J, Klareskog L, Huizinga TW, Alm G, Syvänen AC, Rönnblom L: Association of a haplotype in the promoter region of the interferon regulatory factor 5 gene with rheumatoid arthritis. Arthritis Rheum 2007, 56:2202–2210.

31. Graham RR, Kyogoku C, Sigurdsson S, Vlasova IA, Davies LR, Baechler EC, Plenge RM, Koeuth T, Ortmann WA, Hom G, Bauer JW, Gillett C, Burtt N, Cunninghame Graham DS, Onofrio R, Petri M, Gunnarsson I, Svenungsson E, Rönnblom L, Nordmark G, Gregersen PK, Moser K, Gaffney PM, Criswell LA, Vyse TJ, Syvänen AC, Bohjanen PR, Daly MJ, Behrens TW, Altshuler D: Three functional variants of IFN regulatory factor 5 (IRF5) define risk and protective haplotypes for human lupus. Proc Natl Acad Sci USA 2007, 104:6758–6763.

32. Hom G, Graham RR, Modrek B, Taylor KE, Ortmann W, Garnier S, Lee AT, Chung SA, Ferreira RC, Pant PV, Ballinger DG, Kosoy R, Demirci FY, Kamboh MI, Kao AH, Tian C, Gunnarsson I, Bengtsson AA, Rantapää-Dahlqvist S, Petri M, Manzi S, Seldin MF, Rönnblom L, Syvänen AC, Criswell LA, Gregersen PK, Behrens TW: Association of systemic lupus erythematosus with C8orf13-BLK and ITGAM-ITGAX. N Engl J Med 2008, 358:900–909.

33. Cunninghame Graham DS, Manku H, Wagner S, Reid J, Timms K, Gutin A, Lanchbury JS, Vyse TJ: Association of IRF5 in UK SLE families identifies a variant involved in polyadenylation. Hum Mol Genet 2007, 16:579–591.

34. Graham RR, Kozyrev SV, Baechler EC, Reddy MV, Plenge RM, Bauer JW, Ortmann WA, Koeuth T, González Escribano MF, Argentine and Spanish Collaborative Groups, Pons-Estel B, Petri M, Daly M, Gregersen PK, Martín J, Altshuler D, Behrens TW, Alarcón-Riquelme ME: A common haplotype of interferon regulatory factor 5 (IRF5) regulates splicing and expression and is associated with increased risk of systemic lupus erythematosus. Nat Genet 2006, 38:550–555.

35. Sigurdsson S, Göring HH, Kristjansdottir G, Milani L, Nordmark G, Sandling JK, Eloranta ML, Feng D, Sangster-Guity N, Gunnarsson I, Svenungsson E, Sturfelt G, Jönsen A, Truedsson L, Barnes BJ, Alm G, Rönnblom L, Syvänen AC: Comprehensive evaluation of the genetic variants of interferon regulatory factor 5 (IRF5) reveals a novel 5 bp length polymorphism as strong risk factor for systemic lupus erythematosus. Hum Mol Genet 2008, 17:872–881.

36. Takaoka A, Yanai H, Kondo S, Duncan G, Negishi H, Mizutani T, Kano S, Honda K, Ohba Y, Mak TW, Taniguchi T: Integral role

of IRF-5 in the gene induction programme activated by Toll-like receptors. Nature 2005, 434:243–249.

37. Dieguez-Gonzalez R, Calaza M, Perez-Pampin E, de la Serna AR, Fernandez-Gutierrez B, Castañeda S, Largo R, Joven B, Narvaez J, Navarro F, Marenco JL, Vicario JL, Blanco FJ, Fernandez-Lopez JC, Caliz R, Collado-Escobar MD, Carreño L, Lopez-Longo J, Cañete JD, Gomez-Reino JJ, Gonzalez A: Association of interferon regulatory factor 5 haplotypes, similar to that found in systemic lupus erythematosus, in a large subgroup of patients with rheumatoid arthritis. Arthritis Rheum 2008, 58:1264–1274.

38. Plenge RM, Cotsapas C, Davies L, Price AL, de Bakker PI, Maller J, Peɔer I, Burtt NP, Blumenstiel B, DeFelice M, Parkin M, Barry R, Winslow W, Healy C, Graham RR, Neale BM, Izmailova E, Roubenoff R, Parker AN, Glass R, Karlson EW, Maher N, Hafler DA, Lee DM, Seldin MF, Remmers EF, Lee AT, Padyukov L, Alfredsson L, Coblyn J, et al.: Two independent alleles at 6q23 associated with risk of rheumatoid arthritis. Nat Genet 2007, 39:1477–1482.

39. Plenge RM, Seielstad M, Padyukov L, Lee AT, Remmers EF, Ding B, Liew A, Khalili H, Chandrasekaran A, Davies LR, Li W, Tan AK, Bonnard C, Ong RT, Thalamuthu A, Pettersson S, Liu C, Tian C, Chen WV, Carulli JP, Beckman EM, Altshuler D, Alfredsson L, Criswell LA, Amos CI, Seldin MF, Kastner DL, Klareskog L, Gregersen PK: TRAF1-C5 as a risk locus for rheumatoid arthritis - a genomewide study. N Engl J Med 2007, 357:1199–1209.

40. Komander D, Barford D: Structure of the A20 OTU domain and mechanistic insights into deubiquitination. Biochem J 2008, 409:77–85.

41. Sun SC: Deubiquitylation and regulation of the immune response. Nat Rev Immunol 2008, 8:501–511.

42. Lee EG, Boone DL, Chai S, Libby SL, Chien M, Lodolce JP, Ma A: Failure to regulate TNF-induced NF-kappaB and cell death responses in A20-deficient mice. Science 2000, 289:2350–2354.

43. Graham RR, Cotsapas C, Davies L, Hackett R, Lessard CJ, Leon JM, Burtt NP, Guiducci C, Parkin M, Gates C, Plenge RM, Behrens TW, Wither JE, Rioux JD, Fortin PR, Graham DC, Wong AK, Vyse TJ, Daly MJ, Altshuler D, Moser KL, Gaffney PM: Genetic variants near TNFAIP3 on 6q23 are associated with systemic lupus erythematosus. Nat Genet 2008, 40:1059–1061.

44. Bates JS, Lessard CJ, Leon JM, Nguyen T, Battiest LJ, Rodgers J, Kaufman KM, James JA, Gilkeson GS, Kelly JA, Humphrey MB,

Harley JB, Gray-McGuire C, Moser KL, Gaffney PM: Meta-analysis and imputation identifies a 109 kb risk haplotype spanning TNFAIP3 associated with lupus nephritis and hematologic manifestations. Genes Immun 2009, 10:470–477.

45. Kurreeman FA, Padyukov L, Marques RB, Schrodi SJ, Seddighzadeh M, Stoeken-Rijsbergen G, Helmvan Mil AH, Allaart CF, Verduyn W, Houwing-Duistermaat J, Alfredsson L, Begovich AB, Klareskog L, Huizinga TW, Toes RE: A candidate gene approach identifies the TRAF1/C5 region as a risk factor for rheumatoid arthritis. PLoS Med 2007, 4:e278.

46. Kurreeman FA, Rocha D, Houwing-Duistermaat J, Vrijmoet S, Teixeira VH, Migliorini P, Balsa A, Westhovens R, Barrera P, Alves H, Vaz C, Fernandes M, Pascual-Salcedo D, Michou L, Bombardieri S, Radstake T, van Riel P, Putte L, Lopes-Vaz A, Prum B, Bardin T, Gut I, Cornelis F, Huizinga TW, Petit-Teixeira E, Toes RE, European Consortium on Rheumatoid Arthritis Families: Replication of the tumor necrosis factor receptor-associated factor 1/complement component 5 region as a susceptibility locus for rheumatoid arthritis in a European family-based study. Arthritis Rheum 2008, 58:2670–2674.

47. Barton A, Thomson W, Ke X, Eyre S, Hinks A, Bowes J, Gibbons L, Plant D, Wellcome Trust Case Control Consortium, Wilson AG, Marinou I, Morgan A, Emery P, YEAR consortium, Steer S, Hocking L, Reid DM, Wordsworth P, Harrison P, Worthington J: Re-evaluation of putative rheumatoid arthritis susceptibility genes in the post-genome wide association study era and hypothesis of a key pathway underlying susceptibility. Hum Mol Genet 2008, 17:2274–2279.

48. Carpentier I, Beyaert R: TRAF1 is a TNF inducible regulator of NF-kappaB activation. FEBS Lett 1999, 460:246–250.

49. Tsitsikov EN, Laouini D, Dunn IF, Sannikova TY, Davidson L, Alt FW, Geha RS: TRAF1 is a negative regulator of TNF signaling. enhanced TNF signaling in TRAF1-deficient mice. Immunity 2001, 15:647–657.

50. Wang Y, Rollins SA, Madri JA, Matis LA: Anti-C5 monoclonal antibody therapy prevents collagen-induced arthritis and ameliorates established disease. Proc Natl Acad Sci USA 1995, 92:8955–8959.

51. Wang Y, Kristan J, Hao L, Lenkoski CS, Shen Y, Matis LA: A role for complement in antibody-mediated inflammation: C5-deficient DBA/1 mice are resistant to collagen-induced arthritis. J Immunol 2000, 164:4340–4347.

52. Panoulas VF, Smith JP, Nightingale P, Kitas GD: Association of the TRAF1/C5 locus with increased mortality, particularly from malignancy or sepsis, in patients with rheumatoid arthritis. Arthritis Rheum 2009, 60:39–46.

53. Larson RS, Springer TA: Structure and function of leukocyte integrins. Immunol Rev 1990, 114:181–217.

54. Sohn JH, Bora PS, Suk HJ, Molina H, Kaplan HJ, Bora NS: Tolerance is dependent on complement C3 fragment iC3b binding to antigen-presenting cells. Nat Med 2003, 9:206–212.

55. Ehirchiou D, Xiong Y, Xu G, Chen W, Shi Y, Zhang L: CD11b facilitates the development of peripheral tolerance by suppressing Th17 differentiation. J Exp Med 2007, 204:1519–1524.

56. Han S, Kim-Howard X, Deshmukh H, Kamatani Y, Viswanathan P, Guthridge JM, Thomas K, Kaufman KM, Ojwang J, Rojas-Villarraga A, Baca V, Orozco L, Rhodes B, Choi CB, Gregersen PK, Merrill JT, James JA, Gaffney PM, Moser KL, Jacob CO, Kimberly RP, Harley JB, Bae SC, Anaya JM, Alarcón-Riquelme ME, Matsuda K, Vyse TJ, Nath SK: Evaluation of imputation-based association in and around the integrin-alpha-M (ITGAM) gene and replication of robust association between a non-synonymous functional variant within ITGAM and systemic lupus erythematosus (SLE). Hum Mol Genet 2009, 18:1171–1180.

57. Nath SK, Han S, Kim-Howard X, Kelly JA, Viswanathan P, Gilkeson GS, Chen W, Zhu C, McEver RP, Kimberly RP, Alarcón-Riquelme ME, Vyse TJ, Li QZ, Wakeland EK, Merrill JT, James JA, Kaufman KM, Guthridge JM, Harley JB: A nonsynonymous functional variant in integrin-alpha(M) (encoded by ITGAM) is associated with systemic lupus erythematosus. Nat Genet 2008, 40:152–154.

58. Han S, Kim-Howard X, Deshmukh H, Kamatani Y, Viswanathan P, Guthridge JM, Thomas K, Kaufman KM, Ojwang J, Rojas-Villarraga A, Baca V, Orozco L, Rhodes B, Choi CB, Gregersen PK, Merrill JT, James JA, Gaffney PM, Moser KL, Jacob CO, Kimberly RP, Harley JB, Bae SC, Anaya JM, Alarcón-Riquelme ME, Matsuda K, Vyse TJ, Nath SK: Evaluation of imputation-based association in and around the integrin-alpha-M (ITGAM) gene and replication of robust association between a non-synonymous functional variant within ITGAM and systemic lupus erythematosus (SLE). Hum Mol Genet 2009, 18:1171–180.

59. Willcocks LC, Lyons PA, Clatworthy MR, Robinson JI, Yang W, Newland SA, Plagnol V, McGovern NN, Condliffe AM, Chilvers ER, Adu D, Jolly EC, Watts R, Lau YL, Morgan AW, Nash G, Smith KG: Copy number

of FCGR3B, which is associated with systemic lupus erythematosus, correlates with protein expression and immune complex uptake. J Exp Med 2008, 205:1573–1582.

60.   Fanciulli M, Norsworthy PJ, Petretto E, Dong R, Harper L, Kamesh L, Heward JM, Gough SC, de Smith A, Blakemore AI, Froguel P, Owen CJ, Pearce SH, Teixeira L, Guillevin L, Graham DS, Pusey CD, Cook HT, Vyse TJ, Aitman TJ: FCGR3B copy number variation is associated with susceptibility to systemic, but not organ-specific, autoimmunity. Nat Genet 2007, 39:721–723.

61.   Begovich AB, Carlton VE, Honigberg LA, Schrodi SJ, Chokkalingam AP, Alexander HC, Ardlie KG, Huang Q, Smith AM, Spoerke JM, Conn MT, Chang M, Chang SY, Saiki RK, Catanese JJ, Leong DU, Garcia VE, McAllister LB, Jeffery DA, Lee AT, Batliwalla F, Remmers E, Criswell LA, Seldin MF, Kastner DL, Amos CI, Sninsky JJ, Gregersen PK: A missense single-nucleotide polymorphism in a gene encoding a protein tyrosine phosphatase (PTPN22) is associated with rheumatoid arthritis. Am J Hum Genet 2004, 75:330–337.

62.   Jawaheer D, Seldin MF, Amos CI, Chen WV, Shigeta R, Etzel C, Damle A, Xiao X, Chen D, Lum RF, Monteiro J, Kern M, Criswell LA, Albani S, Nelson JL, Clegg DO, Pope R, Schroeder HW Jr, Bridges SL Jr, Pisetsky DS, Ward R, Kastner DL, Wilder RL, Pincus T, Callahan LF, Flemming D, Wener MH, Gregersen PK, North American Rheumatoid Arthritis Consortium: Screening the genome for rheumatoid arthritis susceptibility genes: a replication study and combined analysis of 512 multicase families. Arthritis Rheum 2003, 48:906–916.

63.   Lee HS, Korman BD, Le JM, Kastner DL, Remmers EF, Gregersen PK, Bae SC: Genetic risk factors for rheumatoid arthritis differ in Caucasian and Korean populations. Arthritis Rheum 2009, 60:364–371.

64.   Ikari K, Momohara S, Inoue E, Tomatsu T, Hara M, Yamanaka H, Kamatani N: Haplotype analysis revealed no association between the PTPN22 gene and RA in a Japanese population. Rheumatology (Oxford) 2006, 45:1345–1348.

65.   Cloutier JF, Veillette A: Cooperative inhibition of T-cell antigen receptor signaling by a complex between a kinase and a phosphatase. J Exp Med 1999, 189:111–121.

66.   Gjörloff-Wingren A, Saxena M, Williams S, Hammi D, Mustelin T: Characterization of TCR-induced receptor-proximal signaling events negatively regulated by the protein tyrosine phosphatase PEP. Eur J Immunol 1999, 29:3845–3854.

67. Vang T, Miletic AV, Arimura Y, Tautz L, Rickert RC, Mustelin T: Protein tyrosine phosphatases in autoimmunity. Annu Rev Immunol2008, 26:29–55.

68. Orrú V, Tsai SJ, Rueda B, Fiorillo E, Stanford SM, Dasgupta J, Hartiala J, Zhao L, Ortego-Centeno N, D›Alfonso S, Italian Collaborative Group, Arnett FC, Wu H, Gonzalez-Gay MA, Tsao BP, Pons-Estel B, Alarcon-Riquelme ME, He Y, Zhang ZY, Allayee H, Chen XS, Martin J, Bottini N: A loss-of-function variant of PTPN22 is associated with reduced risk of systemic lupus erythematosus. Hum Mol Genet2009, 18:569–579.

69. Orozco G, García-Porrúa C, López-Nevot MA, Raya E, González-Gay MA, Martín J: Lack of association between ankylosing spondylitis and a functional polymorphism of PTPN22 proposed as a general susceptibility marker for autoimmunity. Ann Rheum Dis 2006,65:687–688.

70. Watford WT, Hissong BD, Bream JH, Kanno Y, Muul L, O›Shea JJ: Signaling by IL-12 and IL-23 and the immunoregulatory roles of STAT4. Immunol Rev 2004, 202:139–156.

71. Mathur AN, Chang HC, Zisoulis DG, Stritesky GL, Yu Q, O›Malley JT, Kapur R, Levy DE, Kansas GS, Kaplan MH: Stat3 and Stat4 direct development of IL-17-secreting Th cells. J Immunol 2007, 178:4901–4907.

72. Remmers EF, Plenge RM, Lee AT, Graham RR, Hom G, Behrens TW, de Bakker PI, Le JM, Lee HS, Batliwalla F, Li W, Masters SL, Booty MG, Carulli JP, Padyukov L, Alfredsson L, Klareskog L, Chen WV, Amos CI, Criswell LA, Seldin MF, Kastner DL, Gregersen PK: STAT4 and the risk of rheumatoid arthritis and systemic lupus erythematosus. N Engl J Med 2007, 357:977–986.

73. Namjou B, Sestak AL, Armstrong DL, Zidovetzki R, Kelly JA, Jacob N, Ciobanu V, Kaufman KM, Ojwang JO, Ziegler J, Quismorio FP Jr, Reiff A, Myones BL, Guthridge JM, Nath SK, Bruner GR, Mehrian-Shai R, Silverman E, Klein-Gitelman M, McCurdy D, Wagner-Weiner L, Nocton JJ, Putterman C, Bae SC, Kim YJ, Petri M, Reveille JD, Vyse TJ, Gilkeson GS, Kamen DL, et al.: High-density genotyping of STAT4 reveals multiple haplotypic associations with systemic lupus erythematosus in different racial groups. Arthritis Rheum 2009, 60:1085–1095.

74. Kastelein RA, Hunter CA, Cua DJ: Discovery and biology of IL-23 and IL-27: related but functionally distinct regulators of inflammation. Annu Rev Immunol 2007, 25:221–242.

75. Nair RP, Duffin KC, Helms C, Ding J, Stuart PE, Goldgar D, Gudjonsson JE, Li Y, Tejasvi T, Feng BJ, Ruether A, Schreiber S, Weichenthal M,

Gladman D, Rahman P, Schrodi SJ, Prahalad S, Guthery SL, Fischer J, Liao W, Kwok PY, Menter A, Lathrop GM, Wise CA, Begovich AB, Voorhees JJ, Elder JT, Krueger GG, Bowcock AM, Abecasis GR, Collaborative Association Study of Psoriasis: Genome-wide scan reveals association of psoriasis with IL-23 and NF-kappaB pathways. Nat Genet 2009, 41:199–204.

76. Silverberg MS, Cho JH, Rioux JD, McGovern DP, Wu J, Annese V, Achkar JP, Goyette P, Scott R, Xu W, Barmada MM, Klei L, Daly MJ, Abraham C, Bayless TM, Bossa F, Griffiths AM, Ippoliti AF, Lahaie RG, Latiano A, Paré P, Proctor DD, Regueiro MD, Steinhart AH, Targan SR, Schumm LP, Kistner EO, Lee AT, Gregersen PK, Rotter JI, et al.: Ulcerative colitis-risk loci on chromosomes 1p36 and 12q15 found by genome-wide association study. Nat Genet 2009, 41:216–220.

77. Yokoyama K, Su Ih IH, Tezuka T, Yasuda T, Mikoshiba K, Tarakhovsky A, Yamamoto T: BANK regulates BCR-induced calcium mobilization by promoting tyrosine phosphorylation of IP(3) receptor. EMBO J 2002, 21:83–92.

78. Kozyrev SV, Abelson AK, Wojcik J, Zaghlool A, Linga Reddy MV, Sanchez E, Gunnarsson I, Svenungsson E, Sturfelt G, Jönsen A, Truedsson L, Pons-Estel BA, Witte T, D›Alfonso S, Barizzone N, Danieli MG, Gutierrez C, Suarez A, Junker P, Laustrup H, González-Escribano MF, Martin J, Abderrahim H, Alarcón-Riquelme ME: Functional variants in the B-cell gene BANK1 are associated with systemic lupus erythematosus. Nat Genet 2008, 40:211–216.

79. Kurosaki T, Hikida M: Tyrosine kinases and their substrates in B lymphocytes. Immunol Rev 2009, 228:132–148.

80. Hibbs ML, Tarlinton DM, Armes J, Grail D, Hodgson G, Maglitto R, Stacker SA, Dunn AR: Multiple defects in the immune system of Lyn-deficient mice, culminating in autoimmune disease. Cell 1995, 83:301–311.

81. Lu R, Vidal GS, Kelly JA, Delgado-Vega AM, Howard XK, Macwana SR, Dominguez N, Klein W, Burrell C, Harley IT, Kaufman KM, Bruner GR, Moser KL, Gaffney PM, Gilkeson GS, Wakeland EK, Li QZ, Langefeld CD, Marion MC, Divers J, Alarcón GS, Brown EE, Kimberly RP, Edberg JC, Ramsey-Goldman R, Reveille JD, McGwin G Jr, Vilá LM, Petri MA, Bae SC, et al.: Genetic associations of LYN with systemic lupus erythematosus. Genes Immun 2009, 10:397–403.

82. Croft M: The role of TNF superfamily members in T-cell function and diseases. Nat Rev Immunol 2009, 9:271–285.

83. Cunninghame Graham DS, Graham RR, Manku H, Wong AK, Whittaker JC, Gaffney PM, Moser KL, Rioux JD, Altshuler D, Behrens TW, Vyse TJ: Polymorphism at the TNF superfamily gene TNFSF4 confers susceptibility to systemic lupus erythematosus. Nat Genet2008, 40:83–89.

84. Manku H, Graham DS, Vyse TJ: Association of the co-stimulator OX40L with systemic lupus erythematosus. J Mol Med 2009, 87:229–234.

85. Fillatreau S, Gray D, Anderton SM: Not always the bad guys: B cells as regulators of autoimmune pathology. Nat Rev Immunol 2008,8:391–397.

86. Suzuki A, Yamada R, Chang X, Tokuhiro S, Sawada T, Suzuki M, Nagasaki M, Nakayama-Hamada M, Kawaida R, Ono M, Ohtsuki M, Furukawa H, Yoshino S, Yukioka M, Tohma S, Matsubara T, Wakitani S, Teshima R, Nishioka Y, Sekine A, Iida A, Takahashi A, Tsunoda T, Nakamura Y, Yamamoto K: Functional haplotypes of PADI4, encoding citrullinating enzyme peptidylarginine deiminase 4, are associated with rheumatoid arthritis. Nat Genet 2003, 34:395–402.

87. Kang N, Clarke AJ, Nicholson IA, Chard RB: Circulatory arrest for repair of postcoarctation site aneurysm. Ann Thorac Surg 2004,77:2029–2033.

88. Gandjbakhch F, Fajardy I, Ferré B, Dubucquoi S, Flipo RM, Roger N, Solau-Gervais E: A functional haplotype of PADI4 gene in rheumatoid arthritis: positive correlation in a French population. J Rheumatol 2009, 36:881–886.

89. Martinez A, Valdivia A, Pascual-Salcedo D, Lamas JR, Fernández-Arquero M, Balsa A, Fernández-Gutiérrez B, de la Concha EG, Urcelay E:PADI4 polymorphisms are not associated with rheumatoid arthritis in the Spanish population. Rheumatology (Oxford) 2005,44:1263–1266.

90. Caponi L, Petit-Teixeira E, Sebbag M, Bongiorni F, Moscato S, Pratesi F, Pierlot C, Osorio J, Chapuy-Regaud S, Guerrin M, Cornelis F, Serre G, Migliorini P, ECRAF: A family based study shows no association between rheumatoid arthritis and the PADI4 gene in a white French population. Ann Rheum Dis 2005, 64:587–593.

91. Ng PC, Levy S, Huang J, Stockwell TB, Walenz BP, Li K, Axelrod N, Busam DA, Strausberg RL, Venter JC: Genetic variation in an individual human exome. PLoS Genet 2008, 4:e1000160.

92. Jakobsson M, Scholz SW, Scheet P, Gibbs JR, VanLiere JM, Fung HC, Szpiech ZA, Degnan JH, Wang K, Guerreiro R, Bras JM, Schymick JC, Hernandez DG, Traynor BJ, Simon-Sanchez J, Matarin M, Britton A, Leemput J, Rafferty I, Bucan M, Cann HM, Hardy JA, Rosenberg

NA, Singleton AB: Genotype, haplotype and copy-number variation in worldwide human populations. Nature 2008, 451:998–1003.

93. Bhangale TR, Rieder MJ, Nickerson DA: Estimating coverage and power for genetic association studies using near-complete variation data. Nat Genet 2008, 40:841–843.

94. Bentley DR, Balasubramanian S, Swerdlow HP, Smith GP, Milton J, Brown CG, Hall KP, Evers DJ, Barnes CL, Bignell HR, Boutell JM, Bryant J, Carter RJ, Keira Cheetham R, Cox AJ, Ellis DJ, Flat-bush MR, Gormley NA, Humphray SJ, Irving LJ, Karbelashvili MS, Kirk SM, Li H, Liu X, Maisinger KS, Murray LJ, Obradovic B, Ost T, Parkinson ML, Pratt MR, et al.: Accurate whole human genome sequencing using reversible terminator chemistry. Nature 2008, 456:53–59.

95. Kryukov GV, Shpunt A, Stamatoyannopoulos JA, Sunyaev SR: Power of deep, all-exon resequencing for discovery of human trait genes. Proc Natl Acad Sci USA 2009, 106:3871–3876.

96. Chen Y, Zhu J, Lum PY, Yang X, Pinto S, MacNeil DJ, Zhang C, Lamb J, Edwards S, Sieberts SK, Leonardson A, Castellini LW, Wang S, Champy MF, Zhang B, Emilsson V, Doss S, Ghazalpour A, Horvath S, Drake TA, Lusis AJ, Schadt EE: Variations in DNA elucidate molecular networks that cause disease. Nature 2008, 452:429–435.

Chapter 9

STATISTICAL DESIGN OF PERSONALIZED MEDICINE INTERVENTIONS: THE CLARIFICATION OF OPTIMAL ANTICOAGULATION THROUGH GENETICS (COAG) TRIAL

Benjamin French[1] , Jungnam Joo[2], Nancy L Geller[2], Stephen E Kimmel[1], Yves Rosenberg[3], Jeffrey L Anderson[4], Brian F Gage[5], Julie A Johnson[6], Jonas H Ellenberg[1]

[1]Department of Biostatistics and Epidemiology, University of Pennsylvania School of Medicine, 423 Guardian Drive, Philadelphia, Pennsylvania 19104 USA

[2] Office of Biostatistics Research, National Heart, Lung and Blood Institute, 6701 Rockledge Drive MSC 7913, Bethesda, Maryland 20892 USA

[3] Atherothrombosis and Coronary Artery Disease Branch, National Heart, Lung and Blood Institute, 6701 Rockledge Drive MSC 7956, Bethesda, Maryland 20892 USA

[4] JL Sorenson Heart and Lung Center, Intermountain Medical Center, 5121 S Cottonwood St, Murray, Utah 84107 USA

[5] Division of General Medical Sciences, Washington University School of Medicine, 660 S Euclid Ave, St. Louis, Missouri 63110 USA

[6] Department of Pharmacotherapy and Translational Research, University of Florida College of Pharmacy, Box 100486, Gainesville, Florida 32610 USA.

ABSTRACT

Background

There is currently much interest in pharmacogenetics: determining variation in genes that regulate drug effects, with a particular emphasis on improving drug safety and efficacy. The ability to determine such variation motivates the application of personalized drug therapies that utilize a patient's genetic makeup to determine a safe and effective drug at the correct dose. To ascertain whether a genotype-guided drug therapy improves patient care, a personalized medicine intervention may be evaluated within the framework of a randomized controlled trial. The statistical design of this type of personalized medicine intervention requires special considerations: the distribution of relevant allelic variants in the study population; and whether the pharmacogenetic intervention is equally effective across subpopulations defined by allelic variants.

Methods

The statistical design of the Clarification of Optimal Anticoagulation through Genetics (COAG) trial serves as an illustrative example of a personalized medicine intervention that uses each subject's genotype information. The COAG trial is a multicenter, double blind, randomized clinical trial that will compare two approaches to initiation of warfarin therapy: genotype-guided dosing, the initiation of warfarin therapy based on algorithms using clinical information and genotypes for polymorphisms in *CYP2C9* and *VKORC1*; and clinical-guided dosing, the initiation of warfarin therapy based on algorithms using only clinical information.

Results

We determine an absolute minimum detectable difference of 5.49% based on an assumed 60% population prevalence of zero or multiple genetic variants in either *CYP2C9* or *VKORC1* and an assumed 15% relative effectiveness of genotype-guided warfarin initiation for those with zero or multiple genetic variants. Thus we calculate a sample size of 1238 to achieve a power level of 80% for the primary outcome. We show that reasonable departures from these assumptions may decrease statistical power to 65%.

Conclusions

In a personalized medicine intervention, the minimum detectable difference used in sample size calculations is not a known quantity, but rather an unknown quantity that depends on the genetic makeup of the subjects enrolled. Given the possible sensitivity of sample size and power calculations to these key assumptions, we recommend that they be monitored during the conduct of a personalized medicine intervention.

BACKGROUND

Personalized Medicine Interventions

The recent availability of lower-cost genetic testing has motivated medical researchers to determine whether patient care and safety is improved by using a patient's genetic information to initiate and manage drug therapy [1]. To evaluate scientific hypotheses regarding a personalized medicine intervention, a randomized clinical trial can be used to contrast outcomes between subjects randomized to receive genotype-guided drug therapy and those randomized to receive an identical therapy without reference to their genetic characteristics [2]. However, because not all subjects may benefit from the pharmacologic intervention due to their genetic makeup, genotype-guided therapy may not benefit the entire study population. Hence, any putative difference between treatment groups will be attenuated, which may adversely impact key components of the statistical design, such as sample size and statistical power. Therefore, the primary statistical challenge of designing a personalized therapy intervention is to accommodate the potential differential effectiveness of genotype-guided therapy across subpopulations defined by allelic variation.

Although interventions that use a subject's clinical factors, gene expression profile, or perhaps other factors can also be considered as personalized medicine, we restrict our attention to interventions that use genotype. In addition, personalized medicine interventions may be evaluated using several different study designs. For example, in a targeted design, frequently used to evaluate genetic-based therapies for cancer, study eligibility may be restricted to a marker-positive subset of the population anticipated to benefit from therapy based on their genetic characteristics [3]. We focus on untargeted designs, such as those that have been used to evaluate genotype-guided dosing of warfarin, in which all subjects are enrolled regardless of their genetic characteristics.

Genotype-Guided Dosing of Warfarin

Warfarin sodium is the most common oral anticoagulant used for the prevention and treatment of thromboembolism, the formation of a clot in a blood vessel or cardiac chamber that may be carried by the blood stream and obstruct another vessel. Initiation of warfarin therapy is usually based on empiric dosing, which may put patients at an increased risk for either major bleeding complications due to over-anticoagulation or thromboembolic events due to under-anticoagulation. Therefore, initiation of warfarin therapy at an improper dose may be associated with increased costs and higher morbidity [4].

Many patient-specific clinical factors impact warfarin dose-response. In addition, two genes influence warfarin dose: the cytochrome P-450 family 2 subfamily C polypeptide 9 enzyme (*CYP2C9*) gene effects pharmacokinetics, i.e., the effects of the body on the drug; and the vitamin K epoxide reductase complex 1 (*VKORC1*) gene effects pharmacodynamics, i.e., the effects of the drug on the body. Thus, *CYP2C9* variants alter S-warfarin metabolism [5]; *VKORC1* variants alter warfarin response [6]. Both *CYP2C9* and *VKORC1* have proven useful in algorithms to predict the ultimate maintenance dose for optimal warfarin therapy [7]. However, they have not yet been proven to be beneficial in choosing the initial warfarin dose or to impact clinical outcomes.

The goal of this manuscript is to provide practical guidance on the statistical design of a personalized medicine intervention that uses each subject›s genotype information in an untargeted design. The statistical design of the COAG trial serves as an illustrative example. We briefly summarize the clinical rationale and the general study design for the COAG trial. We use power and sample size calculations to illustrate the primary statistical challenge of designing a personalized therapy intervention: to accommodate the potential differential effectiveness of genotype-guided therapy across subpopulations defined by allelic variation. We provide a sensitivity analysis to quantify the extent to which power and sample size calculations may be sensitive to key assumptions required in the statistical design of a personalized medicine intervention. We conclude with general recommendations for the statistical design of personalized medicine interventions.

METHODS

The objective of the COAG trial (clinicaltrials.gov identifier: NCT00839657) is to conduct a multicenter, double blind, randomized clinical trial that compares two approaches to initiation of warfarin therapy:

Genotype-guided dosing, the initiation of warfarin therapy based on algorithms using clinical information and genotypes for polymorphisms in two genes known to influence warfarin response (*CYP2C9* and *VKORC1*); and

Clinical-guided dosing, the initiation of warfarin therapy based on algorithms using only clinical information.

Both approaches will include a baseline dose-initiation algorithm [8] and a dose-revision algorithm [9] applied after four or five days of warfarin therapy. Subsequent doses will be determined using a standard dose-titration algorithm, which is identical for both groups. By comparing the efficacy of genotype-guided dosing to that of clinical-guided dosing, the COAG trial will determine whether the incremental use of genetic information improves stability of anticoagulation during the early treatment period. Future studies could then determine whether such an improvement leads to significantly reduced costs and lower morbidity.

Eligible subjects will be recruited from at least 12 clinical sites in the United States. Clinical and genotype data will be collected on all subjects. Subjects will be randomized to initiate warfarin therapy either using genotype-guided or clinical-guided dosing. All subjects will receive their warfarin on a standard-of-care schedule. Study investigators, clinicians, and subjects will be blinded to treatment assignment and warfarin dose for the first four weeks of the trial. After four weeks of therapy, subjects will be unblinded to dose and followed for up to an additional five months. The Institutional Review Board of all participating institutions approved the COAG trial. Written informed consent will be obtained from all patients who participate in the trial.

The primary outcome of the COAG trial is the percentage of time that participants spend within a therapeutic range for anticoagulation (PTTR) during the first four weeks of therapy. The therapeutic range is defined using the International Normalized Ratio (INR), which reflects the ratio of a patient's prothrombin time to that for a control sample. An INR between 2.0 and 3.0, inclusive, is typically considered to be within the therapeutic range. To calculate the PTTR for each subject, we will use a standard interpolation method that assumes a linear change in INR from one measurement to the next [10]. Figure 1 illustrates the linear interpolation method for a hypothetical subject whose therapeutic INR range is between 2.0 and 3.0, with a corresponding PTTR of 60%.

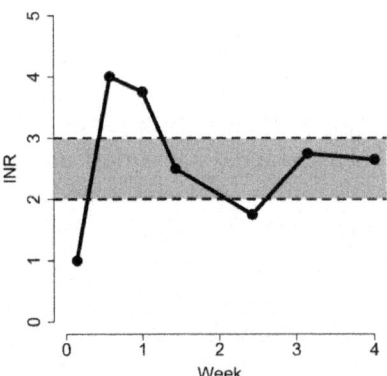

Figure 1: INR measurements (solid circles) and linear interpolations (solid lines) for a hypothetical subject with 60% of time within the therapeutic INR range (shaded region).

Analysis of the primary outcome will be by intention-to-treat [11]. It will not be possible for subjects to switch from their assigned treatment group, but there might be crossovers due to the unavailability of genetic information at the time that the initial dose is dispensed. Every attempt will be made to determine a subject's genotype prior to administration of the initial dose. Given recent technologies, same-day genotyping for warfarin is now possible in practice. In the COAG trial, clinical sites are using one of two genotyping platforms; each has a rapid turnaround time. Both platforms have been FDA approved, have high call and concordance rates, very low failure rates, and the ability to genotype the SNPs needed for the selected dosing algorithms.

For those subjects assigned to the genotype-guided dosing group whose genetic information is not available prior to the initial dose, the initial dose will be determined using the clinical dose-initiation algorithm. Once genetic information becomes available, the dose for these subjects will be determined using the genetic dose-initiation and dose-revision algorithms. The genotype-guided dose-initiation algorithm on day one only uses information on *VKORC1* (not *CYP2C9*) [8]. Therefore, we expect the dose differences on day one to be small relative to the dose differences after the first day. The genotype-guided dose-initiation algorithm on day two, as well as the genotype-guided dose-revision algorithm on days four and five [9], uses information from both *VKORC1* and *CYP2C9*, so that the availability of genetic information by day two will allow the full use of the subject's genetic information to determine their dose for days two through five. We fully expect genotype information to be available on almost all subjects within 24 hours, and certainly by the time of the dose-revision calculations on days four and five.

Randomization

To provide balance in treatment assignment within sites, random assignment to either the genotype-guided or clinical-guided dosing group will be stratified by clinical site. Randomization will also be stratified by race (African American versus not, including Caucasian and Asian American) because race is associated with differential predictive ability of dosing algorithms, with lesser accuracy in African Americans [8], and the dosing algorithms used in the trial predict dose differently among African Americans [9]. In addition, African-American race is associated with the prevalence *CYP2C9* and *VKORC1* variants and is associated with the prevalence of other genetic variants that influence warfarin dose-response. Finally, some clinical sites may recruit a small number of African Americans due to the demographic makeup of their surrounding community.

We will use a block-randomized procedure to assign the treatment groups. Blocking ensures that there will be a balance in the number of patients in each treatment group within each clinical site. Thus, we will use permuted blocks with block sizes of four and six, randomly chosen, which will minimize any imbalances in treatment group assignment. The RANUNI function in SAS 9.2 will be used to generate the randomization numbers within each site for two strata [12].

Sample Size and Statistical Power

A critical element of the statistical design of a randomized clinical trial is to determine a sample size so that a statistical test has adequate power to detect a clinically relevant difference in the primary outcome between treatment groups. The parameters considered in the estimation of sample size include: a minimum detectable difference in the primary outcome between groups; an assumed level of significance for the statistical test of the primary outcome; a measure of variability for the primary outcome in the study population; and the percentage of subjects, if any, expected to drop out of the trial. The sample size parameters for a personalized medicine intervention require additional considerations: the distribution of relevant allelic variants in the study population; and whether the intervention is equally effective across subpopulations defined by allelic variants (e.g., if patients with particular genotypes are not expected to benefit from genotype-guided drug therapy, as we illustrate for warfarin). Due to uncertainly in the distribution of allelic variants and uncertainty in the effectiveness of the intervention across subpopulations, careful attention is required in the design of a personalized medicine to ensure that the study will have adequate power to detect a clinically relevant minimum detectable difference.

In the statistical design of the COAG trial, we focused on the difference in the relative effectiveness of genotype-guided across two genetic subpopulations: those with a single genetic variant versus zero or multiple genetic variants in either *CYP2C9* or *VKORC1*. We viewed the primary outcome of PTTR in each treatment group as a weighted average of PTTR and the corresponding treatment effect (Δ) across subpopulations defined by 1 versus 0, > 1 variants, in which the weights (w) are determined by the populations prevalences that sum to 1:

$$PTTR_C = w_1 \times PTTR_1 + w_{0,>1} \times PTTR_{0,>1;} \tag{1}$$

$$PTTR_G = w_1 \times PTTR_1 \times \Delta_1 + w_{0,>1} \times PTTR_{0,>1} \times \Delta_{0,>1;} \tag{2}$$

where $PTTR_C$ and $PTTR_G$ denote the PTTR in the clinical-guided and genotype-guided dosing groups, respectively. It is straightforward to generalize this approach to more than two subpopulations of interest. Adding additional terms into the weighted average, given the population prevalence and the anticipated treatment effect in each subpopulation, could accommodate more than two subpopulations. Indeed, this approach is generalizeable to any setting in which treatment effects are expected to differ across any number of subpopulations. Specific assumptions are discussed in the following section.

Minimum Detectable Difference

We considered the distribution of *CYP2C9* and *VKORC1* variants and whether genotype-guided dosing of warfarin is equally effective across groups defined by *CYP2C9* and *VKORC1* variants. Current evidence suggests that there will be a subgroup with certain genotypes that will not benefit from genotype-guided dosing [13], most likely because their predicted dose from genotype-guided dosing algorithms will not meaningfully differ from predicted dosing with clinical dosing. We based sample size estimates on the comparison of PTTR between the genotype-guided and clinical-guided dosing groups:

$$PTTR_C = 0.4 \times 73\% + 0.6 \times 61\% = 65.80\%; \qquad (3)$$

$$PTTR_G = 0.4 \times 73\% \times 1 + 0.6 \times 61\% \times 1.15 = 71.29\%; \qquad (4)$$

where:

The proportion in the population who possess a single genetic variant (in either *CYP2C9* or *VKORC1*) and who possess zero or multiple variants is assumed to be 0.4 and 0.6, respectively;

A PTTR of 73% and 61% is assumed for those who possess a single genetic variant and for those who possess zero or multiple variants, respectively;

A 0% relative difference in PTTR is assumed for those with a single genetic variant; and

A 15% relative difference in PTTR for those with zero or multiple variants is assumed to be a clinically relevant difference between the genotype-guided and clinical-guided dosing groups [14].

We assumed that subjects who possess a single genetic variant (in either *CYP2C9* or *VKORC1*) would not benefit from clinical-guided dosing because previous data suggest that the genotype-guided algorithm will predict essentially the same dose as the clinical-guided algorithm. These subjects are expected to attain the same PTTR regardless of their treatment assignment and thus attenuate the mean difference in PTTR between the two groups. To wit, the assumed 15% relative difference in PTTR between the genotype-guided and clinical-guided dosing groups is attenuated to an absolute difference of 5.49% ($PTTR_G$ - $PTTR_C$). Therefore, we assumed an overall minimum detectable difference of 5.49% between groups in the full cohort for sample size calculations. If we had ignored the fact that the intervention is not equally effective across subpopulations defined by genetic variants and assumed a minimum detectable difference of 15%, then the trial would have chosen an inadequately small sample size to achieve adequate power.

The assumed proportion of 0.4 who possess a single genetic variant is based on the Couma-Gen trial [13] and the International Warfarin Pharmacogenetics Consortium (IWPC) [15]. We considered the sensitivity of sample size calculations to a range a population proportions. The assumption that those who possess a single genetic variant will not benefit from genotype-guided dosing, while suggested by the Couma-Gen trial, was not supported in another clinical trial in which all patients benefited from dosing based on *CYP2C9* [16]. In the latter study, the effect of a pharmacogenetic dosing algorithm was similar regardless of the number of *CYP2C9* variants present. Therefore, we believe that our assumptions are conservative.

Level of Significance

To determine the level of significance (α) for the statistical test of PTTR between the genotype-guided and clinical-guided dosing groups, we considered an alpha-allocation approach [17–19]. In this approach, a portion (α_A) of the overall level of significance is used to test the comparison in the full cohort; the remaining portion (α_S) is used to test the comparison in a pre-defined primary subgroup. The alpha-allocation approach facilitates a traditional primary analysis to assess a statistically significant difference between the treatment groups, as well as a predefined primary subgroup analysis that is not relegated to a secondary analysis, as in a traditional analysis.

We defined the primary subgroup based on subjects whose predicted initial dose employing the genetic and clinical dose-initiation algorithms differs by $\geq$ 1.0 mg, a factor known at the time of randomization and therefore not a post-randomization selection. We posited that the subgroup of participants with a larger difference between the predicted initial doses should have a larger separation in PTTR between the two groups. If the improvement in PTTR is related to the magnitude of difference in dosing between the genotype-guided and clinical-guided dosing groups, then the primary subgroup comparison should reflect a larger absolute difference than the 5.49% assumed for the full cohort analysis. We assumed that a clinically relevant absolute difference to detect in the primary subgroup is 9.15%, from a PTTR of 61% to 70.15% in Equation (4), reflecting a 15% relative difference.

We selected $\alpha_A = 0.04$ for the full cohort analysis and $\alpha_S = 0.01$ for the primary subgroup analysis, for an overall type-I error rate of $\alpha = 0.05$. However, allocating alpha so that sum of α_A and α_S is equal to α is a conservative Bonferroni-type correction, which may be unnecessarily conservative if there is a positive correlation between the tests in the full cohort and in the primary subgroup [20, 21]. The correlation between the two tests will be obtained under the null hypothesis when the size of the primary subgroup is known. The correlation will then be incorporated to obtain $\alpha_S > \alpha - \alpha_A$ given that α_A is fixed, so that the overall type-I error rate is controlled at α.

Other assumptions in the computation of sample size were the standard deviation of the PTTR in the study population and the percentage of subjects expected to drop out before reaching the primary endpoint. The within-study variability of PTTR in the literature varied across study designs and populations under study. However, there was a reasonable consistency of variability for the genetic-guided and clinical-guided dosing groups in the studies reviewed. We assumed a standard deviation of 25% based on a study of dose-refinement algorithms in which the standard deviation averaged 23% [22]. We also assumed that 10% of subjects would drop out before reaching the primary endpoint and increased the sample size by dividing the calculated sample size by the square of one minus the drop-out rate [23].

Primary Analysis

The null hypothesis for the primary outcome is that the percent of time that subjects spend within the therapeutic INR range (PTTR) during the first four weeks of therapy is equal between the genotype-guided and clinical-guided dosing groups. We will estimate the difference in mean PTTR between the genotype-guided and clinical-guided dosing groups using a linear regression model, both for the full cohort and for the primary subgroup whose predicted initial dose employing the genetic and clinical dose-initiation algorithms differs by ≥ 1 mg. Inference will be based on a Wald test with a level of significance of 0.05 allocated between the full cohort analysis and the primary subgroup analysis. Because randomization will be stratified by site and race, these variables will be included in the linear regression model. We will perform additional analyses in subgroups defined *a priori* by allelic variation (zero versus a single versus multiple *CYP2C9* or *VKORC1* variants) and by race (African American versus not).

Additional genetic factors may be considered in secondary analyses. Specifically, because *CYP2C9* and *VKORC1* genotypes may not be the only genetic variants that determine optimal warfarin dosing, it is possible that more variants will be identified during the trial. To adjust for additional genetic factors in secondary analyses, we will include them as covariates in a linear regression model if their prevalence differs between the clinical and genetic groups, and also consider possible interactions with *CYP2C9* and/or *VKORC1*.

RESULTS

Table 1 provides the sample size required for the full cohort analysis using a two-sample *t*-test with α_A = 0.04 (two-sided), assuming various proportions with a single genetic variant (0.4, 0.5, and 0.6), estimates for the standard deviation of PTTR (20%, 25%, and 30%), and power levels (80% and 90%), and drop-out rate (10%). A sample size of 1140 would provide 90% power to detect an absolute difference of 5.49% in the full cohort, given that the proportion with a single genetic variant is 0.4 and the standard deviation is 25%. We selected a sample size of 1238 to protect against departures from the assumed proportion with a single genetic variant, study drop-out rate, and standard deviation of PTTR. For example, if the proportion with a single genetic variant is 0.5 and the standard deviation is 25%, then there is 80% power to detect an absolute difference in PTTR of 4.58%. If the proportion with a single genetic variant is 0.4 and standard deviation is 30%, then there is 80% power to detect the assumed 5.49% absolute difference.

Table 1: Sample size estimates for the full cohort analysis; *p* denotes the proportion with a single genetic variant and Δ denotes the corresponding minimum detectable difference

		Standard Deviation of PTTR					
		20%		25%		30%	
		Power		Power		Power	
p	Δ	80%	90%	80%	90%	80%	90%
0.4	5.49%	550	730	860	1140	1238	1642
0.5	4.58%	792	1050	1238	1642	1782	2364
0.6	3.67%	1238	1642	1932	2564	2782	3692

A sample size of 1238 provides sufficient power for the primary subgroup analysis using a two-sample *t*-test with α_S = 0.01 (two-sided). Recall that the size of the primary subgroup is determined by the percentage of subjects whose predicted initial dose employing the genetic and clinical dose-initiation algorithms differs by ≥ 1 mg. If the relative size of the primary subgroup is 50% and the standard deviation of PTTR is 25%, then there is 93.6% power to detect a 9.15% absolute difference. In addition, if the relative size of the primary subgroup is 60% and the standard deviation is 30%, then there is 87.8% power. In fact, the power will be higher because α_S will be increased according to the correlation between the tests in the full cohort and in the primary subgroup.

Sensitivity Analysis

In the statistical design of the COAG trial, there was a concern that the genotype-guided and clinical-guided dosing algorithms may not produce sufficiently differentiable doses between the treatment groups, which may lead to an underestimation of the minimum detectable difference in PTTR between groups. We assumed that any difference between the two groups would arise from the subgroup of patients with either zero or multiple genetic variants. (Recall that the assumed relative difference in the genotype-guided dosing group was 15% for those with zero or multiple variants.) For subjects in this allelic subgroup, if the difference between the two algorithm predictions is negligible or clinically irrelevant, then it is reasonable to expect no difference in PTTR. In this case the PTTR for the genotype-guided dosing group can be expressed as:

$$\text{PTTR}_G = 0.4 \times 73\% + 0.6 \times 61\% \times [1 + (0.15 \times d)], \tag{5}$$

where d is the proportion of subjects with zero or multiple genetic variants in whom there is a clinically meaningful difference between the predicted dose determined by the genotype-guided and clinical-guided dosing algorithms. Hence the expected 15% difference would be diluted by a factor d and it would be more difficult to detect a clinically relevant difference between groups.

To explore the impact of dilution of the treatment effect, we examined the distribution of the differences between the predicted doses among groups defined by allelic variation in the IWPC cohort [15] and calculated the difference between the rounded predicted doses. An absolute dose difference < 1.0 mg per day was defined as the ‹same› predicted dose; an absolute dose difference of $\geq$ 1.0 mg per day was defined as a ‹different› predicted dose. The rationale for the 1.0 mg cut-point is that the average initial dose is 5.0 milligrams; therefore, a 1.0 mg absolute difference represents a clinically relevant 20% difference, on average. Approximately 9% of IWPC participants in the (0, >1) allelic variant group would have received the ‹same› initial dose, i.e., $d = 0.91$. With this dilution of the treatment effect, in order to detect an overall effect size of 5.49% in PTTR, the relative effect size in the (0, > 1) group would need to be 16.5%.

Table 2 provides power estimates for the test of the full cohort analysis for a range of diluted treatment effects corresponding to the parameter d, the proportion of subjects with zero or multiple genetic variants in whom there is a difference between the predicted doses. There is sufficient power when $d > 0.9$.

We are not highly confident in our estimate of how frequently the predicted dose will differ between the two algorithms and therefore have not taken this potential dilution effect into account in our calculations for sample size and power. However, given the potential impact of the dilution effect on the sample size requirements of the study seen in Table 2, we have planned to monitor this factor during the operation of the trial.

Table 2: Power estimates for the full cohort analysis in which the treatment effect is diluted; d denotes the proportion of subjects with zero or multiple genetic variants in whom there is a difference between the predicted initial doses

d	Treatment Effect	Power
0.7	3.84	65%
0.8	4.39	77%
0.9	4.94	86%
1.0 (Undiluted)	5.49	93%

DISCUSSION

In this manuscript we provided practical guidance on the statistical design of a personalized medicine intervention that uses each subject's genotype information in an untargeted design. We used power and sample size calculations to illustrate the primary statistical challenge of designing this type of personalized therapy intervention: to accommodate the potential differential effectiveness of genotype-guided therapy across subpopulations defined by allelic variation. To determine a minimum detectable difference in PTTR between groups, we assumed that 40% of enrolled subjects would have a single genetic variant and would therefore not benefit from genotype-guided warfarin therapy. Hence, the minimum detectable difference used in sample size calculations is not a known quantity, but rather an unknown quantity that depends on the genetic makeup of the subjects enrolled. In addition, the sample size for the primary subgroup analysis depends on the proportion of subjects whose predicted initial dose employing the genetic and the clinical dose-initiation algorithms differs by ≥ 1.0 mg. Due to the importance of these parameters for adequate sample size and statistical power to detect a clinically meaningful difference, they will be monitored during the course of the trial.

As shown in Table 1, the sample size is sensitive to the standard deviation of PTTR. The Data Safety and Monitoring Board (DSMB) may suggest an 'internal pilot study' in which an estimate of the standard deviation will be obtained using the first half of the observed data and the sample size calculations will be updated based on the new estimate [24, 25]. The pre-planned sample size will be assumed to represent a minimum sample size (i.e., the final sample size based on the 'internal pilot study' will not be less than the pre-planned sample size). In this case, the 'internal pilot study' is known as restricted. For restricted designs, the disparity in the type-I error rate in testing the primary hypothesis is negligible [26]. Therefore, it will not be necessary to adjust the type-I error rate of any hypothesis tests regarding the primary outcome. In assessing the need for a sample size increase, data will neither be unblinded nor assessed for the primary outcome. In addition, a sample size adjustment will not impact the overall design of the study. Because the DSMB will not monitor efficacy during the conduct on the COAG trial, there is no conflict between any interim sample size adjustment and interim measures of efficacy.

In our sensitivity analysis, we examined the dilution of the treatment effect due to a clinically irrelevant difference between the predicted doses (employing the genetic and clinical dose-initiation algorithms) for subjects with zero or multiple genetic variants. However, we did not consider the impact of a clinically relevant difference between the predicted doses for subjects with a single variant. In this situation the PTTR for the genotype-guided dosing group can be expressed as:

$$\text{PTTR}_G = 0.4 \times 73\% \times [1 + (0.15 \times d')] + 0.6 \times 61\% \times [1 + (0.15 \times d)], \qquad (6)$$

where d' is the proportion of subjects with a single genetic variant in whom there is a meaningful difference between the predicted doses and d is defined in Equation (5). For example, in the IWPC cohort, approximately 26% of subjects with a single genetic variant would have received a ‹different› initial dose, i.e., $d' = 0.26$. For these subjects, we expect that there would be a difference in PTTR, which would increase the power of the full cohort analysis. Because we were not highly confident in this estimate, we did not examine the increase in power associated with this allelic subgroup. Therefore, our sensitivity analysis is conservative.

An individual's genetic information could be used prior to randomization to identify subjects who are potentially unresponsive to either drug therapy or the pharmacologic intervention, motivating researchers to decide whether to include or exclude those subjects from the trial [27]. For example, in a targeted design, study eligibility may be restricted to subjects who, based on their genetic characteristics, are predicted to be responsive [28]. By excluding potentially unresponsive individuals, a targeted design will require a smaller sample size to detect a statistically significant effect. Conversely, in a traditional (untargeted) design, particularly of an intervention designed to select dose, subjects for whom genetic-based drug therapy is not effective are eligible, because they would still receive drug treatment regardless of their genetic makeup. For example, subjects in the COAG trial would receive warfarin therapy regardless of their *CYP2C9* and *VKORC1* variants. As we have shown with the statistical design of the COAG trial, by including potentially unresponsive subjects, a larger sample size may be required. Cost-benefit considerations regarding the cost of genetic screening for eligibility versus the cost of enrolling potentially unresponsive subjects may be useful to determine which design is more practical in specific applications.

For the COAG trial, we favored including all participants, regardless of their genetic variants. First, the assumption that those who possess a single genetic variant will not benefit from genotype-guided dosing, while suggested by the Couma-Gen trial [13], was not supported in another clinical trial in which all patients benefited from dosing based on *CYP2C9* [16]. Therefore, if we excluded subjects who may not benefit from genotype-guided dosing, we would be unable to evaluate our assumptions. Second, all subjects are genotyped prior to randomization, so that much of the cost is already incurred in screening. Third, including subjects potentially unresponsive to genotype-guided dosing allows the results of the trial to be more generalizable. That is, if the COAG trial indicates that genotype-guided dosing provides increased efficacy compared to clinical-guided dosing, then it motivates consideration of the policy question of whether all patients prescribed warfarin should be genotyped to predict the drug›s efficacy.

We recommend that the statistical design of a personalized medicine intervention that uses each subject›s genotype information, within the framework of a randomized clinical trial, consider the distribution of relevant allelic variants in the study population and whether the intervention is equally effective across subpopulations defined by allelic variants. In the statistical design of the COAG trial, we considered the distribution of *CYP2C9* and *VKORC1* variants and whether genotype-guided dosing of warfarin therapy would provide an equal improvement in efficacy across populations defined by genetic variants. We assumed that subjects with a single genetic variant would not benefit from genotype-guided dosing, thus attenuating the postulated 15% relative difference between the two treatment groups to a 5.49% absolute difference. In our sample size calculations, if we ignored the fact that the genotype-guided dosing is not equally effective across subpopulations defined by genetic variants and assumed a minimum detectable difference of 15%, then the COAG trial would likely have chosen an inadequately small sample size to achieve adequate power. We also recommend that key assumptions regarding sample size and statistical power be monitored during the conduct of the trial, to inform any requisite increase in the sample size needed to detect a clinically relevant difference in the primary outcome between treatment groups. Further research is required to determine whether an interim sample size adjustment based on the observed proportion of allelic variants increases the type-I error rate.

CONCLUSIONS

In summary, we found that sample size and power calculations may be sensitive to key assumptions required in the design of a personalized medicine intervention: the distribution of relevant allelic variants in the study population; and whether the pharmacogenetic intervention is equally effective across subpopulations defined by allelic variants. Given the novelty of pharmacogenetic research, we recommend that these assumptions be monitored during the conduct of a personalized medicine intervention.

ACKNOWLEDGEMENTS

We gratefully acknowledge the National Heart, Lung and Blood Institute (N01 HV88210) and the University of Pennsylvania for supporting this research, and two reviewers for comments that greatly improved the manuscript.

COAG Investigators: Sherif Abdel-Rahman, University of Texas; Robert J Desnick and Jonathan L Halperin, Mount Sinai School of Medicine; Margaret C Fang, University of California, San Francisco; Brian F Gage, Washington University School of Medicine; Richard B Horenstein, University of Maryland School of Medicine; Julie A Johnson, University of Florida; Scott Kaatz, Henry Ford Hospital; Robert D McBane, Mayo Clinic College of Medicine; Emile R Mohler III, Hospital of the University of Pennsylvania; James A S Muldowney III, Vanderbilt University; Scott M. Stevens, Intermountain Medical Center; Steven Yale, Marshfield Clinical Research Foundation.

AUTHORS' CONTRIBUTIONS

All authors made substantial contributions to conception and design. BF and JJ drafted the manuscript. All authors revised the manuscript critically for important intellectual content and approved the final manuscript.

REFERENCES

1.   Terra SG, Johnson JA: Pharmacogenetics, pharmacogenomics, and cardiovascular therapies: The way forward. Am J Cardiovasc Drugs 2002, 2:287–296.

2.   Wang SJ, O›Neill RT, Hung HM: Approaches to evaluation of treatment effect in randomized clinical trials with genomic subset.Pharm Stat 2007, 6:227–244.

3.   Sargent DJ, Conley BA, Allegra C, Collette L: Clinical trial designs for predictive marker validation in cancer treatment trials. J Clin Oncol 2005, 23:2020–2027.

4.   Garcia D, Regan S, Crowther M, Hughes RA, Hylek EM: Warfarin maintenance dosing patterns in clinical practice: Implications for safer anticoagulation in the elderly population. Chest 2005, 127:2049–2056.

5.   Sanderson S, Emery J, Higgins J: CYP2C9 gene variants, drug dose, and bleeding risk in warfarin-treated patients: A HuGEnet systematic review and meta-analysis. Genet Med 2005, 7:97–104.

6.   Rieder MJ, Reiner AP, Gage BF, Nickerson DA, Eby CS, McLeod HL, Blough DK, Thummel KE, Veenstra DL, Rettie AE: Effect of VKORC1haplotypes on transcriptional regulation and warfarin dose. N Engl J Med 2005, 352:2285–2293.

7.   Schelleman H, Chen J, Chen Z, Christie J, Newcomb CW, Brensinger CM, Price M, Whitehead AS, Kealey C, Thorn CF, Samaha FF, Kimmel SE: Dosing algorithms to predict warfarin maintenance dose in Caucasians and African Americans. Clin Pharmacol Ther 2008,84:332–339.

8.   Gage BF, Eby D, Johnson JA, Deych E, Rieder MJ, Ridker PM, Milligan PE, Grice G, Lenzini P, Rettie AE, Aquilante CL, Grosso L, Marsh S, Langaee T, Farnett LE, Voora D, Veenstra DL, Glynn RJ, Barrett A, McLeod HL: Use of pharmacogenetic and clinical factors to predict the therapeutic dose of warfarin. Clin Pharmacol Ther 2008, 84:326–331.

9.   Lenzini P, Wadelius M, Kimmel S, Anderson JL, Jorgensen A, Pirmohamed M, Caldwell MD, Limdi N, Burmester JK, Dowd MB, Angchaisuksiri P, Bass AR, Chen J, Eriksson N, Rane A, Lindh JD, Carlquist JF, Horne BD, Grice G, Milligan PE, Eby C, Shin J, Kim H, Kurnik D, Stein CM, McMillin G, Pendleton RC, Berg RL, Deloukas P, Gage BF: Integration of genetic, clinical, and laboratory data to refine warfarin dosing. Clin Pharmacol Ther 2010, 87:572–578.

10.   Rosendaal FR, Cannegieter SC, van der Meer FJ, Briët E: A method to determine the optimal intensity of oral anticoagulant therapy.Thromb Haemost 1993, 69:236–239.PubMed

11.   Ellenberg JH: Intention-to-treat analysis. In Encyclopedia of Biostatistics. Edited by: Armitage P, Colton T. New York: John Wiley & Sons; 1998:2056–2060.

12.   Fishman GS, Moore LR: A statistical evaluation of multiplicative congruential random number generators with modulus $2^{31} - 1$. J Amer Statist Assoc 1982, 77:129–136.View Article

13. Anderson JL, Horne BD, Stevens SM, Grove AS, Barton S, Nicholas ZP, Kahn SF, May HT, Samuelson KM, Muhlestein JB, Carlquist JF, Couma-Gen Investigators: Randomized trial of genotype-guided versus standard warfarin dosing in patients initiating oral anticoagulation. Circulation 2007, 116:2563–2570.

14. Dolan G, Smith LA, Collins S, Plumb JM: Effect of setting, monitoring intensity and patient experience on anticoagulation control: A systematic review and meta-analysis of the literature. Curr Med Res Opin 2008, 24:1479–1472.View Article

15. The International Warfarin Pharmacogenetics Consortium: Estimation of the warfarin dose with clinical and pharmacogenetic data.N Engl J Med 2009, 360:753–764.View Article

16. Caraco Y, Blotnick S, Muszkat M: CYP2C9 genotype-guided warfarin prescribing enhances the efficacy and safety of anticoagulation: A prospective randomized controlled study. Clin Pharmacol Ther 2008, 83:460–470.

17. Moyé LA: P-value interpretation and alpha allocation in clinical trials. Ann Epidemiol 1998, 8:351–357.

18. Moyé LA, Deswal A: Trials within trials: Confirmatory subgroup analyses in controlled clinical experiments. Control Clin Trials 2001,22:605–619.

19. Coats AJ: CAPRICORN: A story of alpha allocation and beta-blockers in left ventricular dysfunction post-MI. Int J Cardiol 2001,78:109–113.

20. Alosh M, Hugue MF: A flexible strategy for testing subgroups and overall population. Stat Med 2009, 15:3–23.View Article

21. Joo J, Geller NL, French B, Kimmel SE, Rosenberg Y, Ellenberg JE: Prospective alpha allocation in the Clarification of Optimal Anticoagulation through Genetics (COAG) trial. Clin Trials 2010, 7:597–604.

22. Lenzini PA, Grice GR, Milligan PE, Dowd MB, Subherwal S, Deych E, Eby CS, King CR, Porche-Sorbet RM, Murphy CV, Marchand R, Millican EA, Barrack RL, Clohisy JC, Kronquist K, Gatchel SK, Gage BF: Laboratory and clinical outcomes of pharmacogenetic vs. clinical protocols for warfarin initiation in orthopedic patients. J Thromb Haemost 2008, 6:1655–1662.

23. Lachin JM: Introduction to sample size determination and power analysis for clinical trials. Control Clin Trials 1981, 2:93–113.

24. Wittes J, Brittain E: The role of internal pilot studies in increasing the efficiency of clinical trials. Stat Med 1990, 9:65–71.

25. Betensky RA, Tierney C: An examination of methods for sample size recalculation during an experiment. Stat Med 1997, 16:2587–2598.

26. Wittes J, Schabenberger O, Zucker D, Brittain E, Proschan M: Internal pilot studies I: Type I error rate of the naive t-test. Stat Med 1999, 18:3481–91.

27. Simon R: The use of genomics in clinical trial design. Clin Cancer Res 2008, 14:5984–5993.

28. Simon R, Maitouram A: Evaluating the efficiency of targeted designs for randomized clinical trials. Clin Cancer Res 2004, 10:6759–6763.

Chapter 10

FROM BIOPHYSICS TO EVOLUTIONARY GENETICS: STATISTICAL ASPECTS OF GENE REGULATION

Michael Lässig
Institut für Theoretische Physik, Universität zu Köln

ABSTRACT

This is an introductory review on how genes interact to produce biological functions. Transcriptional interactions involve the binding of proteins to regulatory DNA. Specific binding sites can be identified by genomic analysis, and these undergo a stochastic evolution process governed by selection, mutations, and genetic drift. We focus on the links between the biophysical function and the evolution of regulatory elements. In particular, we infer fitness landscapes of binding sites from genomic data, leading to a quantitative evolutionary picture of regulation.

INTRODUCTION

Genomic functions often cannot be understood at the level of single genes but require the study of gene networks. This systems biology credo is nearly commonplace by now. Evidence comes from the comparative analysis of entire genomes: Current estimates put, for example, the number of human genes at around 22000, hardly more than the 14000 of the fruit fly, and not even an order of magnitude higher than the 6000 of baker's yeast. The complexity and diversity of higher animals therefore cannot be explained in terms of their gene numbers. If, however, a biological function requires the concerted action of several genes, and conversely, a gene takes part in several functional contexts, an organism may be defined less by its individual genes but by their interactions. The emerging picture of the genome as a strongly interacting system with many degrees of freedom brings new challenges for experiment and theory, many of which are of a statistical nature. And indeed, this picture continues to make the subject attractive to a growing number of statistical physicists.

Genes encode proteins, and proteins perform functions in the cell. Hence, a gene takes part in a biological function only if it is expressed, i.e., if the protein produced from it is present in the cell. Genes interact by regulation: the protein of one gene can influence the production of protein from another gene. Gene regulation can take place during transcription, the process by which the cell reads the information contained in a gene and copies it to messenger RNA (which is subsequently used to make a functional protein). This is the most fundamental level of interactions between genes: the transcription of one gene may be enhanced or reduced by the expression of other genes. Transcriptional regulation is thus a good starting point for theory. We should keep in mind, however, that it is not the only mode of gene interactions. Especially in eukaryotes, additional regulation mechanims involving histones, chromatin, micro-RNAs etc. become relevant, which are just entering the stage of model building. An excellent introduction to the biology of regulation can be found in [1].

This article is a primer on theoretical aspects of gene interactions, and we limit ourselves to transcriptional regulation. Clearly, the subject has rather diverse aspects:

(1)   Transcription is a biophysical process, which involves the interaction of DNA and proteins. Its regulation takes place through the binding of proteins to DNA at specific loci in the vicinity of the gene to be regulated. Already at this level, this process is rather complex and not yet fully understood. What enables the protein to find one or a few specific functional sites in a genome of up to billions of base pairs, bind there with sufficient strength to influence transcription, and leave again once its task is performed?

(2)   Given that the protein can find its functional sites, can we as well? If that is possible, we can predict the specific gene interactions building regulatory networks from sequence data. The analysis of regulatory DNA is a major topic of research in bioinformatics, with the aim of identifying statistical characteristics of functional loci and of building search algorithms.

(3)   Regulation is also becoming an important part of evolutionary biology [2, 3]. If regulatory networks are to explain the differentiation of higher animals, there must be efficient modes of evolution for the interactions between genes. At the level of regulary DNA, these modes remain largely to be explored. It is clear, however, that the underlying evolutionary dynamics is the basis of a quantitative understanding of regulatory networks.

All three aspects of regulation contribute to a unified theoretical picture. Key concepts such as the biophysical binding energy, the bioinformatic scoring function, and the evolutionary fitness turn out to be rather deeply related. We will focus on these crosslinks between different fields, which are likely to become important for future research. A challenge for an introductory presentation is the diversity of relevant background material, only a rather ecclectic account of which can be presented here. Yet, I hope it transpires even from this short introduction that present quantitative genomics is an area of science shaped by a remarkable confluence of ideas from different disciplines.

BIOPHYSICS OF TRANSCRIPTIONAL REGULATION

The fundamental step in the regulatory interaction between two genes is a binding process: the protein produced by the first gene acts as a transcription factor for the second gene, i.e., it binds to a functional site on the DNA close to the second gene and thereby enhances or suppresses its transcription. Binding sites are short, typically segments of 10 to 15 base pairs in prokaryotes and even shorter segments in eukaryotes. They are primarily located in the cis-regulatory region of a gene, which lies just upstream of its protein-coding sequence and extends over hundreds of base pairs in prokaryotes and over thousands of base pairs in eukaryotes.

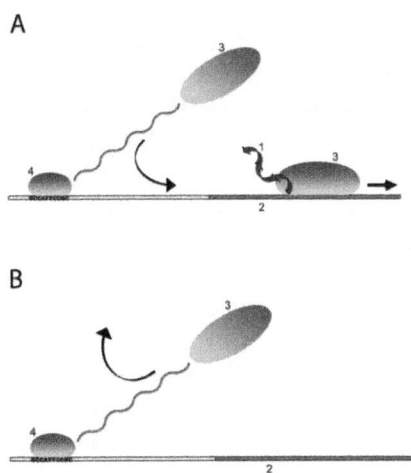

Figure 1: Transcriptional regulation. Transcription is the synthesis of messenger RNA (1) whose genetic code is a copy of the coding DNA (2) of a gene, by means of RNA

polymerase (3). A transcription factor (4) bound to a DNA target site interacts with RNA polymerase molecules, (a) enhancing or (b) reducing the transcription rate of a nearby gene.

The scenario of transcriptional regulation is sketched in Fig. 1. A transcription factor bound to a functional binding site regulates the downstream gene by recruiting or repelling RNA polymerase. This protein-protein interaction catalyzes or suppresses the process of transcription of the gene. All these binding processes should not be understood as on or off; they happen with certain probabilities, which are determined by the binding energies and the numbers of the molecules involved.

Factor-DNA Binding Energies

The interaction of a transcription factor protein with DNA is two-fold: There is a position-unspecific attraction with energy E_u and a specific interaction, whose energy depends on the particular locus where the factor binds. The unspecific part is the electrostatic interaction between the positively charged protein and the negatively charged DNA backbone, while the specific part involves hydrogen bonds between the binding domain of the protein and the nucleotides of the binding locus. A locus is specified by its starting position r and its length ℓ (with relevant values ℓ of order 10). The specific binding energy $E(r)$ depends on ℓ consecutive nucleotides $\mathbf{a} = (a_1,..., a_\ell)$ counted downstream from the starting position, the sequence state or genotype of that locus. Switching between unspecific and specific binding takes place via a conformation change of the factor protein. As a result of these interactions, the factor protein can be in three thermodynamic states as shown in fig. 2: unbound (i.e., freely diffusing), unspecifically bound (i.e., diffusing along the DNA backbone), and specifically bound.

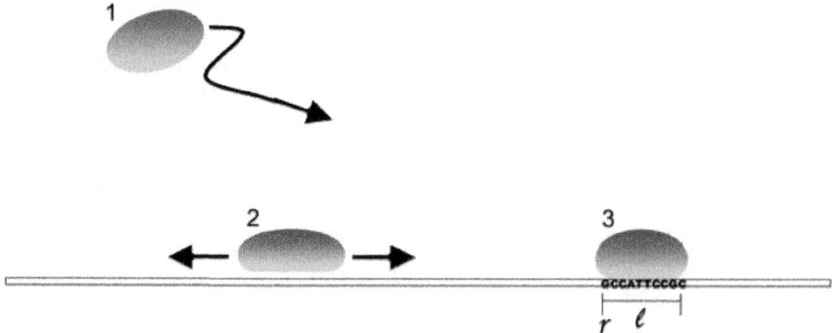

Figure 2: Thermodynamic states of a transcription factor. (1) Unbound state, with three-dimensional diffusion. (2) Unspecific bound state, with one-dimensional diffu-

sion along the DNA backbone. (3) Specific bound state. The binding energy depends on the genotype at the binding locus, which has length ℓ and whose position is specified by the coordinate r.

The biophysics of factor-DNA binding has been established in a series of seminal papers [4–7]. More recently, the characteristics of specific binding have been measured for some bacterial transcription factors [8–12]. These can be summarized as follows:

(a)   The single nucleotides of a binding locus $\mathbf{a} = (a_1, ..., a_\ell)$ give approximately independent contributions to the binding energy,

$$E(\mathbf{a}) = \sum_{i=1}^{\ell} \varepsilon_i(a_i).$$

(1)

(b)   At each position i, there is typically one preferred nucleotide a_i^* with $\varepsilon_i(a_i^*) = \min_a \varepsilon_i(a)$. Hence, there is a unique "ground state" sequence $\mathbf{a}^* = (a_1^*, ..., a_\ell^*)$ with minimal binding energy $(\mathbf{a}^*)$, i.e., with strongest binding.

(c)   ismatches with respect to the minimum-energy sequence involve energy costs $\varepsilon_i(a) - \varepsilon_i(a_i^*) \approx 1 - 3\, k_B T$ per nucleotide.

(d)   There is an energy difference $E_u - E^* \sim 15\, k_B T$ between unspecific and strongest specific binding. Experimental data for the binding energies $\varepsilon_i(a)$ are known only for a few transcription factors.

Approximate values for these energies can also be inferred from nucleotide frequencies in functional binding sites [10]. A promising recent approach is to infer binding energies from large-throughput expression data [13]. For order-of-magnitude estimates, one often uses the so-called two-state approximation [7], which is homogeneous in the nucleotide positions and distinguishes only between match and mismatch:

$$\varepsilon_i(a) - \varepsilon_i(a_i^*) = \begin{cases} \varepsilon & \text{if } a_i \neq a_i^* \\ 0 & \text{if } a_i = a_i^* \end{cases}$$

(2)

with $\varepsilon \approx 2 k_B T$. In this approximation, the binding energy of a sequence $\mathbf{a}$ is simply related to the Hamming distance $d(\mathbf{a}, \mathbf{a}^*)$, i.e., the number of nucleotide mismatches between $\mathbf{a}$ and $\mathbf{a}^*$,

$$E(\mathbf{a}) = E^* + \varepsilon \cdot d(\mathbf{a}, \mathbf{a}^*).$$

(3)

Energy Distribution in the genome

Fig. 3(a) shows the sequence of energy values E(r) found in a segment of the E. coli genome for a specific transcription factor, the cAMP response protein

(CRP) This «energy landscape» looks quite random, i.e., consecutive energy values are approximately uncorrelated. The distribution $W_{dat}(E)$ of energies over the entire noncoding part of the E. coli genome is shown in fig. 3(b). We can compare this with the distribution $W_0(E)$ obtained from a random sequence with the same nucleotide frequencies (i.e., from a scrambled genome). According to eq. (1), the binding energy E is then a sum of independent random variables ε_i, and its distribution becomes approximately Gaussian by the law of large numbers. Fig.3(b) shows that the actual distribution $W_{dat}(E)$ is indeed of the same form as $W_0(E)$ for most energies. However, a closer look at the low-energy tail of the distribution shows that there are significantly more strong binding sites than expected from a random sequence [14–16]. So at least some of them are there not by chance but for a reason.

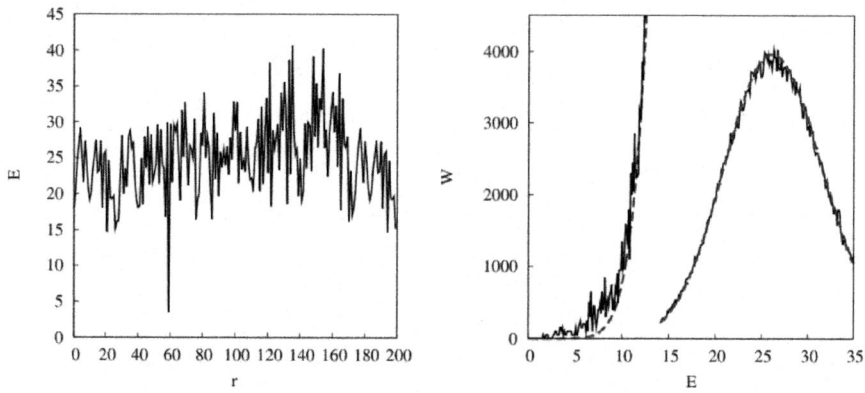

Figure 3: Transcription factor binding energies of the E. coli genome. (a) Energy "land-scape" E(r) for specific binding of the CRP factor at 200 consecutive positions r in an intergenic region, with a binding site at position 59. (b) Count histogram $W_{dat}(E)$ with energy bins of width 0.1 obtained from all intergenic regions, together with the distribution $W_0(E)$ for a random sequence (dashed line, shown with a 30 fold zoom into the region E < 14). From [16].

Search Kinetics

All three thermodynamic modes of a factor molecule – free diffusion, unspecific binding, and specific binding – are important for the search kinetics towards a functional site [4–6]. The unspecific attraction causes the transcription factor to be bound to DNA with a finite probability, i.e., a given molecule spends about equal amounts of time on and off the DNA backbone. Hence, the search process is a mixture of effectively one-dimensional diffusion along the DNA backbone and three-dimensional diffusion in the surrounding medium. This proves more efficient than purely one- or three-dimensional diffusion.

In the 1D mode, the factor diffuses in a flat energy landscape if it is in the conformation of unspecific binding, or in the landscape E(r) if it is in the conformation of specific binding. In this way, it can sample the low-energy part of the landscape E(r) while avoiding its barriers. The main obstacles on its way to a functional site are spurious binding sites, which have a low energy E(r) by chance and act as traps. We lack a completely satisfactory picture of the search kinetics, which is an area of current research [14, 17]. However, this process proves to be remarkably fast. Typical search times are less than a minute, i.e., substantially shorter than typical functional intervals in a cell cycle of at least minutes. Therefore, the regulatory effect of a site is related to its probability of binding a factor molecule at equilibrium, which can be evaluated by standard thermodynamics.

Thermodynamics of Factor Binding

We start with the idealized but instructive problem of a single factor protein interacting with a genome of length $L \gg 1$, which contains a single functional site, while the rest of the sequence is random. Since the protein is bound to the DNA with a probability of about 1/2, we neglect the unbound state for the subsequent probability estimates and study only the bound protein, which is at equilibrium between specific and unspecific binding. At each position r, the likelihood of these two states is given by the Boltzmann factors $\exp[-E(r)/k_B T]$ and $\exp[-E_u/k_B T]$, respectively. Hence, the partition function for a single protein has the form

$$Z = \sum_{r=1}^{L} e^{-E(r)/k_B T} + L e^{-E_u/k_B T}.$$

$$(4)$$

The functional site, which is assumed to be positioned at $r = r_f$, must have a low specific binding energy $E = E(r_f)$. We now single out this position and write

$$Z = e^{-E/k_B T} + \sum_{r \neq r_f} e^{-E(r)/k_B T} + L e^{-E_u/k_B T}$$

$$\approx e^{-E/k_B T} + Z_0,$$

$$(5)$$

where Z_0 is the partition function of a completely random sequence. The probability of the factor being bound specifically at the functional site is then

$$p(E) = \frac{e^{-E/k_B T}}{Z} = \frac{1}{1 + e^{(E - F_0)/k_B T}},$$

$$(6)$$

where $F_0 = -k_B T \log Z_0$ is the free energy for a random genome. Thus, the binding probability depends on the binding energy in a sigmoid way, with a threshold energy $E = F_0$ between strong and weak binding.

This strongly nonlinear dependence is known to physicists as a Fermi function.

It is easy to generalize the thermodynamic formalism to more than one factor molecule. Ignoring the overlap between close sites, each position r can be empty or be occupied either by an unspecifically or by a specifically bound factor. Using a chemical potential σ, the many-factor partition function can hence be written as

$$Z(\sigma) = \prod_{r=1}^{L} Z(\sigma, r),$$

(7)

where

$$Z(\sigma, r) = 1 + e^{\sigma - E(r)/k_B T} + e^{\sigma - E_u/k_B T}$$

(8)

is a sum over the three thermodynamic states at position r: no factor bound, one factor bound specifically or unspecifically. The chemical potential σ is determined by the number of factor molecules, n, via the relation $n = (d/d\sigma) \log Z(\sigma)$. For actual transcription factor numbers, which are of order $1 - 10^4$, this relation is well approximated by [14]

$$\sigma = \frac{F_0}{k_B T} + \log n.$$

(9)

The functional site is now occupied by a specifically bound factor with probability

$$p(E) = \frac{e^{\sigma - E/k_B T}}{Z(\sigma, r_f)} = \frac{1}{1 + e^{(E - F_0)/k_B T - \log n}}.$$

(10)

The binding probability – and hence the effects of the functional site on the regulated gene – are thus determined by the binding energy, the number of factor molecules, and on the genomic background (via the free energy F_0). The dependence p(E) is a Fermi function with threshold energy $E = F_0 + k_B T \log n$, which is shifted with respect to the single-molecule case. Clearly, p is also a Fermi function of log n at fixed binding energy, with a threshold at $\log n = (E - F_0)/k_B T$. If there is more than one functional site in the genome, the calculation remains unaffected as long as their number is much smaller than n.

Sensitivity and Genomic Design of Regulation

The regulatory machinery can be very efficient: in bacteria, it has been shown that single factor molecules can have regulatory effects. We can use eq. (6) to enquire how the cell can reach this high level of sensitivity, following mostly ref. [14]. We assume a minimal genome which has a single functional site of maximum binding strength E^* and is otherwise random. If a single factor molecule is to affect regulation, its binding to the functional site must not be overwhelmed by the remainder of the genome. This leads to a criterion on the signal-to-noise ratio of regulatory interactions,

$$F_0 \gtrsim E^*, \tag{11}$$

which in turn imposes a number of constraints on the design of regulatory DNA:

(a)   In a random genome, there must be at most a number of order one minimum-energy binding sites. Estimating the probability to find such a site at a given position as $(1/4)^\ell$, we obtain the condition

$$L(1/4)^\ell \lesssim 1. \tag{12}$$

This gives a lower bound on the site length, $\ell \gtrsim \log L/\log 4$. For a bacterial genome ($L \sim 10^6$), we obtain $\ell \gtrsim 10$, which gives the right length of functional binding sites. However, this bound is not fulfilled in eukaryotes. Indeed, eukaryotic genomes use a different design with groups of adjacent binding sites.

(b)   For each minimum-energy site, there are ℓ suboptimal sites of Hamming distance 1 from the minimum-energy sequence. These must not suppress the binding to the minimum-energy site, i.e.,

$$\exp(-E^*/k_B T) \gtrsim \ell \exp[-(E^* + \varepsilon)/k_B T] \tag{13}$$

in the two-state approximation. This gives a lower bound on the binding energy per nucleotide, $\varepsilon/k_B T \gtrsim \log \ell \approx 2 - 3$.

(c)   Finally, the unspecific binding in the entire genome must not suppress the specific binding to a minimum-energy site, i.e.,

$$\exp(-E^*/k_B T) \gtrsim L \exp(-E_u/k_B T). \tag{14}$$

This produces a lower bound on the energy gap between unspecific and optimal specific binding, $(E_u - E^*)/k_B T \gtrsim \log L \approx 15$.

Quite remarkably, these bounds are fulfilled as approximate equalities in bacteria. Hence, the machinery of transcriptional regulation operates just at the treshold of single-molecule sensitivity, i.e, $F_0 \approx E^*$.

Programmability and Evolvability of Regulatory Networks

Of course, not every regulatory interaction is equally sensitive. To switch genes on or off, the cell uses the dependencies of the binding probability both on factor numbers and on binding energies. During the cell cycle, the level of n can vary over several orders of magnitude, say, between a few and tens of thousands of molecules. At a given value of n, the effects on the regulated genes differ since their functional sites have different values of E. The binding energies can change on evolutionary time scales by mutations of the site sequence, which leads to regulatory differences between individuals and, ultimately, between species. Both parameters are thus necessary to encode pathways in regulatory networks. This is most flexible if minimum-energy sites are indeed sensitive to a single factor molecule as discussed above. Differentialprogrammability as a network design principle [14] thus favors complicated molecular structures with longer binding sites and larger binding energies. However, this competes with the evolvability of the system by a stochastic evolution process [18]. We have seen that the single-molecule sensitivity is just marginally reached in bacteria. This indicates that the actual machinery may result from a compromise between programmability and evolvability: binding sites are just complicated enough to work. It also indicates that genomic structures can only be understood from their evolution. This aspect will be developed further below, after we have introduced sequence analysis aspects in the next section.

Bioinformatics of Regulatory DNA

Predicting regulatory interactions between genes is clearly a key problem in bioinformatics, which is as important as the analysis of individual genes and proteins. It is not surprising that this problem is very difficult since, as we have discussed in the previous section, targeting regulatory input in a large genome is a tremendous signal-to-noise problem even for the cell itself. Its solution via the analysis of regulatory DNA requires finding statistical criteria to distinguish between functional binding sites and background sequence. A general introduction to the relevant sequence statistics can be found in ref. [19].

Markov Model for Background Sequence

We begin by specifying a stochastic model for the nonfunctional segments of intergenic DNA. These are assumed to be Markov sequences with uniform single-nucleotide frequencies $p_0(a)$ (a = A, C, G, T). Hence, the probability of finding a given sequence has the factorized form

$$P_0(a_1, ..., a_k) = \prod_{i=1}^{k} p_0(a_i).$$

(15)

This assumption should not be taken too literally. The term "nonfunctional" refers to binding of a particular transcription factor. Intergenic DNA contains plenty of non-random elements with other functions (e.g., binding sites for other factors) or without known function (such as repeat elements). The salient point is, however, that most of intergenic DNA is well approximated by a Markov sequence with respect to binding of a given transcription factor. To make this more precise, we project the distribution $P_0(\mathbf{a})$ for segments of length ℓ onto the binding energy E as independent variable. Denoting the projected distribution for simplicity with the same letter P_0, we have

$$P_0(E) = \sum_{\mathbf{a}} P_0(\mathbf{a}) \delta(E - E(\mathbf{a})).$$

(16)

This distribution is close to the actual genomic distribution $W_{dat}(E)$ for most values of E, as we have seen in fig. 3. It is possible to improve the background model by introducing small frequency couplings between neigboring letters [15, 16].

Probabilistic Model for Functional Sites

The sequences $\mathbf{a} = (a_1, ..., a_\ell)$ at functional sites of a given transcription factor are assumed to be drawn from a different distribution $Q(\mathbf{a})$. We write this distribution in the form

$Q(\mathbf{a}) = P_0(\mathbf{a}) \exp [S(\mathbf{a})].$ (17)

The quantity $S(\mathbf{a})$, which is called the relative log likelihood score of the distributions P_0 and Q, will turn out to have an important evolutionary meaning as well.

The single-nucleotide distribution $q_i(a)$ at a given position i within functional loci is obtained by summing the full distribution Q over all other positions

$$q_i(a) = \sum_{a_1, ..., a_{i-1}, a_{i+1}, ..., a_\ell} Q(\mathbf{a}).$$

(18)

The set of these marginal distributions, $q_i(a)$ ($i = 1, ..., \ell$; a = A, C, G, T) is called the position weight matrix for binding sites of a given factor [20]. If the score function is additive in the nucleotide positions, $S(\mathbf{a}) = \sum_{i=1}^{\ell} s_i(a_i)$, the Q distribution has a factorized form, $Q(\mathbf{a}) = \prod_{i=1}^{\ell} q_i(a_i)$ with

$q_i(a) = p_0(a) \exp [s_i(a)].$ (19)

This additivity assumption is made in most of the existing literature since the position weight matrix (18) can be inferred from a sample of known functional site sequences, which in turn determines directly the single nucleotide scores (19). This scoring is the basis for a number of site prediction methods in single species and by cross-species analysis; see, e.g., refs. [20–24].

Here we treat functional sites as coherent statistical units and do not make the assumption of additivity of the score function [16]. As will be discussed in the next section, functionality imposes correlations between the nucleotide frequencies within a functional site, preventing factorization of the Q distribution. Of course, it is not possible to reconstruct the full distribution $Q(\mathbf{a})$, which lives on a 4^{ℓ}-dimensional sequence space, from a limited sample of experimentally known functional sites. However, we can again project this distribution onto the binding energy as independent variable, $Q(E) = \sum_{\mathbf{a}} Q(\mathbf{a}) \delta(E - E(\mathbf{a}))$. Since all regulatory effects of a functional site depend on its sequence a only via the binding energy, we can also write the score as a function of the energy, $S(\mathbf{a}) = S(E(\mathbf{a}))$ (this will become obvious in the next section). Hence, the relationship (17) has the same form for the projected distributions,

$$Q(E) = P_0(E) \exp [S(E)]. \tag{20}$$

Bayesian Model for Genomic loci

Assuming that functional loci are distributed randomly with a small probability λ, we now combine the models for background sequence and for functional sites into a model for the full distribution of sequences $\mathbf{a}$ in intergenic DNA,

$$W(\mathbf{a}) = (1 - \lambda) P_0(\mathbf{a}) + \lambda Q(\mathbf{a}). \tag{21}$$

(At the moment, we are ignoring the possible overlap between functional sites). In the language of statistics, this is a probabilistic model with hidden variables. The output of this model consists of pairs (m, $\mathbf{a}$): First, the model variable m is randomly drawn, labelling a locus as nonfunctional (m = 0) with probability $1 - \lambda$ or as functional (m = 1) with probability λ. Then the sequence is drawn from the corresponding distribution $P_0(\mathbf{a})$ or $Q(\mathbf{a})$. However, only the sequence counts $\mathbf{a}$ are available data. The "hidden" variable m can be inferred from the data in a probabilistic way using Bayes' formula, which expresses the joint probability distribution of data and model in terms of its conditional and its marginal distributions

$$\text{prob}(\mathbf{a}, m) = \text{prob}(\mathbf{a}|m) \, \text{prob}(m) = \text{prob}(m|\mathbf{a}) \, \text{prob}(\mathbf{a}) \tag{23}$$

with $\text{prob}(\mathbf{a}) = \sum_m \text{prob}(\mathbf{a}|m)\text{prob}(m)$. We can solve for the conditional probability of the model for given data $\mathbf{a}$,

$$prob(m \mid \mathbf{a}) = \frac{prob(\mathbf{a} \mid m)prob(m)}{\sum_m prob(\mathbf{a} \mid m)prob(m)}.$$

$$(23)$$

For the probability of functionality, $\rho_f(\mathbf{a}) = prob(m = 1|\mathbf{a})$, this formula reads

$$\rho_f(\mathbf{a}) = \frac{\lambda Q(\mathbf{a})}{W(\mathbf{a})} = \frac{1}{1 + \exp\left[-S(\mathbf{a}) + \log\frac{1-\lambda}{\lambda}\right]}.$$

$$(24)$$

The dependence on S has again the form of a Fermi function. Its threshold value $S = \log[(1 - \lambda)/\lambda]$ separates sequences that are more likely to be functional or more likely to be background.

The full Bayesian model (21) can again be projected onto the energy variable,

$$W(E) = (1 - \lambda)P_0(E) + \lambda Q(E).$$

In this form, it can be tested against genomic data [16]. To plot the distributions P_0, Q, and W as functions of E, we use eq. (1) with an energy matrix $\varepsilon_i(a) = \varepsilon_0 \log[q_i(a)/p_0(a)]$ estimated from the position weight matrix up to an overall constant ε_0[10]. For our example of the CRP transcription factor, the distribution Q(E) can be estimated from the about 50 known binding sites in the E. coli genome. Using this Q distribution and a probability of functionality $\lambda \approx 6 \times 10^{-4}$, the full distribution W(E) produces an excellent fit of the count histogram $W_{dat}(E)$ over the entire range of energies; see fig. 4(a). The log likelihood score function $S(E) = \log[Q(E)/P_0(E)]$ is shown in fig. 4(b), shifted such that the curve has its zero at a point $E_s \approx 13$ beyond which binding becomes negligible.

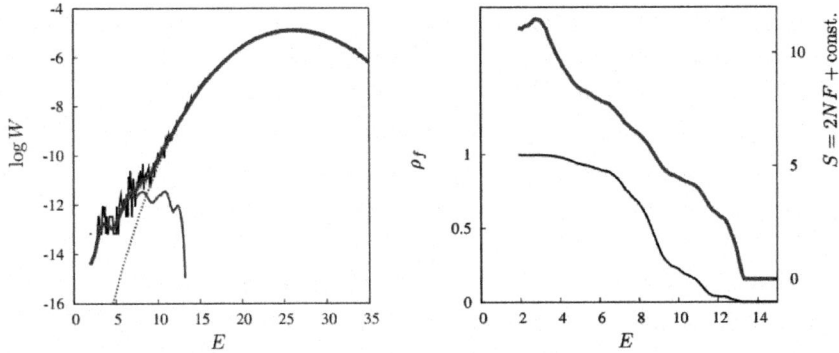

Figure 4: Bayesian model for regulatory DNA and score function. (a) Energy count histogram $W_{dat}(E)$ for CRP sites in E. coli as in fig. 3 (log scale), model distribution W(E) (thick line), and its decomposition (25) into background component $(1 - \lambda)P_0(E)$ (thin dashed line) and component $\lambda Q(E)$ ($E < E_s \approx 13$) of functional sites (thin solid line).

(b) Log-likelihood score $S(E) = \log [Q(E)/P_0(E)]$ (shifted by a constant, thick line) and probability of functionality $\rho_f(E)$ (thin line). From [16].

The resulting probability of functionality $\rho f(E)$ as given by eq. (24) is also shown in fig. 4(b). This indicates the dilemma for the prediction of individual binding sites based on sequence data from a single species. Many functional sites have energies in the "twilight" region between the ensembles λQ and $(1 - \lambda)P_0$, where ρf takes values around 1/2. Hence, depending on the energy cutoff chosen, any prediction is torn between many false negatives or many false positives.

Dynamic Programming and Sequence Analysis

It is straightforward to generalize the Bayesian approach to longer segments of intergenic DNA, which are covered by an unknown number s of non-overlapping functional sites as shown in fig. 5[22]. The hidden variables are now the sequence of left initial positions $\mathbf{r}_f \equiv (r_1,..., r_s)$ of the functional sites (with the no-overlap constraint $r_{v+1} \geq r_v + \ell$ for $v = 1,..., s - 1$). The full sequence distribution in a segment of length L has the form

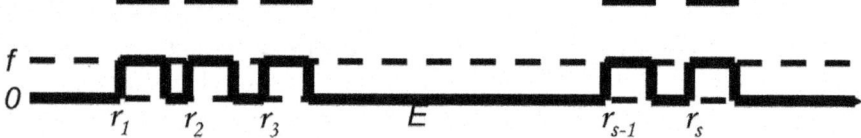

Figure 5: Analysis of regulatory sequences. A configuration of s nonoverlapping binding sites is given by the sequence of left initial positions $\mathbf{r}_f = (r_1,..., r_s)$ (with $r_{v+1} - r_v \geq \ell$ for $v = 1, 2,..., s - 1$). It can be associated with a path $m(r)$ which takes the values $m = 1$ at the nucleotide positions of binding sites and $m = 0$ elsewhere. Dynamic programming algorithms based on a Bayesian model (27) of genomic sequences assign to each site configuration a probability of occurence $\rho (r|a_1,..., a_L)$ for given sequence data $a_1,..., a_L$; see eq. (29).

$$W_L(a_1, ..., a_L) = Z^{-1} \sum_{\mathbf{r}_f} \tilde{\lambda} W_L(a_1, ..., a_L \mid \mathbf{r}_f),$$

$$(26)$$

where Z is a normalization factor, $\lambda^{\tilde{\lambda}} = \lambda + O(\lambda^2)$ is a weight factor for each functional locus (the negligible correction terms originate from the no-overlap constraint), and $W_L(a_1,..., a_L|\mathbf{r}_f)$ is the sequence distribution for given positions of functional loci,

$$W_L(a_1, \ldots, a_L \mid r_f) =$$

$$p_0(a_1) \cdots p_0(a_{r_1 - 1}) \prod_{\nu=1}^{s} Q(a_{r_\nu}, \ldots, a_{r_\nu + \ell - 1}) p_0(a_{r_\nu + \ell}) \cdots p_0(a_{r_{\nu+1} - 1}) =$$

$$p_0(a_1) \cdots p_0(a_L) \exp\left[\sum_{\nu=1}^{s} S(a_{r_\nu}, \ldots, a_{r_\nu + \ell - 1}) \right]$$

$$(27)$$

with $r_{s+1} = L + 1$. The sum over sequences r_f of arbitrary length s seems formidable at first, but W_L is easy to compute from the recursion

$$W_r(a_1, \ldots, a_r) = (1 - \lambda\hat{})p_0(a_r)W_{r-1}(a_1, \ldots, a_{r-1}) + \lambda\tilde{}Q(a_{r-\ell+1}, \ldots, a_r) W_{r-\ell}(a_1, \ldots, a_{r-\ell})$$

$$(28)$$

with the initial condition $W_0 = 1$ and $\lambda\hat{}^{\hat{\lambda}} = \lambda\tilde{}^{\tilde{\lambda}} + O(\lambda\tilde{}^{\tilde{\lambda}\,2})$. This type of recursion relation is usually called a dynamic programming algorithm in computer science. In physics, it is known as a transfer matrix, and the sum (27) is recognized as the corresponding discrete path integral in imaginary time r, if we interpret r_f as encoding a path m(r) that takes the value m = 1 at the nucleotide positions $r_\nu, \ldots, r_\nu + \ell - 1$ ($\nu = 1, \ldots, s$) within functional loci and m = 0 otherwise (see fig. 5). Both concepts prove very useful also in more general problems of sequence alignment.

In analogy to (24), the probability of a set r_f of functional loci for given sequence data is

$$p(r_f \mid a_1, \ldots, a_L) = \frac{W_L(a_1, \ldots, a_L \mid r_f)}{W_L(a_1, \ldots, a_L)}.$$

$$(29)$$

The most likely set $r*f^{r_f^*}$ can be obtained by the following "backward" algorithm: Given the sequence ($W_1, \ldots, W_L$) obtained from the «forward» recursion (28), we can decide for every point r whether it is more likely to be a background position or the endpoint of a functional locus, ignoring all sequence information from positions > r. This depends on whether the leading contribution to W_r comes from the first or second term on the r.h.s. of (28) and defines the local optimum model m*(r). The global optimum set of functional loci respecting the no-overlap constraint is then $r*f^{r_f^*} = \{r \mid b(r) = 1\}$, where b(r) is given by the recursion $b(r) = \ell$ if $b(r+1) \leq 1$ & m*(r) = 1 and b(r) = max(b(r+1) - 1, 0) otherwise, with the initial condition b(L + 1) = 0.

The Bayesian model can easily be extended to sequences containing several types of binding sites, which bind different transcription factors and are distinguished by their Q distributions. Dynamic programming algorithms can thus predict the likely coverage of a sequence with binding sites of known type [22]. This is the first step in extending the statistical analysis from single binding sites to entire regions of regulatory DNA. Indeed, models of this kind have been applied successfully to predict regulatory elements in eukaryotes, which typically consist of functional groups of adjacent binding sites. In the algorithms currently used, however, the scoring in (27) is strictly additive for groups of non-overlapping binding sites: it does not take into account dependencies between the sites within one functional group or overlapping sites within one sequence.

Evolution of regulatory DNA

In the statistical picture developed so far, background sequences and functional sites are reduced to ensembles P_0 and Q. This picture is incomplete in two ways. On one hand, it is quite disconnected from the biophysical aspects discussed before: the specific function of binding sites hardly enters the standard formalism of position weight matrices. On the other hand, there is not yet any notion of time and dynamics. Sequences change by various mutation processes, and the observed sequence ensembles derive from this evolutionary dynamics. The evolution of functional loci is fundamentally different from that of background sequence: it is subject to natural selection, that is, the fitness of an organism depends on its genotype **a** at a functional locus via the effects on the regulated gene. At this point, the biophysics of binding enters the evolution of functional sequences [25–27]. Moreover, it becomes clear that the statistical framework has to be extended from individual sequences to distributions of genotypes in a population. In this section, we develop an evolutionary picture of regulatory DNA, from which we obtain expressions for the sequence ensembles P_0, Q, and the score function S. The next four paragraphs are a self-contained introduction to the underlying concepts of population genetics.

Deterministic Population Dynamics and Fitness

We start by describing the evolution of a large population, which contains individuals of different genotypes **a**. Each genotype is assumed to produce a specific phenotype, which may influence the reproductive success of the individuals carrying it. With respect to factor binding, the phenotype can be associated with the binding energy E(**a**), since presumably all organismic effects of a locus depend on its genotype only via the binding energy. However, the discussion in the following paragraphs is more general.

We first assume that the subpopulations of a given genotype reproduce separately, i.e., there neither transitions between genotypes through mutations nor (in a sexually reproducing population) mixing through genomic recombination. Writing the dynamics of the subpopulations in the form of simple growth laws,

$$\frac{d}{dt} N_a(t) = F_a(t) N_a(t),$$
(30)

defines the (Malthusian) fitness $F_a(t)$ of each genotype. For notational simplicity, we now limit ourselves to the case of just two genotypes **a** and **b**, where (30) can be written as growth laws for the total population size $N(t) = N_a(t) + N_b(t)$ and for the population fraction $x(t) = N_b(t)/N(t)$ of genotype **b**,

$$\frac{d}{dt} N(t) = \bar{F}(t) N(t),$$
(31)

$$\frac{d}{dt} x(t) = \Delta F_{ab}(t) x(t)[1 - x(t)],$$
(32)

with $\bar{F}(t) \equiv [1 - x(t)]F_a(t) + x(t)F_b(t)$ and $\Delta F_{ab}(t) \equiv F_b(t) - F_a(t)$. This decomposition is useful since the overall growth rate $\bar{F}(t)$ is often strongly time-dependent due to external conditions (e.g., seasonality), while fitness differences, which reflect intrinsic properties of the phenotypes, are more stable. Different genotypes coexisting in a population frequently produce the same or very similar phenotypes and thus have equal fitness ($\Delta F_{ab} = 0$).

Assuming ΔF_{ab} to be constant over the time of observation, the solution of eq. (32) is the evolutionary trajectory

$$x(t) = \frac{x_0 \exp[\Delta F_{ab}(t - t_0)]}{1 - x_0(\exp[\Delta F_{ab}(t - t_0)] - 1)}$$
(33)

with the initial condition $x(t_0) = x_0$, shown in fig. 6(a). For $\Delta F_{ab} \neq 0$, the fixed points of this dynamics are the monomorphic population states $x = 0$, and $x = 1$, of which $x = 1$ is stable for $\Delta F_{ab} > 1$ and $x = 0$ for $\Delta F_{ab} < 1$. The approach to the stationary state takes place on a characteristic time scale $\tau_d = 1/\Delta F_{ab}$.

In the important case of neutral evolution ($\Delta F_{ab} = 0$), the evolutionary outcome remains indefinite. These results, which can readily be generalized to more than two phenotypes, are a simple version of Fisher›s fundamental theorem of natural selection: any population with initially coexisting phenotypes of different fitness will evolve towards a state where only the fittest phenotype is present.

A

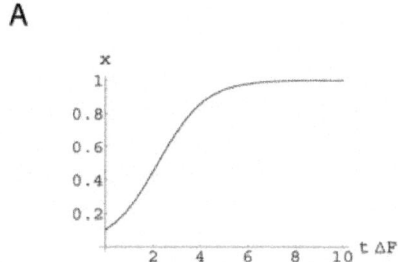

B

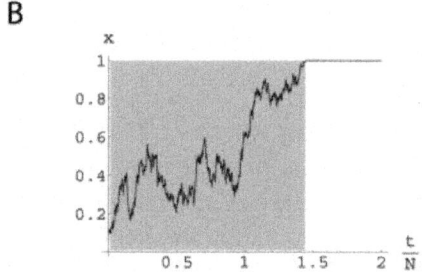

C

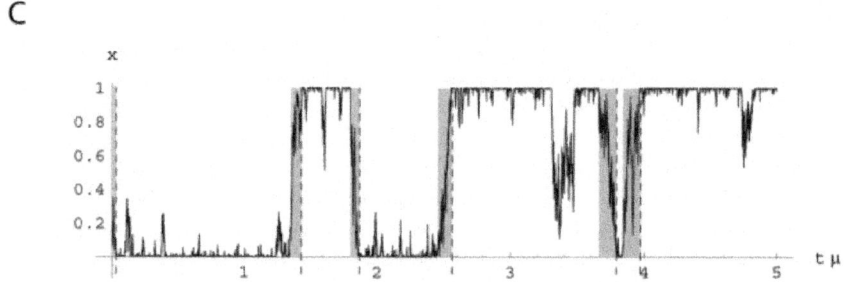

Figure 6: Evolution of genotype composition x (t). (a) Deterministic evolution with fitness difference $\Delta F_{ab} > 0$, leading to certain fixation of genotype b (time is shown in units of $\tau_d = 1/\Delta F_{ab}$). (b) Stochastic evolution with selection and genetic drift, leading to fixation of one of the genotypes. The time to fixation (grey shading) is of order τ_s ($N \Delta F_{ab} = 0.5$, time is shown in units of N). (c) Stochastic evolution with selection, genetic drift, and mutations in the regime $N\mu \ll 1$, leading to a substitution dynamics with rates $u_{a \to b}$ and $u_{b \to a}$ given by (49). Substitution events are marked by dashed lines. The typical time between initial mutation and fixation (grey shading) for a given substitution, τ_s, is much shorter than the time between subsequent substitutions, $1/u_{a \to b}$ resp. $1/u_{b \to a}$ ($N \Delta F_{ab} = 0.5$, $N\mu = 0.05$, time is shown in units of $1/\mu$).

Fisher's theorem seems to prove the popularized Darwinian notion of the "survival of the fittest". However, it rests on very restrictive assumptions that are never fulfilled in a natural population. The deterministic growth law (32) neglects mutations and recombinations, as well as the reproductive fluctuations present in any population due to its finite number of individuals. These other evolutionary forces have to be incorporated in our theoretical picture before we can even define fitness as a measurable quantity and before the theory can address the important case of neutral evolution.

Stochastic Dynamics and Genetic Drift

Stochastic fluctuations of the reproduction process in a large but finite population have been studied extensively in population genetics, see [28, 29]. They are called genetic drift, an unfortunate name which may falsely suggest a deterministic effect. To take these fluctuations into account, we replace eq. (30) by a stochastic growth law,

$$\frac{d}{dt} N_a(t) = F_a(t) N_a(t) + \chi_a(t),$$

(34)

where $\chi_a(t)$ are Gaussian random variables with $\overline{\chi_a(t)} = 0$ and

$$\overline{\chi_a(t)\chi_b(t')} = N_a(t)\delta(t - t')\delta_{a,b}.$$

(35)

This form of noise is simply due to the law of large numbers, and the continuum dynamics (34) emerges as an effective large-N description for a plethora of discrete evolution models, which are defined at the level of individuals and have finite generation times. In the application to real populations, N has to be interpreted as the so-called effective population size, which can be inferred from genome data and is in general smaller than the actual population size.

In the case of two genotypes, eq. (34) can again be projected onto the population fraction x,

$$\frac{d}{dt} x(t) = \Delta F_{ab}(t) x(t)[1 - x(t)] + \chi_x(t),$$

(36)

where $\chi_x(t) = (\partial x/\partial N_a)\chi_a(t) + (\partial x/\partial N_b)\chi_b(t)$ are Gaussian random variables with zero mean and

$$\overline{\chi_x(t)\chi_x(t')} = \frac{x(1 - x)}{N}\delta(t - t').$$

(37)

This dynamics produces stochastic evolutionary trajectories x(t) as shown in fig. 6(b). To capture their statistics, we convert the Langevin equation (36) into a Fokker-Planck equation for the probability distribution of the genotype composition [28,30],

$$\frac{\partial}{\partial t}\mathcal{P}(x,t) = \frac{1}{2N}\frac{\partial^2}{\partial x^2}x(1-x)\mathcal{P}(x,t) - \Delta F_{ab}(t)\frac{\partial}{\partial x}x(1-x)\mathcal{P}(x,t).$$

(38)

The mathematical subtlety of this equation lies in the x-dependent diffusion "constant" $x(1-x)/2N$, which reflects the multiplicative nature of the reproduction process. As a consequence, the two monomorphic population states $x = 0$ and $x = 1$ are also fixed points also of the stochastic dynamics. Any evolutionary trajectory $x(t)$ will eventually lead to one of these states with probability 1; this is called the fixation of the corresponding genotype in the population. In other words, the Fokker-Planck equation (38) describes diffusion in the interval (0, 1) with absorbing boundaries. There is a family of stationary states

$$\mathcal{P}(x) = (1-\varphi)\delta(x) + \varphi\delta(1-x),$$

(39)

parametrized by the fixation probability φ of genotype **b**. The value of φ depends on the initial condition x_0 and can be computed by solving the backward diffusion equation

$$\frac{\partial}{\partial t}\mathcal{P}(x,t \mid x_0,t_0) = x_0(1-x_0)\left(\frac{1}{2N}\frac{\partial^2}{\partial x_0^2} + \Delta F_{ab}(t)\frac{\partial}{\partial x_0}\right)\mathcal{P}(x,t \mid x_0,t_0).$$

(40)

For time-independent ΔF_{ab}, the stationary solution $\varphi(x_0) = \lim_{t\to\infty} P^{\mathcal{P}}(x = 1, t|x_0, t_0)$ has the form [28, 30]

$$\phi(x_0, \Delta F_{ab}, N) = \frac{1 - \exp(-2N\Delta F_{ab}x_0)}{1 - \exp(-2N\Delta F_{ab})},$$

(41)

which for near-neutral evolution ($N\Delta F_{ab} \ll 1$) reduces to

$$\varphi(x_0, 0, N) = x_0 + N\Delta F_{ab}x_0(1-x_0) + ..$$

(42)

The characteristic time τ_s of the stochastic dynamics interpolates between the diffusive scale N and the deterministic scale: $\tau_s \approx \min(N, \tau_d)$. It determines the typical time of the evolution process up to fixation, shown shaded in fig. 6(b).

Hence, the stochastic population dynamics depends no longer only on the fitness difference of the genotypes as in the deterministic case, but also on the initial state of the population and the the population size. Yet, our evolutionary picture is still incomplete. Population states with coexisting genotypes enter the dynamics as initial conditions, but since mutations are neglected, the model does not explain how this coexistence is generated and maintained.

Mutation Processes and Evolutionary Equilibria

At the level of an individual, mutations are rare stochastic genotype changes $a \rightarrow b$, which take place with rates $\mu_{a \rightarrow b}$, often coupled to the reproduction process. (These rates are all of the same order of magnitude, in estimates we therefore omit the indices.) We include mutations into the population dynamics (34) by their systematic effect on the genotype subpopulations,

$$\frac{d}{dt} N_a(t) = F_a(t) N_a(t) + \sum_b [\mu_{b \rightarrow a} N_b(t) - \mu_{a \rightarrow b} N_a(t)] + \chi_a(t),$$
(43)

while their stochastic effect (whose variance is of order $N\mu$) is neglected since it is small against the reproductive sampling noise $\chi_a(t)$. In the case of two different genotypes, this dynamics can again be projected onto the variable x,

$$\frac{d}{dt} x(t) = \Delta F_{ab}(t) x(t)[1 - x(t)] + \mu_{a \rightarrow b}[1 - x(t)] - \mu_{b \rightarrow a} x(t) + \chi_x(t),$$
(44)

which leads to the Fokker-Planck equation [31]

$$\frac{\partial}{\partial t} P(x,t) = \frac{1}{N} \frac{\partial^2}{\partial x^2} x(1 - x) P(x,t) - \Delta F_{ab}(t) \frac{\partial}{\partial x} x(1 - x) P(x,t)$$

$$- \mu_{a \rightarrow b} \frac{\partial}{\partial x} (1 - x) P(x,t) + \mu_{b \rightarrow a} \frac{\partial}{\partial x} x P(x,t).$$
(45)

For time-independent ΔF_{ab}, this equation has a single stable stationary state,

$$P(x) = \frac{1}{Z} x^{-1 + N\mu_{a \rightarrow b}} (1 - x)^{-1 + N\mu_{b \rightarrow a}} \exp(2N\Delta F_{ab} x)$$
(46)

with a normalization constant Z that can be expressed in terms of Bessel and Gamma functions [32].

Substitution Dynamics

Here we are interested in the stochastic evolution (45) and its equilibrium state (46) for $N\mu \ll 1$, which is the relevant dynamical regime for nuclear DNA in eukaryotes and in most prokaryotes (but not in viral systems). In this regime, the mutation term in (45) is small against the diffusion term except for values of x close to the boundaries 0 or 1. In this region, the continuum approximation of eq. (45) is no longer valid, and (46) has to be replaced by a stationary solution $P^{\mathcal{P}}_d(N_a)$ of the underlying discrete evolution model, which gives the probability that the population contains N_a individuals of genotype a (with $N_a = N - N_b = 0, 1, ..., N$). The discrete solution is easily shown to have the singularity $P_d(0) \simeq (N\mu_{a \rightarrow b})^{-1} P_d(1)$. This singularity is correctly captured if we use the approximation $P_d(N_a) \simeq \int_{N_a/N}^{(N_a+1)/N} dx \, \mathcal{P}(x)$ for all N_a (except at the other boundary, where there is a similar singularity $\mathcal{P}_d(N) \simeq (N\mu_{b\to a})^{-1} P_d(N-1)$) [33].

From this solution, we read off the following characteristics of the evolutionary dynamics at equilibrium, which are illustrated by the trajectory of fig. 6(c)[32]:

(a)   For sufficiently small values of μ, the population remains monomorphic for most of the time. Using the shorthands $Q(\mathbf{a}) = \mathcal{P}_d(N_a = 0)$ and $Q(\mathbf{b}) = \mathcal{P}_d(N_a = N)$, we have

$$Q(\mathbf{a}) + Q(\mathbf{b}) = 1 - O(\mu N \log N). \tag{47}$$

(b)   The ratio of probabilities for the two monomorphic population states is given by the ratio of "forward" and "backward" mutation rate, the fitness difference, and the effective population size:

$$\frac{Q(\mathbf{b})}{Q(\mathbf{a})} = \frac{\mu_{\mathbf{a}\to\mathbf{b}}}{\mu_{\mathbf{b}\to\mathbf{a}}} \exp(2N\Delta F_{\mathbf{ab}}) + O(N_\mu). \tag{48}$$

(c)   The monomorphic population states $x = 0$ and $x = 1$ are unstable due to mutations even at arbitrarily small values of μ, which cause occasional transitions of the entire population from genotype $\mathbf{a}$ to $\mathbf{b}$, and vice versa. These so-called substitutions are marked by dashed lines in fig. 6(c). The substitution rate $u_{\mathbf{a}\to\mathbf{b}}$ can be evaluated as the product of creating a single mutant of genotype $\mathbf{b}$ in an initially monomorphic $\mathbf{a}$ population, $N\mu_{\mathbf{a}\to\mathbf{b}}$ $N\mu_{\mathbf{a}\to\mathbf{b}}$, and its probability of fixation, φ ($x_0 = 1/N$, $\Delta F_{\mathbf{ab}}$, N). The time between initial mutation and fixation (shown by grey shading in fig. 6(c)) is still of order τ_s and thus much shorter than the time scale $1/\mu$, on which mutation effects become important. Hence, the fixation probability φ is given to leading order by (41), which has been derived for $\mu = 0$. Together we have [28, 30]

$$u_{\mathbf{a}\to\mathbf{b}} = N\mu_{\mathbf{a}\to\mathbf{b}} \frac{1 - \exp(-2\Delta F_{\mathbf{ab}})}{1 - \exp(-2N\Delta F_{\mathbf{ab}})}. \tag{49}$$

Hence, the substitution rate $u_{\mathbf{a}\to\mathbf{b}}$ is enhanced over $\mu_{\mathbf{a}\to\mathbf{b}}$ for $\Delta F_{\mathbf{ab}} > 0$ and suppressed for $\Delta F_{\mathbf{ab}} < 0$, as shown in Fig. 7. For weak selection ($N|\Delta F_{\mathbf{ab}}| \ll 1$), eq. (49) becomes

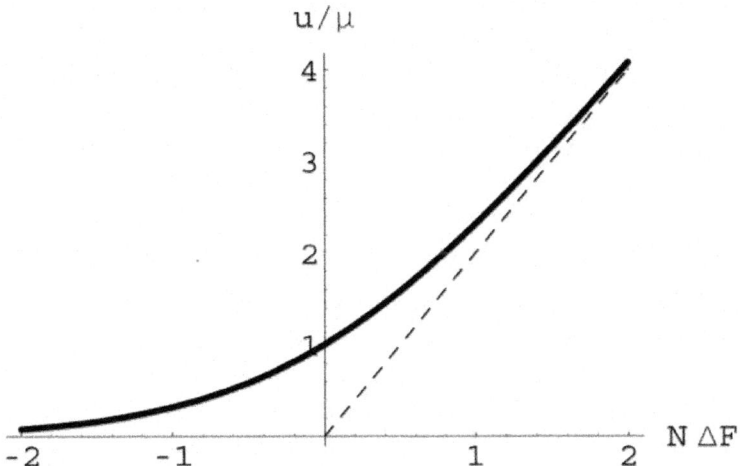

Figure 7: Substitution rate in a population versus mutation rate in an individual. The ratio of these rates, $u_{a\to b}/\mu_{a\to b}$, depends on the product $N\,\Delta F_{ab}$ of effective population size and fitness difference between the genotypes (in the relevant regime $N \gg 1$, $\Delta F_{ab} \ll 1$, $N\,\Delta F_{ab}$ finite). The substitution rate $u_{a\to b}$ is equal to μ_{ab} for neutral mutations ($\Delta F_{ab} = 0$), reduced for deleterious mutations ($\Delta F_{ab} < 0$), and enhanced for advantageous mutations ($\Delta F_{ab} > 0$).

$$u_{a\to b} = \mu_{a\to b}(1 + N\Delta F_{ab} + ...).$$

This reproduces Kimura's famous original result: for neutral evolution, the substitution rate equals the mutation rate in an individual, independently of the population size. For this reason, the rates $\mu_{a\to b}$ are referred to as neutral mutation rates. For strong selection ($N|\Delta F_{ab}| \gg 1 \gg |\Delta F_{ab}|$), eq. (49) takes the asymptotic forms

$$u_{a\to b} = \mu_{a\to b}\begin{cases} 2N\,|\,\Delta F_{ab}\,|\,\exp(2N\Delta F_{ab}) & (2N\Delta F_{ab} \ll 1), \\ 2N\Delta F_{ab} & (2N\Delta F_{ab} \gg 1). \end{cases} \tag{51}$$

The backward substitution rate $u_{b\to a}$ is given by a formula similar to (49) with $\Delta F_{ba} = -\Delta F_{ab}$. Forward and backward substitution rate have the simple ratio

$$\frac{u_{a\to b}}{u_{b\to a}} = \frac{\mu_{a\to b}}{\mu_{b\to a}}\exp(2N\Delta F_{ab}) \tag{52}$$

for $N \gg 1$. Comparing with (48), we obtain the consistency condition

$$\frac{u_{a\to b}}{u_{b\to a}} = \frac{Q(b)}{Q(a)}. \tag{53}$$

Hence, for sufficiently small mutation rates ($\mu N \log N \ll 1$), a simple picture emerges: The evolution of a population can be described as a sequence of transitions between monomorphic genotype states (substitutions). The substitution rate u is determined by the corresponding mutation rate in an individual, the fitness difference between the genotypes, and the effective population size.

Neutral Dynamics in Sequence Space, Sequence Entropy

This evolutionary picture can be generalized to multiple genotypes, for example, the 4^ℓ dimensional sequence space of genomic loci $\mathbf{a} = (a_1,..., a_\ell)$. Transitions between different sequence states are point mutations $\mathbf{a} \rightarrow \mathbf{b}$, which change exactly one nucleotide. (We neglect here insertion and deletion processes, which change the length of the sequence). We first discuss neutral evolution, where the substitution rate $u_{a \rightarrow b}$ equals the mutation rate in an individual, $\mu_{a \rightarrow b}$, for all elementary transitions $\mathbf{a} \rightarrow \mathbf{b}$. Bona fide neutral mutation rates can be inferred from DNA sequence alignments of sufficiently close species, recent insights have also come from studying repeat elements.

We assume the neutral dynamics has an equilibrium distribution $P_0(\mathbf{a})$ which obeys detailed balance, i.e., the relation

$$\frac{\mu_{a \rightarrow b}}{\mu_{b \rightarrow a}} = \frac{P_0(\mathbf{b})}{P_0(\mathbf{a})}$$

(54)

holds for each pair of sequence states linked by an elementary transition process $\mathbf{a} \rightarrow \mathbf{b}$. This says that the probability current at equilibrium, $\mu_{a \rightarrow b} P_0(\mathbf{a})$ - $\mu_{b \rightarrow a} P_0(\mathbf{b})$, vanishes for each elementary transition. Clearly, any distribution $P_0(\mathbf{a})$ satisfying the conditions (54) is stationary under the dynamics with rates $\mu_{a \rightarrow b}$, but not every such dynamics has a stationary distribution which satisfies (54) (the simplest counterexample involving three states and a circular probability current $\mathbf{a} \rightarrow \mathbf{b} \rightarrow \mathbf{c}$ at stationarity).

However, as will be verified below, detailed balance is a good approximation for the genomic substitution dynamics at least in prokaryotes. (There are known violations at CpG islands in eukaryotes [34]). In the simplest type of models, every nucleotide a mutates independently of all other positions with uniform rates $\mu_{a \rightarrow b}$ (i.e., $\mu_{a \rightarrow b} = \mu_{a \rightarrow b}$ for any two sequences $\mathbf{a}= (..., a ,...)$ and $\mathbf{b} = (..., b, ...)$ differing by exactly one nucleotide). This produces a factorized equilibrium distribution $P_0(\mathbf{a})$ of the form (15).

We can project the equilibrium distribution onto a measurable quantity as independent variable. For binding site sequences, a convenient choice is the binding energy E, and the projected distribution $P_0(E)$ has the form (16). Hence we can define the sequence entropy [35]

$S_0(E) = \log P_0(E),$ (55)

which counts the log density of sequence states **a** at energy E, weighed by the distribution $P_0(\mathbf{a})$.

Dynamics Under selection, the Score-Fitness Relation

The dynamics of substitutions can be studied in the same way for evolution under selection, which is specified at the level of genotypes by an arbitrary fitness function $F(\mathbf{a})$ [18, 36]. This generalizes the results of [37] for a model with selection acting independently at different nucleotide positions,

$F(\mathbf{a}) = \sum_{i=1}^{\ell} f_i(a_i)$. For each elementary transition $\mathbf{a} \to \mathbf{b}$, the substitution rate $u_{a \to b}$ is determined by the neutral rate $\mu_{a \to b}$, the fitness difference ΔF_{ab}, and the effective population size N according to (49). Given the detailed balance (54) of neutral evolution and the relation (52) between forward and backward rates, it then follows immediately that the evolutionary dynamics under selection also obeys detailed balance, as given by (53) with an equilibrium distribution $Q(\mathbf{a})$ of the form (48). Thus we have [18, 36]:

The equilibrium distribution $Q(\mathbf{a})$ of fixed genotypes generated by a substitution dynamics (49) with fitness function $F(\mathbf{a})$ is related to its neutral counterpart $P_0(\mathbf{a})$ by

$Q(\mathbf{a}) = P_0(\mathbf{a}) \exp [2NF (\mathbf{a}) + \text{const.}],$ (56)

with the constant given by normalization.

We can project eq. (56) onto the fitness as independent variable. Defining the distribution $Q(F) \equiv \sum_a Q(\mathbf{a})\delta (F (\mathbf{a}) - F)$, similarly $P_0(F)$, and the sequence entropy $S_0(F) = \log P_0(F)$, the projected identity takes the form

$Q(F) = \exp [2NF + S_0(F) + \text{const.}]$ (57)

For binding site sequences, we have a similar projection on the binding energy, $Q(E) = \exp [2NF(E) + S_0(E) + \text{const.}]$, since all genotypes with the same "phenotype" E have the same fitness, i.e., the same score S. The projected identities express the equilibrium distribution under selection in terms of fitness and sequence entropy, reflecting the balance between stochasticity (genetic drift) and selection [18]. For strong selection, the exponent $2NF - S_0$ is dominated by the fitness term, and $Q(F)$ takes appreciable values only at points of near-maximal fitness, i.e., where $F_{max} - F \lesssim 1/2N$. For moderate selection, there is a nontrivial balance between both terms, and for weak selection, the Q distribution can be approximated by its neutral counterpart $P_0 = \exp(S_0)$. Clearly, the roles of fitness and sequence entropy are formally analogous to those of energy and entropy in statistical physics of thermodynamic systems, if 2N is identified with the inverse temperature $1/k_B T$. Some consequences of this analogy are discussed in ref. [38].

The dynamics of substitutions establishes a rather general evolutionary grounding of genome statistics, if we identify the equilibrium distributions $P_0(\mathbf{a})$ and $Q(\mathbf{a})$ with the genomic distributions discussed in the previous section, as already anticipated by our notation. Comparing eqs. (56) and (17) gives a relation between fitness and score [16, 18]:

The log-likelihood score $S(\mathbf{a}) = \log [Q(\mathbf{a})/P_0(\mathbf{a})]$ equals the fitness function multiplied by twice the effective population size up to a constant,

$$S(\mathbf{a}) = 2NF(\mathbf{a}) + const..\qquad(58)$$

This relation allows us to use sequence data of a given genome to infer quantitative patterns of its evolution. We now discuss specific consequences for the evolution of regulatory DNA; an application to protein evolution can be found in ref. [37].

Measuring Selection for Binding Sites

We first give a precise definition of functionality for regulatory (and other) elements: A binding locus is functional if the genotype at that locus is under selection (for binding of the corresponding factor). Nonfunctional loci have evolutionarily neutral genotypes. This definition asks whether binding at a given locus makes a difference to the organism or not. It is weaker than that of a functional binding site, which is a functional locus with a sequence **a** that is likely to actually bind the factor. A functional locus can lose its binding sequence due to deleterious mutations, leading to suboptimal fitness of the organism. Conversely, a nonfunctional locus can have by chance a sequence which does bind the factor: this is a spurious binding site without consequences for the organism. To measure the selection on functional sites in silico, we apply the identity (58) to the genomic distributions $P_0(\mathbf{a})$ and $Q(\mathbf{a})$. (Assuming equilibrium for most loci seems to be justified for our example of CRP binding sites in E. coli since we find very similar distributions in the distant bacterial species Salmonella typhimurium, and the factor protein itself is highly conserved between these species.) After projection onto the energy, the fitness landscape 2NF(E) for CRP binding sites is thus given by fig. 4(b)[16]. The fitness is constant in the no-binding region ($E \gtrsim E_s \approx 13$) since the evolution is always neutral in that region. This constant is set to 0 in our normalization, i.e., F(E) measures the fitness gain of functional sites due to factor binding. Loci with strong binding are also under strong selection, with effective fitness values 2NF of order 10. Genetic drift counteracts selection, producing also loci with weaker binding and reduced effective fitness. This fitness "landscape" is thus qualitatively of the form predicted from the underlying biophysics [18, 25]. Of course, it should be kept in mind that this landscape results from averaging over a family of binding sites, which may have a spectrum of individual selection coefficients and selected binding strengths.

Nucleotide Frequency Correlations

A further consequence of (57) is the generic occurence of nucleotide frequency correlations within functional loci [18]. If the fitness function $F(\mathbf{a})$ is not additive in the nucleotide positions, nucleotide frequencies are correlated in selected genotypes even if they are independent under neutral evolution. This happens quite generically since selection acts on the entire genotype $\mathbf{a}$ as a functional unit and not on its single nucleotides. For binding sites, fitness effects follow from the expression level of the regulated gene, which depends on the sequence $\mathbf{a}$ via the binding probability of the corresponding transcription factor. While the binding energy is often approximately additive in the nucleotide positions as given by (1), the binding probability (10) is a strongly nonlinear function of the energy. This introduces correlations between nucleotide frequencies at any two positions within functional loci, preventing factorization of the distribution $Q(\mathbf{a})$.

Stationary Evolution of Binding Sites

Functional loci with a substantial level of selection (as found for the CRP binding sites in E. coli) evolve in a way quite different from background sequence. This is quantified in fig. 8(a), which shows pairs of binding energies (E_1, E_2) for experimentally verified CRP binding sites in E. coli and the corresponding sites regulating orthologous genes in S. typhimurium [16, 27]. The evolutionary distance t between the two species and characteristics of the neutral mutation process can be inferred from alignments of background sequence. The "phenotypic" evolution of CRP binding is quantified by the energy transition probabilities $G_0(E_2|E_1)$ under neutral evolution and $G_f(E_2|E_1)$ under stationary selection [16]. These are readily obtained by simulating the substitution dynamics over a time interval t for given initial value E_1, both with neutral rates $\mu_{a \to b}$ and with rates $u_{a \to b}$ given by (49) and the fitness function $2NF(E)$ measured in E. coli. The resulting conditional expectation values $\langle G_0(E_2|E_1) \rangle$ and $\langle G_f(E_2|E_1) \rangle$ for the binding energy in S. typhimurium are also shown in fig.8(a). The data conform to the selection model, showing a substantially stronger conservation of binding energy than expected for neutral evolution [16, 27, 39].

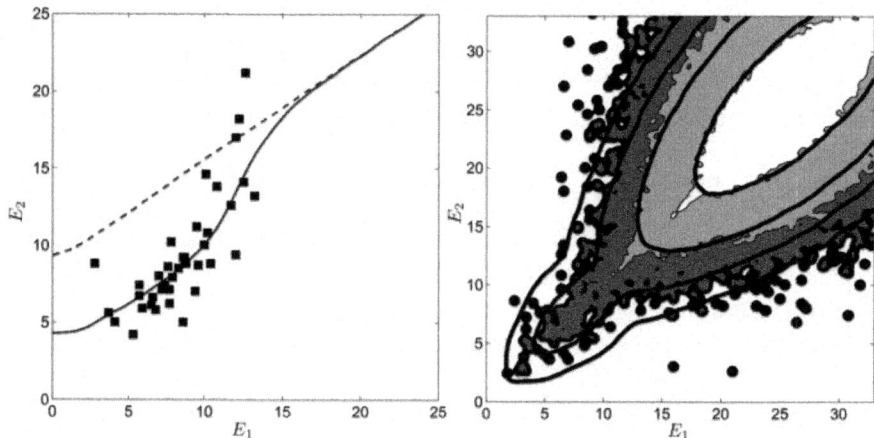

Figure 8: Evolution of binding sites. (a) Binding energy pairs (E_1, E_2) for 32 experimentally verified CRP binding sites in E. coli from the DPInteract database [57] and their aligned orthologs in S. typhimurium (dots). Conditional expectation value for the binding energy in S. typhimurium under neutral evolution, $\langle G_0(E_2|E_1) \rangle$ (dashed line), and under selection, $\langle G_f(E_2|E_1) \rangle$ (solid line). (b) Distribution of energy pair counts $W_{dat}(E_1, E_2)$ (filled contours), compared to the distribution $W(E_1, E_2)$ given by the Bayesian model (62). The symmetry of these distributions under exchange of E_1 and E_2 reflects detailed balance of the substitution dynamics. From [16,39].

We can now build a probabilistic model for cross-species comparisons [16]. It is based on the joint distributions of energy pairs

$$P_0(E_1, E_2) = G_0(E_2|E_1) \, P_0(E_1) \tag{59}$$

under neutral evolution and

$$Q(E_1, E_2) = G_f(E_2|E_1) \, Q(E_1) \tag{60}$$

under stationary selection, which are determined by the corresponding distributions in one species and the energy transition probabilities. Detailed balance of the substitution dynamics implies

$$\frac{P_0(E_2)}{P_0(E_1)} = \frac{G_0(E_2 \mid E_1)}{G_0(E_1 \mid E_2)} \quad \text{and} \quad \frac{Q(E_2)}{Q(E_1)} = \frac{G_f(E_2 \mid E_1)}{G_f(E_1 \mid E_2)}, \tag{61}$$

i.e., the joint distributions $P_0(E_1, E_2)$ and $Q(E_1, E_2)$ must be symmetric functions of their arguments. These distributions combine into a model for pairs of aligned loci, which generalizes the single-species model (25) and takes the form

$$W(E_1, E_2) = (1 - \lambda)P_0(E_1, E_2) + \lambda Q(E_1, E_2). \tag{62}$$

(This model can be extended further to include non-stationary selection.) The distribution $W(E_1, E_2)$ with a fraction of functionality $\lambda = 0.0018$ is in excellent agreement with the count distribution $W_{dat}(E_1, E_2)$ obtained from E. coli and S. typhimurium, as shown in fig. 8(b). The symmetry of W_{dat} is thus consistent with the underlying assumption of detailed balance. Analogous Bayesian models can be defined for more than two species related by a phylogeny. This approach has been applied to binding site prediction in bacteria [16]; a related study of several species of funghi has been reported in ref. [40].

Adaptive Evolution of Binding Sites

What does this picture say about the adaptive evolution of transcriptional regulation in response to a newly arising selection pressure? The evolution from a genotype with marginal binding $(E(\mathbf{a}) \approx E_s)$ to strong binding requires only about three uphill point mutations in the fitness landscape of fig. 4(b), i.e., there is an effective fitness gain $2N\Delta F \approx 3$ per mutation. Hence, according to (51), the rate of uphill substitutions per locus is enhanced by a factor $2N\Delta F \cdot d(\mathbf{a}, \mathbf{a}^*)$ at least of order 10 over the neutral point mutation rate per nucleotide. At the same time, the downhill rate is strongly suppressed. This shows that the adaptive formation of a binding site from background sequence can indeed be a rapid mode of regulatory evolution, due to the substantial level of selection [18].

However, this mode is only efficient if adaptation can set in immediately after the selection pressure is established. In larger regulatory regions, the exact position of a binding site is often not important. We assume the initial genome contains a set of $\tilde{L}$ shadow sites, i.e., positions $r_1,...,\ r_{\tilde{L}}$ where a given sequence $\mathbf{a}$ would have the same regulatory effect. If one of these shadow sites has already a genotype with marginal binding, it acts as a "seed" for the onset of adaptation [41]. On the other hand, if all shadow sites of the initial genome have energy $E > E_s$, there is typically a substantial waiting time of neutral evolution before one of them reaches the threshold energy E_s. Assuming the initial genome to be entirely background sequence, it will contain at least one such seed if $\int_{E<E_s} P_0(E)dE \gtrsim 1/\tilde{L}$, which is a joint condition on $\tilde{L}$ and the site length ℓ: the shadow regulatory region must be long enough and binding sites must be short enough. The example shows that the evolvability of regulation imposes constraints on genome architecture [18]. Adaptive point substitution may thus be a feasible mode for the formation of a single binding site, but will hardly explain the groups of adjacent sites characteristic of eukaryotic promoters. These may originate from repeat duplication by slippage, which

has recently been shown to be an efficient source of sequence innovation in intergenic regions of Drosophila.

Towards a dynamical picture of the genome

The relationship $S = 2NF + const.$ between score and fitness is a cornerstone of the theoretical picture developed so far, which links its population genetic, bioinformatic and biophysical arches. It relates a key evolutionary variable with the statistics of genomic frequency counts. The physical binding energy is an appropriate phenotypic variable on which fitness and score depend, because molecular function is determined by binding interactions.

We have discussed this picture for transcription factor binding sites, but it can be applied more generally to functional elements in genomes. It relates the statistics of these elements in one genome with their evolutionary dynamics, which is observed in cross-species comparisons. This dynamics is shaped by selection: The components of functional elements are coupled by a common fitness function; such fitness interactions are called epistasis. Hence, functional correlations lead to evolutionary correlations. These can be traced in the Q distribution over fixed genomes of a functional element. A more detailed statistical analysis using the statistics of polymorphisms within a population is briefly sketched below.

Thus, the picture of the genome as a system with multiple interactions has a fundamental dynamical significance. This is important since it allows us to trace functional modules from evolutionary patterns. We conclude the article with a brief outlook on functional integration of regulatory sequences at various and its dynamical implications.

Evolutionary Interactions between Sites

Regulatory function is often determined not by single binding sites, but jointly by a group of sites in the same regulatory region [42]. An important mechanism is binding cooperativity, i.e., the formation of a protein complex between two (or more) factors bound to their corresponding DNA sites. The binding energy of this complex has the form $E = E_1 + E_2 + \Delta E_{12}$, where E_1 and E_2 are the energies of the factors bound individually and $\Delta E_{12} < 0$ is the energy gain due to the protein-protein interaction, which is of the order of a few $k_B T$. Cooperative binding has a number of functional effects [1]:

(a)   It increases the signal-to-noise ratio for the targeting of regulatory input to a specific gene, which is important in larger eukaryotic genomes, where single spurious binding sites are abundant in background sequence.

(b)   It sharpens the response of the binding probability to variations in the factor concentrations around their threshold value. This follows from the thermodynamics of two factors, which is a straightforward generalization of the case of a single factor discussed above.

(c)   It implements logical connections between regulatory input signals to a given gene. The simplest example is an AND connection between two factors, where the regulated gene is affected only if both factors are simultaneously present. This happens if the binding energies and factor concentrations are such that individual binding is weak but joint binding is strong. Larger groups of binding sites can encode a whole repertoire of more complicated logical functions [43].

Regulatory modules with several jointly acting binding sites are frequently found in eukaryotes. The functional coupling of sites in a module translates into interactions between these sites in their sequence evolution. The genomic functional element, i.e., the subset of the regulatory region on which selection acts, is the module as a whole. Its fitness $F(E_1, E_2, \Delta E_{12}, ...)$ is a joint function of the binding energies as the relevant phenotypic variables [18, 25]. The evolutionary dynamics under this selection allows for a large number of compensatory changes, i.e., pairs of correlated substitutions changing two binding energies such that the fitness remains constant. These lead to nucleotide frequency correlations between different sites. Such compensatory changes have indeed been observed in experiments on Drosophila promoters [44].

Site-Shadow Interactions

In larger regulatory regions, there is a number of shadow sites where a binding sequence **a** would have a similar regulatory effect as at the functional sites present. In that case, the genomic functional element contains not only the functional binding sites but also the shadow sites. Once a functional site has disappeared due to deleterious mutations, a shadow site can turn functional by adaptive evolution as described in the last section. The resulting evolutionary dynamics leads to sequence turnover with the actual binding sites present at different but functionally equivalent positions [36]. Substantial sequence turnover has been observed in a number of case studies [44–49]. Also the number of actual sites is subject to evolutionary variation since the same regulatory effect, i.e., the same fitness, can be distributed over fewer stronger or more weaker sites. With increasing number $L^{\tilde{L}}$ of shadow positions, one expects that the number of actual sites grows while individual sites get weaker [36].

Gene Interactions

Evolutionary interactions are not limited to regulatory elements for the same gene. An example are gene duplications and the subsequent evolution of the daughter genes. Selection acts jointly on this pair of genes [50], which have initially identical functions, eventually leading to either loss of one of them or to subfunctionalization, which has been argued to be an important mode of genome evolution in eukaryotes [51, 52]. This process can take place by regulation, i.e., via a correlated distribution of the regulatory elements on the daughter genes. More generally, the evolution of genes in a regulatory network is correlated if their functions are coupled either in series (i.e., one gene acts on the other) or in parallel (i.e., they are part of alternative pathways for the same function). Although some regulatory networks in model organisms – e.g. the embryonic development in the sea urchin [53] – have been studied in detail, we lack a coherent view of their functional evolution to date.

Interactions and Time-Dependent Selection

The functional integration of regulation at multiple levels and the resulting fitness interactions (epistasis) imply that the selection at one genomic site is influenced by changes at other sites. A recent analysis of single-nucleotide polymorphisms and substitutions in Drosophila provides indeed evidence on a genome-wide scale that selection is time-dependent: at individual loci, changes in the direction of selection occur at nearly the rate of neutral evolution [54, 55]. At the same time, selection is sufficiently strong so that the adaptive response can keep up with the rate of selection changes. This rate is faster in non-coding DNA, which points towards the role of regulation in the adaptive differentiation between species. Genomic evolution emerges as a complex stochastic process, shaped jointly be the driving force of time-dependent selection, fitness interactions between sites, and the ongoing background of near-neutral changes. Much more remains to be learned about the interplay of these evolutionary forces: in a large and strongly coupled system, one external signal can trigger an avalanche of subsequent compensatory responses. This dynamics seems now within reach of genomic sequence analysis.

Evolutionary Innovations

Under stationary selection, functional elements are more conserved than background sequence, and the score-fitness relation quantifies the amount of conservation. But evolution is, of course, not limited to conservation. On one hand, there is typically a multitude of different genotypes yielding the same molecular function, and the evolutionary dynamics continuously plays with these alternatives. On the other hand, organisms face long-term changes of their environment, which lead to new selection pressures and a response by adaptive evolution of new functions. If regulation is to account for a large part of the diversification in higher eukaryotes, loss or gain of regulatory function should be an important mode of molecular evolution. Changes in regulatory DNA leading to new functions of gene networks have been observed [56], and it is possible to extend the statistical models described in the previous section to include evolutionary gain or loss of function of individual binding sites [16]. On a broader scale, time-dependent selection and fitness couplings appear act as a major driving forces of genomic change, triggering avalanches of evolutionary innovation. Understanding this molecular basis of innovations is a major challenge for theory and experiment in the coming years. It will profoundly change our dynamical view of the genome.

ACKNOWLEDGEMENTS

This article has been published as part of BMC Bioinformatics Volume 8 Supplement 6, 2007: Otto Warburg International Summer School and Workshop on Networks and Regulation. The full contents of the supplement are available online athttp://www.biomedcentral.com/1471-2105/8?issue=S6

REFERENCES

1. Ptashne M, Gann A: Genes and Signals. 2002, Cold Spring Harbor Laboratory Press

2. Tautz D: Evolution of transcriptional regulation. Curr Opin Genet Dev. 2000, 10 (5): 575-579. 10.1016/S0959-437X(00)00130-1.

3. Wray GA, Hahn MW, Abouheif E, Balhoff JP, Pizer M, Rockman MV, Romano LA: The evolution of transcriptional regulation in eukaryotes. Mol Biol Evol. 2003, 20 (9): 1377-1419. 10.1093/molbev/msg140. [http://dx.doi.org/10.1093molbev/msg140]

4. Berg OG, Winter RB, von Hippel PH: Diffusion-driven mechanisms of protein translocation on nucleic acids. 1. Models and theory. Biochemistry. 1981, 20 (24): 6929-6948. 10.1021/bi00527a028.

5. Winter RB, von Hippel PH: Diffusion-driven mechanisms of protein translocation on nucleic acids. 2. The Escherichia coli repressor-operator interaction: equilibrium measurements. Biochemistry. 1981, 20 (24): 6948-6960. 10.1021/bi00527a029.

6. Winter RB, Berg OG, von Hippel PH: Diffusion-driven mechanisms of protein translocation on nucleic acids. 3. The Escherichia coli lac

repressor-operator interaction: kinetic measurements and conclusions. Biochemistry. 1981, 20 (24): 6961-6977. 10.1021/bi00527a030.

7.   von Hippel PH, Berg OG: On the specificity of DNA-protein interactions. Proc Natl Acad Sci USA. 1986, 83 (6): 1608-1612. 10.1073/pnas.83.6.1608.

8.   Sarai A, Takeda Y: Lambda repressor recognizes the approximately 2-fold symmetric half-operator sequences asymmetrically. Proc Natl Acad Sci USA. 1989, 86 (17): 6513-6517. 10.1073/pnas.86.17.6513.

9.   Fields DS, He Y, Al-Uzri AY, Stormo GD: Quantitative specificity of the Mnt repressor. J Mol Biol. 1997, 271 (2): 178-194. 10.1006/jmbi.1997.1171.

10.   Stormo GD, Fields DS: Specificity, free energy and information content in protein-DNA interactions. Trends Biochem Sci. 1998, 23 (3): 109-113. 10.1016/S0968-0004(98)01187-6.

11.   Oda M, Furukawa K, Ogata K, Sarai A, Nakamura H: Thermodynamics of specific and non-specific DNA binding by the c-Myb DNA-binding domain. J Mol Biol. 1998, 276 (3): 571-590. 10.1006/jmbi.1997.1564. [http://dx.doi.org/10.1006/jmbi.1997.1564]

12.   Omagari K, Yoshimura H, Takano M, Hao D, Ohmori M, Sarai A, Suyama A: Systematic single base-pair substitution analysis of DNA binding by the cAMP receptor protein in cyanobacterium Synechocystis sp. PCC 6803. FEBS Lett. 2004, 563 (1–3): 55-58. 10.1016/S0014-5793(04)00248-0. [http://dx.doi.org/10.1016/S0014-5793(04)00248-0]

13.   Foat BC, Houshmandi SS, Olivas WM, Bussemaker HJ: Profiling condition-specific, genome-wide regulation of mRNA stability in yeast. Proc Natl Acad Sci USA. 2005, 102 (49): 17675-17680. 10.1073/pnas.0503803102. [http://dx.doi.org/10.1073/pnas.0503803102]

14.   Gerland U, Moroz JD, Hwa T: Physical constraints and functional characteristics of transcription factor-DNA interaction. Proc Natl Acad Sci USA. 2002, 99 (19): 12015-12020. 10.1073/pnas.192693599. [http://dx.doi.org/10.1073/pnas.192693599]

15.   Djordjevic M, Sengupta AM, Shraiman BI: A biophysical approach to transcription factor binding site discovery. Genome Res. 2003, 13 (11): 2381-2390. 10.1101/gr.1271603. [http://dx.doi.org/10.1101/gr.1271603]

16.   Mustonen V, Lässig M: Evolutionary population genetics of promoters: predicting binding sites and functional phylogenies. Proc Natl Acad Sci USA. 2005, 102 (44): 15936-41. 10.1073/pnas.0505537102. [http://dx.doi.org/10.1073/pnas.0505537102]

17. Slutzky M, Mirny L: Kinetics of protein-DNA interaction: facilitated target location in a sequence-dependent potential. Biophys J. 2004, 87 (6): 4021-4035. 10.1529/biophysj.104.050765.

18. Berg J, Willmann S, Lässig M: Adaptive evolution of transcription factor binding sites. BMC Evol Biol. 2004, 4: 42-10.1186/1471-2148-4-42. [http://dx.doi.org/10.1186/1471-2148-4-42]

19. Durbin R, Eddy SR, Krogh A, Mitchison G: Biological Sequence Analysis. 1998, Cambridge University Press

20. Stormo GD, Hartzell GW: Identifying protein-binding sites from unaligned DNA fragments. Proc Natl Acad Sci USA. 1989, 86 (4): 1183-1187. 10.1073/pnas.86.4.1183.

21. Hertz GZ, Stormo GD: Identifying DNA and protein patterns with statistically significant alignments of multiple sequences. Bioinformatics. 1999, 15 (7–8): 563-577. 10.1093/bioinformatics/15.7.563.

22. Rajewsky N, Socci ND, Zapotocky M, Siggia ED: The evolution of DNA regulatory regions for proteo-gamma bacteria by interspecies comparisons. Genome Res. 2002, 12 (2): 298-308. 10.1101/gr.207502. Article published online before print in January 2002. [http://www.genome.org/cgi/content/full/12/2/298]

23. van Nimwegen E, Zavolan M, Rajewsky N, Siggia ED: Probabilistic clustering of sequences: inferring new bacterial regulons by comparative genomics. Proc Natl Acad Sci USA. 2002, 99 (11): 7323-7328. 10.1073/pnas.112690399. [http://dx.doi. org/10.1073/pnas.112690399]

24. Lenhard B, Sandelin A, Mendoza L, Engström P, Jareborg N, Wasserman WW: Identification of conserved regulatory elements by comparative genome analysis. J Biol. 2003, 2 (2): 13-10.1186/1475-4924-2-13. [http://dx.doi.org/10.1186/1475-4924-2-13]

25. Gerland U, Hwa T: On the selection and evolution of regulatory DNA motifs. J Mol Evol. 2002, 55 (4): 386-400. 10.1007/s00239-002-2335-z. [http://dx.doi.org/10.1007/s00239-002-2335-z]

26. Moses AM, Chiang DY, Kellis M, Lander ES, Eisen MB: Position specific variation in the rate of evolution in transcription factor binding sites. BMC Evol Biol. 2003, 3: 19-10.1186/1471-2148-3-19. [http://dx.doi.org/10.1186/1471-2148-3-19]

27. Brown CT, Callan CG: Evolutionary comparisons suggest many novel cAMP response protein binding sites in Escherichia coli. Proc Natl Acad Sci USA. 2004, 101 (8): 2404-2409. 10.1073/pnas.0308628100.

28. Kimura M, Crow J: An Introduction to Population Genetics Theory. 1970, Harper & Row, New York

29. Kimura M: The Neutral Theory of Molecular Evolution. 1983, Cambridge University Press

30. Kimura M: On the probability of fixation of mutant genes in a population. Genetics. 1962, 47: 713-719.

31. Kimura M, Ohta T: The Average Number of Generations until Fixation of a Mutant Gene in a Finite Population. Genetics. 1969, 61 (3): 763-771.

32. Rouzine IM, Rodrigo A, Coffin JM: Transition between stochastic evolution and deterministic evolution in the presence of selection: general theory and application to virology. Microbiol Mol Biol Rev. 2001, 65: 151-185. 10.1128/MMBR.65.1.151-185.2001. [http://dx.doi.org/10.1128/MMBR.65.1.151-185.2001]

33. Grün D, Lässig M: to be published.

34. Arndt PF, Hwa T: Identification and measurement of neighbor-dependent nucleotide substitution processes. Bioinformatics. 2005, 21 (10): 2322-2328. 10.1093/bioinformatics/bti376. [http://dx.doi.org/10.1093/bioinformatics/bti376]

35. Peliti L: Quasispecies evolution in general mean-field landscapes. Europhys Lett. 2002, 57: 745-51. 10.1209/epl/i2002-00526-5.

36. Berg J, Lässig M: Stochastic evolution of transcription factor binding sites. Biophysics (Moscow). 2003, 48 (Suppl 1): S36-S44.

37. Halpern AL, Bruno WJ: Evolutionary distances for protein-coding sequences: modeling site-specific residue frequencies. Mol Biol Evol. 1998, 15 (7): 910-917.

38. Sella G, Hirsh AE: The application of statistical physics to evolutionary biology. Proc Natl Acad Sci USA. 2005, 102 (27): 9541-9546. 10.1073/pnas.0501865102. [http://dx.doi.org/10.1073/pnas.0501865102]

39. Mustonen V, Lässig M: to be published.

40. Moses AM, Chiang DY, Pollard DA, Iyer VN, Eisen MB: MONKEY: identifying conserved transcription-factor binding sites in multiple alignments using a binding site-specific evolutionary model. Genome Biol. 2004, 5 (12): R98-10.1186/gb-2004-5-12-r98. [http://dx.doi.org/10.1186/gb-2004-5-12-r98]

41. MacArthur S, Brookfield JFY: Expected rates and modes of evolution of enhancer sequences. Mol Biol Evol. 2004, 21 (6): 1064-1073. 10.1093/molbev/msh105. [http://dx.doi.org/10.1093/molbev/msh105]

42. Arnosti DN: Analysis and function of transcriptional regulatory elements: insights from Drosophila. Annu Rev Entomol. 2003, 48: 579-602.

10.1146/annurev.ento.48.091801.112749.   [http://dx.doi.org/10.1146/annurev.ento.48.091801.112749]

43.   Buchler NE, Gerland U, Hwa T: On schemes of combinatorial transcription logic. Proc Natl Acad Sci USA. 2003, 100 (9): 5136-5141. 10.1073/pnas.0930314100. [http://dx.doi.org/10.1073/pnas.0930314100]

44.   Ludwig MZ, Bergman C, Patel NH, Kreitman M: Evidence for stabilizing selection in a eukaryotic enhancer element. Nature. 2000, 403 (6769): 564-567. 10.1038/35000615. [http://dx.doi.org/10.1038/35000615]

45.   Ludwig MZ, Patel NH, Kreitman M: Functional analysis of eve stripe 2 enhancer evolution in Drosophila: rules governing conservation and change. Development. 1998, 125 (5): 949-958.

46.   McGregor AP, Shaw PJ, Hancock JM, Bopp D, Hediger M, Wratten NS, Dover GA: Rapid restructuring of bicoid-dependent hunchback promoters within and between Dipteran species: implications for molecular coevolution. Evol Dev. 2001, 3 (6): 397-407. 10.1046/j.1525-142X.2001.01043.x.

47.   Dermitzakis ET, Bergman CM, Clark AG: Tracing the evolutionary history of Drosophila regulatory regions with models that identify transcription factor binding sites. Mol Biol Evol. 2003, 20 (5): 703-14. 10.1093/molbev/msg077. [http://dx.doi.org/10.1093/molbev/msg077]

48.   Scemama JL, Hunter M, McCallum J, Prince V, Stellwag E: Evolutionary divergence of vertebrate Hoxb2 expression patterns and transcriptional regulatory loci. J Exp Zool. 2002, 294 (3): 285-99. 10.1002/jez.90009. [http://dx.doi.org/10.1002/jez.90009]

49.   Costas J, Casares F, Vieira J: Turnover of binding sites for transcription factors involved in early Drosophila development. Gene. 2003, 310: 215-20. 10.1016/S0378-1119(03)00556-0.

50.   Wagner A: Selection and gene duplication: a view from the genome. Genome Biol. 2002, 3 (5): reviews1012-10.1186/gb-2002-3-5-reviews1012.

51.   Lynch M, Conery JS: The evolutionary demography of duplicate genes. J Struct Funct Genomics. 2003, 3 (1–4): 35-44. 10.1023/A:1022696612931.

52.   Lynch M, Conery JS: The origins of genome complexity. Science. 2003, 302 (5649): 1401-4. 10.1126/science.1089370. [http://dx.doi.org/10.1126/science.1089370]

53.   Davidson E: A view from the genome: spatial control of transcription in sea urchin development. Curr Opin Genet Dev. 1999, 9 (5): 530-41. 10.1016/S0959-437X(99)00013-1.

54. Mustonen V, Lässig M: Adaptations to fluctuating selection in Drosophila. Proc Natl Acad Sci USA. 2007, 104 (7): 2277-82. 10.1073/pnas.0607105104. [http://dx.doi.org/10.1073/pnas.0607105104]

55. Mustonen V, Lässig M: Sequence evolution under quenched selection fluctuations. preprint. 2006

56. Gasch AP, Moses AM, Chiang DY, Fraser HB, Berardini M, Eisen MB: Conservation and evolution of cis-regulatory systems in ascomycete fungi. PLoS Biol. 2004, 2 (12): e398-10.1371/journal.pbio.0020398. [http://dx.doi.org/10.1371/journal.pbio.0020398]

57. Robison K, McGuire AM, Church GM: A comprehensive library of DNA-binding site matrices for 55 proteins applied to the complete Escherichia coli K-12 genome. J Mol Biol. 1998, 284 (2): 241-254. 10.1006/jmbi.1998.2160. [http://dx.doi.org/10.1006/jmbi.1998.2160]

Chapter 11

STATISTICAL ANALYSIS IN GENETIC STUDIES OF MENTAL ILLNESSES

Heping Zhang

Yale School of Public Health, Yale University, 60 College Street, New Heaven, Connecticut 06520-8034, USA

ABSTRACT

Identifying the risk factors for mental illnesses is of significant public health importance. Diagnosis, stigma associated with mental illnesses, comorbidity, and complex etiologies, among others, make it very challenging to study mental disorders. Genetic studies of mental illnesses date back at least a century ago, beginning with descriptive studies based on Mendelian laws of inheritance. A variety of study designs including twin studies, family studies, linkage analysis, and more recently, genomewide association studies have been employed to study the genetics of mental illnesses, or complex diseases in general. In this paper, I will present the challenges and methods from a statistical perspective and focus on genetic association studies.

INTRODUCTION

Mental illnesses affect the health and well-being of all populations and all ages. Schizophrenia—a chronic, severe, and disabling brain disorder—is one of these mental illnesses, affecting about 1.1 percent of the U.S. population age 18 and older in a given year. People with schizophrenia sometimes hear voices others do not hear, believe that others are broadcasting their thoughts to the world, or become convinced that others are plotting to harm them. These experiences can make them fearful and withdrawn and cause difficulties when these people try to have relationships with others (http://www.nimh.nih.gov). Emil Kraepelin (1856– 1926) described "Dementia Praecox" as an inherited disorder in his influential "Textbook of Psychiatry" (1899). Dementia Praecox, coined "schizophrenia," was first used by Arnold Pick (1851–1924)—a professor of psychiatry at the German branch of Charles University in Prague— to describe a patient with a psychotic disorder resembling hebephrenia in

1891. Nearly a century ago, Cannon and Rosanoff (1911) made an attempt to understand whether there are any forms of nervous and mental diseases that are transmitted from generation to generation in concordance with Mendelian laws.

They examined the families of 11 neuropathetic patients, which are now referred to as probands in pedigrees. Using Mendelian laws as their theoretical expectation, they concluded that the neuropathetic make-up is recessive to normal. Although the report was indeed "preliminary," a few things are noteworthy. First, they noted that "any form of insanity or even all the forms of hereditary insanity do not constitute an independent hereditary character." This raised an early sign of the complexity associated with studying mental disorders compared to the characterization of the disorders and their comorbidity. Here, comorbidity refers to more than one disease condition in the same patient. Second, they remarked "should larger accumulations of such data in the future give similar results, we shall be able" to confirm their result.

The requirement for more samples and replication is another challenge in studies of complex diseases. Last, but not the least, while they said "let us test,..., the hypothesis ..." they did not mean a statistical test. However, the idea of the χ^2-test is evident. Despite this early work, it was not until the 1960s that the researchers began to use scientifically rigorous designs and methods to study the inheritance of mental illnesses. For example, the key idea in adoption studies lies in the belief that any links between an adopted child and the biological parents are attributable to genetics, and any links between that child and adoptive parents can be attributed to environment (Plomin et al., 1997).

This enables us to separate the confounding environment (i.e., a family) from genetic contribution. Consequently, there are two strategies in adoption studies. One approach compares the risk of developing schizophrenia in the adopted children of schizophrenic parents to the risk of adopted children whose parents do not have schizophrenia. Several studies including Heston (1966), Rosenthal (1972) and Tienari (1991) used this approach to study schizophrenia. Each study found an elevated risk in adopted-away children of schizophrenic parents, supporting the role of genetics in the transmission of schizophrenia. The origin of this approach is the schizophrenic parents. Another approach backtracks from adopted children who have developed schizophrenia and compares the risks of schizophrenia in their adoptive and biological families. Kety, Rosenthal and Wender (1978) and others found that the risk was significantly higher in the biological relatives than in the adoptive families, again underscoring the role of genetics as a risk factor. While these schizophrenia adoption studies are in- fluential in understanding the role of genetics in mental disorders, the majority of the genetic factors associated with mental disorders are based on family and twin studies. By comparing the concordance in the risk between identical (monozygotic) and fraternal (dizygotic) twins, twin studies arguably provide the most compelling results about genetic and environmental effects.

For example, the concordance in monozygotic twins for Tourette's syndrome, a complex disorder characterized by repetitive, sudden and involuntary movements or noises called tics, was reported to be about 50% whereas it is less than 10% in dizygotic twins. Twin studies are most helpful in demonstrating the magnitude of genetic effect, but they do not provide insight into the inheritance pattern of a condition. Thus, family studies can offer information that twin studies cannot. Thus, Cannon and Rosanoff (1911) employed a small-scale, simple family study. Using the Mendelian laws, not only might we find evidence of genetics, but also infer the mode of transmission, as Cannon and Rosanoff (1911) concluded for the heredity of insanity. Although twin and family studies continue to be useful for understanding the genetics of complex diseases, different studies are needed to locate a specific gene on a chromosome that may underlie the disease. Gene mapping in humans through linkage analysis emerged in the 1930s, but it was Morton (1955) who laid the foundation for the methodology. It was only during the 1970s and 1980s, when the Elston-Steward (1971) algorithm was developed and implemented (Ott, 1974), that the method thrived as a common tool of genetic studies.

These initial and subsequent developments allowed for linkage analyses of multiple markers simultaneously. In light of the sheer number of genes and that we do not know which specific gene we are looking for, we typically genotype 300 to 400 "landmarks" that cover the 22 pairs of autosomes and the X chromosome. By inferring the transmission patterns of these markers, then linking them to the disease status, we can obtain information about the most probable region where the gene of interest resides. While linkage studies have had some successes (e.g., BRCA1), they have generated many more premature excitements. In the late 1980s, two particular studies attracted significant public attention after they reported that bipolar affective disorders were linked to DNA markers on chromosome 11, and that a susceptibility locus for schizophrenia was located on chromosome 5. Unfortunately, these findings were not replicated. Replications in genetic studies of mental disorders do not come easily. For example, Abelson et al. (2005) identified mutations involving the SLITRK1 gene (13q31.1) in a small number of people with Tourette's syndrome. However, most people with Tourette's syndrome do not have a mutation in the SLITRK1 gene. Because the mutations were reported in so few people with this condition, the association of the SLITRK1 gene with this disorder could not be confirmed.

In fact, Scharf et al. (2008) reported a lack of the association between SLITRK1var321 and Tourette's syndrome in a large family-based sample. Various reasons have been suggested to explain the difficulties detouring progress in genetic studies using linkage analysis. A key concept underlying linkage analysis is the recombination fraction, which re- flects the distance between any two markers, such as a DNA marker and the disease locus. There may be limited information in the data, however, diminishing the power of the linkage study. Furthermore, complex diseases are polygenic, involving multiple genes (Carter and Chung, 1980). Linkage analyses, however, are generally under the assumption of one major gene. Additionally, heterogeneity in the diagnosis and comorbidity of mental illnesses make linkage analysis considerably more difficult, if even possible at all. Many investigators have adopted association analyses to take advantage of the advent of high-throughput genotyping technologies. Recent efforts have identified genes that contribute to a number of complex human traits using the ultra-dense genetic markers (Arking et al., 2006; Klein et al., 2005; Duerr et al., 2006; Chen et al., 2007). Trios (one affected offspring and two parents) have been an effective design for association studies, particularly with the development of the elegant transmission/disequilibrium test (TDT) (Spielman, McGinnis and Ewens, 1993).

The central idea of this test is that each affected child serves as his or her own matched case and control. This acts to control for all potential confounding issues and examines alleles that both are and are not transmitted from the parents. In the absence of association between the affective status and the gene, the distributions of the transmitted and non-transmitted alleles are expected to be the same. Deviations in distribution as evaluated by a χ^2-test indicate the existence of association. Trios are the simplest example of nuclear family, but when other siblings are available, the trio design is not cost-effective. As a result, family-based association tests (FBAT) including sibships (Spielman and Ewens, 1998; Horvath and Laird, 1998; Knapp, 1999), nuclear families (Weinberg, 1999; Lunetta et al., 2000; Rabinowitz and Laird, 2000) and general pedigrees (Martin, Monks, Warren and Kaplan, 2000) have been developed. Another restriction in the use of trios is the requirement of defining the affective status of a disease. Consequently, association tests have been proposed for quantitative traits (Allison, 1997; Rabinowitz, 1997), traits with distribution belonging to an exponential family (Liu, Tritchler and Bull, 2002), ordinal traits (Zhang, Wang and Ye, 2006; Wang, Ye and Zhang, 2006) and multiple traits (Lange et al., 2003; Zhang, Liu and Wang, 2010).

Since the early success in identifying the complement factor H polymorphism in age-related macular degeneration (Klein et al., 2005), case-control association studies have intensified, and many genetic variants have been identified and catalogued (Hindorff et al., 2009). Despite the enormous investment, the intense attention to the genetics of diseases, the rapid improvement in technology, and the increasingly large sample sizes in many studies, it remains challenging to identify disease genes, especially those underlying mental illnesses. Some of the common genetic variants that have been identified for complex diseases only account for a small portion of the genetic risk, which may vary across populations (Goldstein, 2009).

For example, Kopp et al. (2008) and Kao et al. (2008) identified several variations in the MYH9 gene as major contributors to excess risk of kidney disease among AfricanAmericans. They found that 60 percent of AfricanAmericans carry the risk variants as opposed to 4 percent of white Americans. Technology will continue to improve and the amount of genetic data will increase. The purpose of this article is to review some of the progress from a statistical perspective and discuss some of the potential challenges. Obviously, it would take volumes or series to do justice to all of the work in statistical genetics. Instead of taking on that impossible task, this article is oriented toward the publications directly related to my own recent work.

METHODS

Since 1952, the American Psychiatric Association has published four editions of the Diagnostic and Statistical Manual of mental disorders (DSM) and plans to release its fifth edition in 2013. While widely used, the use and development of the DSM has not gone without controversy and criticism. Unlike diseases for which the diagnoses are well accepted by physicians and patients, such as cancer, the diagnosis of mental disorders must reflect biological factors (e.g., gender and racial disparities), non-biological factors such as culture that are not specific to one person, and it also must reflect the natural variation within the same person.

Ordinal Traits

It is clear from the above discussion that a simple dichotomous diagnosis (e.g., yes or no), or a well distributed continuous trait, is unlikely to characterize the state of mental disorders. In fact, the questions used in the diagnosis of mental disorders, such as DSM-IV, are usually posed in terms of severity or frequency and hence in an ordinal scale. Statistical methods for genetic analysis are well established for both quantitative (continuous) and binary traits (see, e.g., Blackwelder and Elston, 1985; Goldgar, 1990; Schork, 1993; Amos, 1994; Risch and Zhang, 1995; Kruglyak et al., 1996; Blangero and Almasy, 1997; Ott, 1999). While there has been some progress in the analysis of ordinal traits (e.g., Heath et al., 2002; Steinke, Borish and Rosenwasser, 2003; Vergne et al., 2003; Zhang, Feng and Zhu, 2003; Feng, Leckman and Zhang, 2004; Zhang, Liu and Wang, 2010), especially in plant science (Rao and Xu, 1998; Xu and Xu, 2006), insufficient attention has been paid to addressing the unique challenges of analyzing ordinal traits. Some researchers have recognized that it is difficult to conduct genetic analyses of ordinal traits because such traits cannot be directly characterized by a linear function of genetic and environmental effects (Rao and Xu, 1998). To fill in this methodological gap, we have made a systematic effort to develop statistical methods for segregation analysis (Zhang, Feng and Zhu, 2003), linkage analysis (Feng, Leckman and Zhang, 2004) and association analysis (Zhang, Wang and Ye, 2006) of ordinal traits (for family studies and case-control studies).

Analysis of Family Data

Long before the era of genomics, researchers collected data in families, also called pedigrees as illustrated in Figure 1. Although the ascertainment process for families varies, Figure 1 depicts a representative three-generation pedigree. The proband is the first person who enters into the study according to defined inclusion and exclusion criteria: such criteria are related to the disease of interest. Other members of the proband's family are included and directly or indirectly assessed, depending on the circumstance. The key idea in analyzing family data is that if a gene is a major driving force behind a disease, a trace in the concordance of diseases in family members would reflect the transmission pattern of a gene under the Mendelian laws. This is the fundamental concept that Cannon and Romanoff (1911) employed. This type of analysis is referred to as . The Elston–Stewart (1971) algorithm set up the quintessential framework to analyze data from general pedigrees through a technique called peeling. The main complication in analyzing pedigree data is the complex relationship among family members, making it difficult to express the likelihood function in an easily computed form. The peeling algorithm makes use of the conditional independence embedded in the pedigree resulting from the Mendelian laws, and so peels off the complete likelihood function into smaller pieces before putting them back together.

Other methods have been relatively recently developed using the concept of latent random variables (Hopper, 1989; Babiker and Cuzick, 1994; Li and Thompson, 1997; Siegmund and McKnight, 1998; Zhang and Merikangas, 2000), which are closely related to the classic ousiotype models of Cannings, Thompson and Skolnick (1978) in pedigree analysis. The basic idea is to use latent variables to represent the contribution of unobserved factors including a major gene, residual genetic factors and common environmental factors. As discussed by Zhang and Merikangas (2000), the computation involving pedigrees is similar to the peeling algorithm. Advantages of using latent variable based models are that the interactions between underlying genetic effects and the observed covariates (e.g., demographic variables) can be considered. Additionally, more relevant to this article, we can accommodate ordinal traits in the latent variable framework.

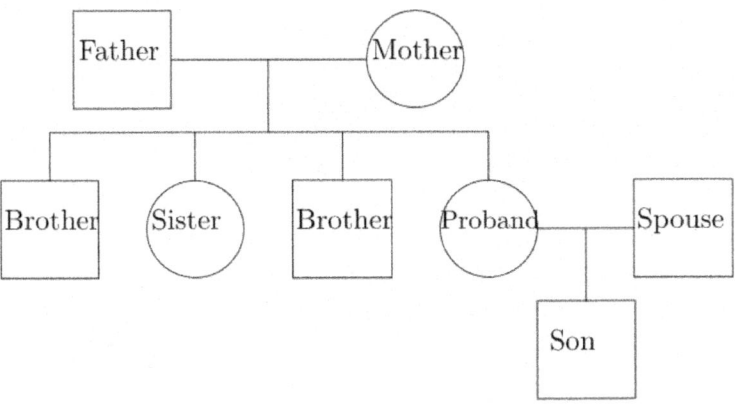

Figure 1: A three-generation pedigree.

A Latent Variable Model

We follow the notation of Zhang and Merikangas (2000) and Zhang, Feng and Zhu (2003). Consider a trait, Y, that takes an ordinal value of 0, 1,...,K. Let x be a p-vector of covariates that is also available for each study subject. Three types of latent random variables Ui 1,Ui 2 and Ui 3 are introduced within family i to represent, respectively, (a) common, unmeasured environmental factors; (b) genetic susceptibility of the family founders (a founder refers to a subject whose parents are not a part of the observed pedigree, e.g., father, mother and spouse in Figure 1); and (c) the transmission of susceptibility genes from a parent to an offspring. The concept of latent variables is straightforward, but the interesting and difficult part lies in the specification of their distributions. They need to be interpretable and convenient. The following are the assumptions that we found useful:

- U_1^i follows Bernoulli distributions $P\{U_1^i = 1\} = 1 - P\{U_1^i = 0\} = \theta_1$, where θ_1 is an unknown parameter.
- $U_2^i = (U_{2,1}^i, U_{2,2}^i, \ldots, U_{2,2n_i-1}^i, U_{2,2n_i}^i)'$, where n_i is the size of pedigree i. Here, $P\{U_{2,2j-1}^i = 1\} = 1 - P\{U_{2,2j}^i = 0\} = \theta_2$ when $U_{2,2j-1}^i$ and $U_{2,2j}^i$ are the U_2^i-variables of a founder.
- $U_3^i = (U_{3,1}^i, \ldots, U_{3,s_i}^i)'$. According to the Mendelian laws, $P\{U_{3,j}^i = 1\} = P\{U_{3,j}^i = 0\} = \frac{1}{2}$, $j = 1, \ldots, s_i$, and s_i is the number of parent-offspring pairs in family i. U_3^i facilitates the transmission of U_2^i-variables from the founders to the offspring. For example, if a parent of subject j has U_2^i-variables, $U_{2,2k-1}^i$ and $U_{2,2k}^i$, and the U_3^i-variable for this

parent-offspring pair is $U_{3,l}^i$, then one of subject j's U_2^i-variables is $U_{2,2j-1}^i = U_{2,2k-1}^i U_{3,l}^i + U_{2,2k}^i(1 - U_{3,l}^i)$.

- All latent variables are independent.

U_1^i is a simple "switch" indicating the presence or absence of a shared environment factor within family i. U_2^i is assigned independently to each of founders who are the source for any gene to enter into a family, and thus mimics the transmission of a single major susceptibility locus with alleles A and a of frequencies θ_2 and $1 - \theta_2$, respectively. Conditional on all of the latent variables, denoted by U^i, within family i, the probability distribution for member j is assumed to be

$$P\{Y_j^i \leq k|U^i\} = \frac{\exp(\mathbf{x}_j'\beta + \alpha_k + \mathbf{a}_j'\gamma)}{1 + \exp(\mathbf{x}_j'\beta + \alpha_k + \mathbf{a}_j'\gamma)},$$
$$k = 0, \ldots, K - 1, \tag{2.1}$$

where $\mathbf{a}_j^i = (U_1^i, U_{2,2j-1}^i + U_{2,2j}^i, U_{2,2j-1}^i U_{2,2j}^i)^T$, and β and γ are p- and 3-vectors of parameters. The α_k is the trait level dependent intercept, $k = 0, \ldots, K - 1$. As Zhang, Feng and Zhu (2003) pointed out, the β parameters measure the strength of association between the trait and the covariates, conditional on the latent variables. The γ parameters indicate the familial and genetic contributions to the trait. The mode of inheritance can be inferred from γ. For example, $\gamma_2 = 0$ and $\gamma_3 \neq 0$ suggests a recessive effect.

The likelihood function can be derived from (2.1). Due to the presence of latent variables, the EM algorithm (Dempster, Laird and Rubin, 1977) is the most convenient choice for parameter estimation (Guo and Thompson, 1992; Zhang and Merikangas, 2000; Zhang, Feng and Zhu, 2003). Although Zhang and Merikangas (2000) and Zhang, Feng and Zhu (2003) presented an effective solution (e.g., a modified likelihood), we should note that the lack of concavity in the likelihood function makes it a challenging task to find the maximum likelihood estimates of the model parameters. In addition, the θ's and γ's are not fully identifiable. The identifiability issue not only causes computational problems, but also presents theoretical challenges in statistical inference. Another important, yet understudied, issue is the validation of the assumptions on the distributions of the latent variables. But, how useful is the latent variable model (2.1)? First, it provides a regression framework to assess familial aggregation and genetic contribution, and possibly interactions between measured covariates and latent factors. Using data from a family study of substance use (Merikangas et al., 1998), Zhang and Merikangas (2000) were able to present extremely significant evidence of familial aggregation p-value $< 10^{-9}$ for alcohol dependence. This study additionally demonstrated that transmission does not follow a major locus pattern. In retrospect, their findings predicted the difficulty of identifying major genes associated with alcoholism. In addition, Zhang and Merikangas (2000) presented simulation examples to delineate when the absence of latent variables in (2.1) affects the estimates of the effects by the measured covariates. For example, hypothetically, if the greater presence of females in a family has an impact on the well-being of the family, ignoring the familial latent variables is likely to result in a biased estimate of the sex difference.

Not only is it important to include the latent factors, but also it is important to adjust for covariates. To further illustrate this point, Zhang, Feng and Zhu (2003) reported the following simulation. Ten thousand data sets were generated from model (2.1) with $\theta1 = 0.3$, β chosen from 0, 1, 5 or 10, γ_1 from 0, 1 or 2, $\alpha_0 = -1$ and $\alpha1 = 1$. To focus on the difference of having or not having covariates, they set $\gamma_2 = \gamma_3 = 0$. Each data set consists of 200 families with 7 family members (similar to Figure 1). One covariate x was generated as follows. For family i, U^i_1 were generated according to whether a random number ri1 from the uniform (0, 1) was greater than 0.3 or not. For member j in family i, an independent random number r_{ij2} from the uniform (0, 1) was generated. Then, $x_{ij} = 0.9r_{ij2} + 0.2r_{i1}$.

To evaluate the performance of the test statistic, the covariate was deliberately ignored in the test. When $\beta = 0$, the covariate played no role in the data generating process. The row corresponding to $\beta = 0$ in Table 1 displays the p-value (the column corresponding to $\gamma_1 = 0$) and the power for two values of γ_1 (1 or 2).

Table 1: The probability estimates of rejecting $\gamma_1 = 0$ at the significance level of 0.05. The covariate is omitted from the testing despite the fact that its coefficient β may not be zero

	$\gamma_1 = 0$	$\gamma_1 = 1$	$\gamma_1 = 2$
$\beta = 0$	0.0494	0.9503	1.0
$\beta = 1$	0.0534	0.9843	1.0
$\beta = 5$	0.1667	0.9971	1.0
$\beta = 10$	0.3828	0.9890	1.0

When β 0, the covariate plays a role in the data generating model. The data in Table 1 reveal the consequence of ignoring the covariate, which is more severe when the effect of the covariate is greater.

Linkage Analysis

While linkage analysis has a long history, it only became a common practice after the availability of several convenient computing programs (Ott, 1974; Kruglyak et al., 1996; Almasy and Blangero, 1998). For statisticians, some of the common terminologies in linkage analysis are puzzling, including the so-called LOD-score method and nonparametric method. Morton (1955) first introduced the term "LODscore." LOD stands for "the logarithm (base 10) of odds." The "odds" is a probability ratio, or likelihood ratio, of the probability under an alternative hypothesis to the probability under the null hypothesis. The LODscore method is essentially a log-likelihood ratio test with two fundamental differences: (a) the use of the base 10 logarithm versus the natural logarithm; (b) the log-likelihood ratio statistic has a multiplier of 2 conforming to a χ^2 distribution under certain regularity conditions. Specifically, the LOD-score is the log(base 10)-ratio of the likelihood when the recombination fraction is less than 1/2 (i.e., two loci are not on the same chromosome, or called unlinked), to the likelihood when the recombination fraction is 1/2 (no linkage).

The recombination fraction is the frequency that a chromosomal crossover occurs between two loci (or genes) during meiosis; 1% of combination frequency is termed the distance of one centimorgan (cM) in a genetic linkage map. Because the LOD-score is in base 10, a score of 3 indicates 1000 to 1 odds in favor of the linkage, which is the conventional threshold for declaring the evidence for linkage. If we convert a LOD-score of 3 into the standard log-likelihood ratio statistic, it yields a p-value of 2×10^{-4} under χ^2_1. By Bonnferoni correction, it corresponds to a genomewide p-value of 0.05 for 250 markers. This number is in the range for the number of microsatellites used in typical linkage studies. In order to compute the LOD-score, we first need a number of parameters that determine the likelihood for a given recombination fraction. Then use the maximum likelihood over the recombination fraction for the likelihood under the alternative hypothesis. The parameters that are required include the mode of inheritance, penetrance, and disease allele frequency.

These parameters are generally unknown and difficult to estimate for complex diseases including mental illnesses. For example, using segregation analysis (see Section 2.1.1) Pauls and Leckman (1986) examined specific genetic hypotheses about the mode of transmission of Gilles de la Tourette's syndrome, by performing segregation analyses in 30 nuclear families (two-generation pedigrees). They concluded that Tourette's syndrome is inherited as an autosomal dominant trait (one copy of the abnormal allele is sufficient to cause the disease). The penetrance (the probability of having the disease for a given genotype) was reported at 0.71 in males and 1.0 in females with at least one abnormal allele. After several decades of research, no major genetic variant has been identified for Tourette's syndrome, and most likely this syndrome involves multiple genes, interacting with environmental factors. This reality makes it difficult to infer the mode of inheritance, penetrance, and disease allele frequency, and conceptually, this may not make sense for complex diseases (non-Mendelian inheritance). This difficulty is somewhat alleviated since the LOD-score method has been found to work reasonably well (e.g., Abreu, Greenberg and Hodge, 1999) under various parameter settings. There have been some efforts to improve the robustness of the method (Gastwirth, 1966, 1985; Whittemore, 1996). See Zheng et al. (2009) for a thorough review. Existing methods do not extend to the case of ordinal traits.

The effectiveness of the robust methods remains to be studied. Naturally, nonparametric linkage methods have been developed to avoid specification of the genetic model parameters. In statistics, "nonparametric" methods typically refers to distribution-free methods such as rank-based tests and methods based on the empirical distribution. In linkage analysis, however, "nonparametric" does not mean "distribution-free," but instead refers to the replacement of true genetic model parameters with the parameters of inheritance of markers, hypothesized to be close to the disease locus. Thus, with nonparametric linkage methods, we still need to compute the likelihood. Two core algorithms are used to compute the likelihood: the Elston–Steward algorithm (1971) and the Lander–Green (1987) algorithm.

As previously discussed, the Elston–Steward algorithm (1971) is a peeling algorithm that makes the computation in a large pedigree feasible by splitting it into small pieces. This algorithm was implemented in early versions of linkage analysis programs (e.g., LIPED and LINKAGE); computational time increased linearly in family size, but exponentially with the number of loci. More recent programs (e.g., GENEHUNTER) use the Lander– Green (1987) algorithm that has first-order complexity in the number of loci, but unfortunately exponential in the family size. Although Markov chain Monte Carlo methods have been used to accommodate linkage analysis of large families and a large number of markers (Guo and Thompson, 1992), in practice, one may have to break large pedigrees apart in order to run programs such as GENEHUNTER. We should note that there had not been a linkage analysis program to handle ordinal traits until the release of LOT (Zhang et al., 2008). Typically, the methods for linkage analysis can be divided into two main steps; only the second step involves the trait (Kruglyak et al., 1996). The first step infers how genetic information travels in a family as represented by the so-called "inheritance vector." We will use the pedigree in Figure 1 to illustrate this concept.

The two parents and spouse are the founders of the family, meaning that their parents are not in the current pedigree. The four siblings and the child are nonfounders. The inheritance pattern at marker locus t is completely described by an inheritance vector $v(t) = (v_1, v_2, v_3, v_4, ..., v_9, v_{10})$. In other words, we devote two elements for every nonfounder. The founders are not included because they are the sources of the genes in the family and the inheritance vector is conditional on their genes. The paired elements describe the outcomes of the paternal and maternal meioses transmitted to the nonfounders. Specifically, $v_{2j-1} = 1$ or 2 according to whether the grand paternal or grand maternal allele is transmitted in the paternal meiosis to the j th nonfounder. v_{2j} carries the similar information for the corresponding maternal meiosis, namely, $v_{2j} = 3$ or 4 according to whether the grand paternal or grand maternal allele was transmitted in the maternal meiosis to the j th nonfounder.

In practice, the genetic markers do not always allow us to determine the true inheritance vector. In this case, the inheritance distribution is the conditional probability distribution over the possible inheritance vectors that conform with the alleles observed at t, which we denote by $p\{v(t) = w\}$ for all inheritance vectors $w \in V$; here V is the set of all possible inheritance vectors. In the absence of any genotypic information, all inheritance vectors are equally likely according to Mendel's first law; the probability distribution is uniform. For segregation analysis, we employed latent variables to reflect the "imaginative" genetic effects in (2.1). In linkage analysis, we have genetic markers that flow through the inheritance vector.

Thus, we can still use (2.1) for linkage analysis except that a_{ij} should be $(U_1^i, U_{2,v_{2j-1}}^i + U_{2,v_{2j}}^i)$. On one hand, we have a reduced number of latent variables. On the other hand, many of the latent variables depend on each other through the inheritance vectors. The computation of the likelihood would be summed over all inheritance vectors w in V, in addition to the probability space of the remaining independent latent variables. Because of this connection and distinction, the challenges in the linkage analysis of ordinal traits are, to a great extent, similar to those in segregation analysis of ordinal traits, for example, the asymptotic mixture of χ^2- distributions and the need to introduce the penalized likelihood (Liang and Rathouz, 1999; Zhang, Feng and Zhu, 2003).

Association test

As discussed above, linkage analysis focuses on testing the position of a marker, although it has been difficult to replicate findings in linkage studies of mental disorders. An association analysis, however, tests whether a genetic variant, including particular allele or genotype of a marker and a haplotype in several markers, is associated with a trait. Some study cohorts recruited for linkage studies have been re-genotyped for genomewide association analyses. For binary or quantitative traits, many methods have been developed and implemented. Two commonly used programs are PLINK (Purcell et al., 2007) and FBAT (Rabinowitz and Laird, 2000). To analyze an ordinal trait, Zhang, Wang and Ye (2006) introduced the following proportional odds model:

$$\text{logit}\{P(y_{ij} \le k | G_{ij})\} = \alpha_k + \beta c_{ij},$$

(2.2)

where $\alpha_0,...,\alpha_{K-1}$ are non-descending level parameters, β is the genetic effect. The genetic factor c_{ij} can be chosen to reflect the underlying mode of inheritance such as the number of the risk allele. Under model (2.2), the null hypothesis is H_0: $\beta = 0$. The score statistic is

$$S = \sum_{i,j} [R^+(y_{ij}) - R^-(y_{ij})] A_{ij},$$

(2.3)

where $R^+(y_{ij})$ and $R^-(y_{ij})$ are the counts of offspring in the entire sample whose trait values are greater or less than y_{ij}, respectively, and A_{ij} is the number of copies of transmitted alleles at the marker locus. Thus, Zhang, Wang and Ye (2006) proposed the following OTDT test based on the score statistic:

$$\frac{[S - E(S|Y)]^2}{\text{Var}(S|Y)},$$

which follows a χ_1^2 -distribution asymptotically. For a case-control study, $R^+(y_{ij})$ and $R^-(y_{ij})$ are the numbers of subjects whose trait values are greater or less than y_{ij}, respectively. If we rewrite the statistic in (2.3) in a general form as $\sum_{i,j} w_{ij} A_{ij}$, this yields the classic TDT when $w_{ij} = 1$ and the QTDT (Rabinowitz, 1997) when $w_{ij} = y_{ij} - \bar{y}$, where $\bar{y}$ is the average of all y_{ij} 's. In other words, all of these tests are a weighted function of the number of transmitted alleles at the marker locus, and the choice of the weights depends on the property of the trait. With this observation, after the proper weights are computed, the existing FBAT software can be used to test the association between any trait and alleles at a marker locus.

In the following, we describe a unified method to choose weights for any kind of trait. It is straightforward to categorize a quantitative trait into any reasonable number of categories (such as deciles) and induce an ordinal scaled trait. This would allow the use of the O-TDT for a quantitative trait. In their simulation studies, Zhang, Wang and Ye (2006) demonstrated that this strategy has comparable power to the QTDT for quantitative traits. This is due to the fact that the number of categories is enough to capture most of the information in the data (e.g., following Cochran's rule; Cochran, 1977). The advantage is that the ordinal scaled test is not affected by the nonnormal distribution of a quantitative trait, and so, the unified approach is robust. One limitation of the test proposed by Zhang, Wang and Ye (2006) is that it does not adjust for covariates. Environmental factors or covariates, such as gender and age, may confound the association of interest. In a subsequent work, Wang, Ye and Zhang (2006) generalized model (2.2) to include covariates as follows:

$$\text{logit}\{P(y_{ij} \leq k|G_{ij}), z_{ij}\} = \alpha_k + \beta c_{ij} + \delta' z_{ij}, \tag{2.4}$$

where z_{ij} denotes the covariates and δ is the vector of the corresponding coefficients. Consequently, the score statistic becomes

$$S = \sum_{i,j} [\hat{\gamma}(y_{ij}, z_{ij}) - \hat{\gamma}(y_{ij} - 1, z_{ij})] A_{ij}, \tag{2.5}$$

where

$$\hat{\gamma}(k, z) = \frac{\exp(\hat{\alpha}_k + \hat{\delta}' z_{ij})}{1 + \exp(\hat{\alpha}_k + \hat{\delta}' z_{ij})},$$

which is the estimated probability of having a trait value no greater than k. Thus, the weight function in (2.5) is the difference between the probability of having a trait value greater than y_{ij} and the probability of having a trait value less than yij . Not surprisingly, this is in essence the same as the weight function in (2.3) where we used counts instead of frequency (or probability). It is important to note that association analysis does not directly equate to a causal relationship. In well-designed genetic association studies, an observed association is expected to result from either a causal functional variant of a gene, or the linkage disequilibrium between the marker and a susceptibility gene. In population-based case-control studies, there are typically attempts to match cases and controls by important demographic and/or baseline information.

It is not wise to over-match subjects. Alternatively, we can collect potentially important environmental variables and consider them in the association analysis. We can also use principal component analysis on the genotypes to explore whether there are "clusters" in the study cohorts that are not appropriately reflected in the environmental variables. In family-based studies, the association tends to be conditional on parental genotypes and all phenotypes.

Unique Challenges in Analyzing Ordinal Traits

Understanding the genetic mechanisms for complex diseases is challenging regardless of whether we analyze binary, ordinal or continuous traits. Any challenges that exist for analyzing binary and continuous traits remain for ordinal traits. What are the unique challenges in analyzing ordinal traits? The key difference is that there is not a simple distribution function for ordinal traits. For continuous traits, the assumption is that the traits can somehow be treated under normality, by transformation if needed. For binary traits, through a link function (e.g., logit) we only need to deal with a Bernoulli distribution. However, for ordinal traits, the two typical approaches are (a) to assume a reliability variable or a continuous latent variable or (b) to assume a proportional odds model as we presented above. The first challenge is in the estimation. The likelihood function is complicated, and based on the numerical results, it has multiple local maxima. In addition, due to identifiability (or near-identifiability), the likelihood function may be relatively flat. Combinations of the EM and other algorithms can provide practical solutions, but finding a more efficient algorithm is an open problem.

The second challenge is in the inference. When latent variables or mixture distributions are used, some of the commonly assumed regularity conditions do not hold. One solution is to use a penalized likelihood function (Zhang, Feng and Zhu, 2003) that prevents the parameters from being near the singularity points. Finally, model diagnostics are difficult. For example, how do we know the latent variable-based model or the proportional odds model provides an adequate fit to the data? Although the models and methods presented above do not address this and other questions, they provide a foundation for further research and improvement.

Comorbidity

The methods described above only deal with a single trait. However, comorbidity is the rule rather than the exception in studies of mental and behavioral disorders. For example, a patient may suffer from both anxiety and depression (Li and Burmeister, 2009), and the same patient may also be addicted to nicotine, alcohol, or other substances (Merikangas et al., 1998; True et al., 1999). From a data analysis perspective, we need to consider how important it is to accommodate multiple diseases/traits. In a real-data example, Chen et al. (2011) analyzed a data set from the Study of Addiction: Genetics and Environment (SAGE). By simply considering addiction to at least two of the six substances (addiction to nicotine, alcohol, marijuana, cocaine, opiates or other drugs), we were able to identify the PKNOX2 gene that reached genomewide signifi- cance level among European-origin females. Interestingly, the PKNOX2 gene has been previously identi- fied as one of the cis-regulated genes for alcohol addiction in mice (Mulligan et al., 2006). To further delineate the benefit of considering multivariate traits, Zhu and Zhang (2009) conducted comprehensive simulation studies, considered the correlations of 0.2 and −0.2 among three quantitative traits, and demonstrated that testing correlated traits jointly is more powerful than testing a single trait at a time. Using generalized estimation equation, Lange et al. (2003) developed a family-based association test for multivariate quantitative traits (FBAT-GEE). Recently, Zhang, Liu and Wang (2010) constructed a nonparametric test based on the generalized Kendall's tau to accommodate any combination of dichotomous, ordinal, and quantitative traits.

Kendall's tau

Kendall's τ is a rank-based correlation between two variables. It contracts the probability of observing the two variables in the same order in two observations with the probability of observing the two variables in the opposite order. Specifically, for a sample of n observations $(X_1, Y_1),...,(X_n,Y_n)$, two observations (X_i,Y_i) and (X_j,Y_j) are called concordant if $(X_i - X_j)(Y_i - Y_j) > 0$ and discordant if $(X_i - X_j)(Y_i - Y_j) < 0$. Then Kendall's τ is based on the difference between the numbers of concordant pairs and discordant pairs. We introduce a kernel function,

$$\phi((X_i, Y_i), (X_j, Y_j))$$

$$= \text{sign}\{(X_i - X_j)(Y_i - Y_j)\}$$

$$= \begin{cases} 1, & \text{if } (X_i - X_j)(Y_i - Y_j) > 0, \\ -1, & \text{if } (X_i - X_j)(Y_i - Y_j) < 0, \\ 0, & \text{if } (X_i - X_j)(Y_i - Y_j) = 0, \end{cases}$$

and define a U-statistic

$$U = \binom{n}{2}^{-1} \sum_{i<j} \phi((X_i, Y_i), (X_j, Y_j)). \tag{2.6}$$

Then, Kendall's τ is

$$\tau = \frac{U}{\sqrt{\text{Var}_0(U)}},$$

(2.7)

where $\text{Var}_0(U)$ is the variance of U under the null hypothesis of no correlation between X and Y, and equal to $n(n-1)(2n+5)/18$ if X and Y are continuous variables (Hollander and Wolfe, 1999).

Generalized Kendall's tau

To test the association between genetic markers and comorbidity, Zhang, Liu and Wang (2010) generalized Kendall's tau as follows. For individuals i and j, let Ti and T_j be their vectors of traits, respectively. Then, a trait kernel is defined as

$$F_{ij} = (f_1(T_i^{(1)} - T_j^{(1)}), \ldots, f_p(T_i^{(p)} - T_j^{(p)}))',$$

where function $fk(\cdot)$ is the identity function for a quantitative or binary trait (Rabinowitz, 1997), or the sign function for an ordinal trait (Zhang, Wang and Ye, 2006).

Also, recall that, as in Section 2.1.4, c is the number of any chosen allele for marker genotype and let C_i refer to the C for the ith subject. Then, Zhang, Liu and Wang (2010) defined a marker kernel as

$$D_{ij} = c_i - c_j.$$

Their U-statistic is defined as

$$U = \binom{n}{2}^{-1} \sum_{i<j} D_{ij} F_{ij}.$$

(2.8)

The association test statistic, or generalized Kendall's tau, is U Var^{-1}_0 (U) U, where Var0(U) is the variance of U under the null hypothesis that there is no association between marker alleles and any linked locus that influences the trait T . The test statistic follows an asymptotic χ^2-distribution under the null hypothesis. Obviously, the statistic in (2.8) does not incorporate covariate effects. This is relatively straightforward for a single trait as was done in (2.4). Here, the traits can be a hybrid of different traits. An alternative is to impose different weights for each pair of samples in the statistic (2.8) according to the information of their covariates. The weight, denoted by $w(z_i, z_j)$ for the pair (i, j), reflects the relative importance attributed by the covariates when we derive the statistic. Zhu, Jiang and Zhang (2010) examined the following weight function.

Write $z = (z^{co}, z^{ca})'$ with $z^{co} = (z^{(1)}, \ldots, z^{(l_1)})'$ for the continuous covariates and $z^{ca} = (z^{(l_1+1)}, \ldots, z^{(l)})'$ for the categorical covariates. They defined the weight function $w(z_i, z_j)$ as

$$w(z_i, z_j) = W(\|z_i^{co} - z_j^{co}\|) I(z_i^{ca} = z_j^{ca}), \tag{2.9}$$

where $W(\cdot)$ is a positive and decreasing function, for example, $W(u) = \exp(-u2/2h2)$, and $I(\cdot)$ is the indicator function. Then a weighted test statistic is given by

$$S = \binom{n}{2}^{-1} \sum_{i<j} D_{ij} F_{ij} w(z_i, z_j). \tag{2.10}$$

Zhang, Liu and Wang (2010) and Zhu, Jiang and Zhang (2010) showed that under the null hypothesis, the test statistic S (weighted or not) has the following asymptotic distribution conditional on all phenotypes and parental genotypes:

$$\text{Var}_0^{-1/2}(S)[S - E_0(S)] \xrightarrow{d} N(0, I_p),$$

where

$$E_0(S) = \frac{2}{n-1} \sum_{i=1}^{n} \bar{u}_i E_0(C_i | M_i^{pa}),$$

$$\text{Var}_0(S) = \frac{4}{(n-1)^2}$$
$$\cdot \sum_{i=1}^{n} \sum_{j=1}^{n} \bar{u}_i \bar{u}'_j \text{Cov}_0(C_i, C_j | M_i^{pa}, M_j^{pa}).$$

Consequently, the following test statistic

$$\chi_{\text{tau}}^2 = [S - E_0(S)]' \text{Var}_0^{-1}(S)[S - E_0(S)]$$

converges to χ_p^2 in distribution under the null hypothesis provided that Var0(S) is full rank. In a case-control study, we do not have the markers from parents and hence the conditional expectations are replaced with the unconditional ones. Thus, the key difference in the test statistics between family studies and population studies lies in the conditioning on the parental markers. The conditioning on the parental markers gives the family studies a major advantage in removing the effect of population admixture, but family studies tend to be more difficult and expensive to carry out. Under the alternative hypothesis, the test statistic χ_{tau}^2 can be written as a weighted sum of non-central

$\chi_1^2 = \sum_{i=1}^{p} e_i \chi_1^2(\phi_i)$, where $e_1 \geq \cdots \geq e_p$ are the nonnegative eigenvalues of $\Sigma_1^{1/2} \Sigma_0^{-1} \Sigma_1^{1/2}$.

$\phi_i = \mu^2_{Ri}$ and μR_i is the ith component of $\mu_p = Q\Sigma_1^{-1/2}\mu$, where Q is an orthonormal matrix such that $Q\Sigma_1^{1/2}\Sigma_0^{-1}\Sigma_1^{1/2}Q' = \text{diag}(e_1, \ldots, e_p)$. μ is the difference in the means of S under the alternative and null hypotheses. Using the approximation theory of Pearson (1959), Solomon and Stephens (1977) and Liu, Tang and Zhang (2009), we can find a certain degree of freedom l and non-centrally parametric υ such that the distribution of χ^2_{tau} can be closely approximated by $\chi^2_1(\upsilon)$. Through simulation studies, Zhu et al. con- firmed that this approximation is accurate enough for power calculation.

It is noteworthy that the weight function in (2.9) is restrictive with respect to categorical covariates, especially so for ordinal covariates. The use of genomic propensity score can give rise to an alternative weight function. Specifically, for a di-allelic marker G (e.g., SNP), the genomic propensity score is the conditional probability $p_g(z) = P(G = g|Z = z)$. This probability can be fitted by a logistic regression model or proportional odds model depending on whether G is chosen as an allele type or genotype. In the latter choice, the model also depends on the mode of inheritance. In the current genomewide association studies, we usually only have genotypes and cannot distinguish the phases of individual alleles. Thus, we have to construct genomic propensity scores by considering various modes of inheritance. Once the genomic propensity score is estimated, it can be treated as a numerical covariate and then we can use (2.9) again.

Examples

Zhang, Liu and Wang (2010) reanalyzed a data set from the Collaborative Study on the Genetics of Alcoholism (COGA) (Begleiter, 1995; Edenberg et al., 2005). The data came from a multicenter (9 sites) consortium that recruited study participants by requiring every proband to meet two alcohol dependence diagnostic criteria based on DSM-IVR (American Psychiatric Association, 1994). The firstdegree relatives of the probands were invited into the study. Zhang, Liu and Wang (2010) included a total of 1614 individuals from 143 families. They considered three phenotypes: (1) alcohol DX-DSM3R+Feighner; (2) maximum number of drinks in a 24-hour period; and (3) the response to "spent so much time drinking, had little time for anything else." Using the first phenotype alone, the p-value of the association between a peak marker D7S679 on chromosome 7 and the trait was 0.0019. However, when the three traits are analyzed together, D7S679 remains the peak marker, and the p-value is reduced to 0.00055, demonstrating the possibility that the other two phenotypes enhanced the association signal. If the other two phenotypes are analyzed alone, the analysis did not lead to anything worthy of further attention.

In the analysis cited above, the association was assessed without considering covariates. In a follow-up analysis, Zhu, Jiang and Zhang (2010) considered two important covariates: age at interview and sex. When these two covariates were controlled for, the p-value of the association between the peak marker D7S679 and the three phenotypes went down further to 0.000313.

DISCUSSION

Studying comorbidity is a significant issue in mental and behavioral research, dating back to a century ago (Cannon and Rosanoff, 1911). This is challenging due to a lack of statistical methods that accommodate the complexity of comorbidity. While dealing with comorbidity in genetic studies is the focus of this review, it is achieved through gradual development, and accumulation of methods.

Various challenges are dealt with along the way. Although I focused on the analysis of ordinal traits and applications in mental health, the presented methods are closely related to robust and rank-based methods for binary and quantitative traits. Furthermore, ordinal traits arise in studies of diseases besides mental illnesses, such as cancer (specifically, different stages). From the statistical perspective, the methods that are presented here have broad applications beyond genetic association studies.

From college admissions, to job searches, to scientific investigations, we make inferences based on multidimensional data. It is important and imperative to consider and develop inferential tools for multivariate outcomes, particularly when the outcomes are discrete. There is extensive literature on the statistical analysis of multivariate normal variables as well as on nonparametric tests for a single variable of nonnormal distribution. However, few options are available for the inference when we have multiple nonnormally distributed variables and potential hybrids of continuous and discrete variables. To overcome this challenge, I presented several useful statistical techniques such as the rank-based U-statistics and the kernel-based weighted statistics to accommodate the mix of continuous and discrete outcomes and the presence of important covariates.

ACKNOWLEDGMENTS

This work is supported in part by National Institute on Drug Abuse Grant R01DA016750. The author wishes to thank Professor David Madigan for his encouragement on this review. He also thanks Dr. Gang Zheng and Professor Joseph Gastwirth for helpful discussions and comments, and Jennnifer Brennan for careful reading and comments.

REFERENCES

1. Abelson, J. F., Kwan, K. Y., O'Roak, B. J., Baek, D. Y., Stillman, A. A., Morgan, T. M., Mathews, C. A., Pauls, D. L., Rasin, M.-R., Gunel, M., Davis, N. R., Ercan-Sencicek, A. G., Guez, D. H., Spertus, J. A., Leckman, J. F., Leon S. Dure, t., Kurlan, R., Singer, H. S., Gilbert, D. L., Farhi, A., Louvi, A., Lifton, R. P., Sestan, N. and State, M. W. (2005). Sequence variants in SLITRK1 are associated with Tourette's syndrome. *Science* 310 317–320.

2. Abreu, P. C., Greenberg, D. A. and Hodge, S. E. (1999). Direct power comparisons between simple lod scores and NPL scores for linkage analysis in complex diseases. *Am. J. Hum. Genet.* 65 847–857.

3. Allison, D. B. (1997). Transmission-disequilibrium tests for quantitative traits. *Am. J. Hum. Genet.* 60 676–690.

4. Almasy, L. and Blangero, J. (1998). Multipoint quantitative-trait linkage analysis in general pedigrees. *Am. J. Hum. Genet.* 62 1198–1121.

5. American Psychiatric Association (1994). *Diagnostic and Statistical Manual of Mental Disorders*, 4th ed. American Psychiatric Association Press, Washington, DC.

6. Amos, C. I. (1994). Robust variance-components approach for assess-ing genetic linkage in pedigrees. *Am. J. Hum. Genet.* 54 535–543.

7. Arking, D. E., Pfeufer, A., Post, W., Kao, W. H. L., Newton-Cheh, C., Ikeda, M., West, K., Kashuk, C., Akyol, M., Perz, S., Jalilzadeh, S., Illig, T., Gieger, C., Guo, C.-Y., Larson, M. G., Wichmann, H. E., Marbán, E., O'Donnell, C. J., Hirschhorn, J. N., Kääb, S., Spooner, P. M., Meitinger, T. and Chakravarti, A. (2006). A common genetic variant in the NOS1 regulator NOS1AP modulates cardiac repolarization. *Nat. Genet.* 38 644–651.

8. Babiker, A. and Cuzick, J. (1994). A simple frailty model for family studies with covariates. *Stat. Med.* 13 1679–1692.

9. Begleiter, e. a. H. (1995). The collaborative study on the genetics of alcoholism. *Alcohol Health Res. World* 19 228–236.

10. Blackwelder, W. C. and Elston, R. C. (1985). A comparison of sib-pair linkage tests for disease susceptibility loci. *Genet. Epidemiol.* 285–97.

11. Blangero, J. and Almasy, L. (1997). Multipoint oligogenic linkage analysis of quantitative traits. *Genet. Epidemiol.* 14 959–964.

12. Cannings, C., Thompson, E. A. and Skolnick, M. H. (1978). Probability functions on complex pedigrees. *Adv. Appl. Probab.* 10 26–61.

13. Cannon, G. L. and Rosanoff, A. J. (1911). Preliminary report of a study of heredity in insanity in the light of the Mendelian laws. Reprinted from *J. Nervous and Mental Disorders* 38 272–279.

14. Carter, C. L. and Chung, C. S. (1980). Segregation analysis of schizophrenia under a mixed genetic model. *Hum. Hered.* 30 350–356.

15. Chen, X., Liu, C. T., Zhang, M. Z. and Zhang, H. P. (2007). A forest-based approach to identifying gene and gene–gene interactions.*Proc. Natl. Acad. Sci. USA* 104 19199–19203.

16. Chen, X., Cho, K., Singer, B. H. and Zhang, H. P. (2011). The nuclear transcription factor PKNOX2 is a candidate gene for substance dependence in European-origin women. *PLoS ONE* 6 e16002.

17. Cochran, W. G. (1977). *Sampling Techniques*, 3rd ed. Wiley, New York.

18. Dempster, A. P., Laird, N. M. and Rubin, D. B. (1977). Maximum likelihood from incomplete data via the EM algorithm (with discussion).*J. Roy. Statist. Soc. Ser. B* 39 1–38.

19. Duerr, R. H., Taylor, K. D., Brant, S. R., Rioux, J. D., Silverberg, M. S., Daly, M. J., Steinhart, A. H., Abraham, C., Regueiro, M. and Griffiths, A. et al. (2006). A genomewide association study identifies IL23R as an inflammatory bowel disease gene. *Science* 314 1461–1463.

20. Edenberg, H. J., Bierut, L. J., Boyce, P., Cao, M., Cawley, S., Chiles, R. and Doheny, K. F. (2005). Description of the data from the Collaborative Study on the Genetics of Alcoholism (COGA) and single-nucleotide polymorphism genotyping for Genetic Analysis Workshop 14. *BMC Genetics* 6 S2.

21. Elston, R. C. and Steward, J. (1971). A general model for the analysis of pedigree data. *Hum. Hered.* 21 523–542.

22. Feng, R., Leckman, J. and Zhang, H. P. (2004). Linkage analysis of ordinal traits for pedigree data. *Proc. Natl. Acad. Sci. USA* 10116739–16744.

23. Gastwirth, J. L. (1966). On robust procedures. *J. Amer. Statist. Assoc.* 61 929–948.

24. Gastwirth, J. L. (1985). The use of maximin efficiency robust tests in combining contingency tables and survival analysis. *J. Amer. Statist. Assoc.* 80 381–384.

25. Goldgar, D. E. (1990). Multipoint analysis of human quantitative genetic variation. *Am. J. Hum. Genet.* 47 957–967.

26. Goldstein, D. B. (2009). Common genetic variation and human traits. *N. Eng. J. Med.* 360 1696–1698.

27. Guo, S. W. and Thompson, E. A. (1992). A Monte Carlo method for combined segregation and linkage analysis. *Am. J. Hum. Genet.* 511111–1126.

28. Heath, A. C., Todorov, A. A., Nelson, E. C., Madden, P. A. F., Bucholz, K. K. and Martin, N. G. (2002). Gene–environment interaction effects on behavioral variation and risk of complex disorders: The example of alcoholism and other psychiatric disorders. *Twin Research* 5 30–37.

29. Heston, L. L. (1966). Psychiatric disorders in foster home reared children of schizophrenic mothers. *Bristish J. Psychiatry* 112 819–825.

30. Hindorff, L. A., Sethupathy, P., Junkinsa, H. A., Ramosa, E. M., Mehtac, J. P., Collinsb, F. S. and Manolioa, T. A. (2009). Potential etiologic and functional implications of genome-wide association loci for human diseases and traits. *Proc. Natl. Acad. Sci. USA* 1069362–9367.

31. Hollander, M. and Wolfe, D. A. (1999). *Nonparametric Statistical Methods*, 2nd ed. Wiley, New York.

32. Hopper, J. L. (1989). Modelling sibship environment in the regressive logistic model for familial disease. *Genet. Epidemiol.* 6 235–240.

33. Horvath, S. and Laird, N. M. (1998). A discordant-sibship test for disequilibrium and linkage: No need for parental data. *Am. J. Hum. Genet.* 63 1886–1897.

34. Kao, W. H. et al. (2009). MYH9 is associated with nondiabetic end-stage renal disease in African-Americans. *Nat. Genet.* 40 1185–1192.

35. Kety, S. S., Rosenthal, D. and Wender, P. (1978). Genetic relationships within the schizophrenia spectrum: Evidence from adoption studies. In *Critical Issues in Psychiatric Diagnosis* (R. L. Spitzer and D. F. Klein, eds.) 213–223. Raven Press, New York.

36. Klein, R. J., Zeiss, C., Chew, E. Y., Tsai, J.-Y., Sackler, R. S., Haynes, C., Henning, A. K., SanGiovanni, J. P., Mane, S. M., Mayne, S. T., Bracken, M. B., Ferris, F. L., Ott, J., Barnstable, C. and Hoh, J. (2005). Complement factor H polymorphism in age-related macular degeneration. *Science* 308 385–389.

37. Knapp, M. (1999). Using exact *P* values to compare the power between the reconstruction-combined transmission/disequilibrium test and the sib transmission/disequilibrium test. *Am. J. Hum. Genet.* 65 1208–1210.

38. Kopp, J. B., Smith, M. W., Nelson, G. W., Johnson, R. C., Freedman, B. I., Bowden, D. W., Oleksyk, T., McKenzie, L. M., Kajiyama, H., Ahuja, T. S., Berns, J. S., Briggs, W., Cho, M. E., Dart, R. A., Kimmel, P. L., Korbet, S. M., Michel, D. M., Mokrzycki, M. H., Schelling, J. R., Simon, E., Trachtman, H., Vlahov, D. and Winkler, C. A. (2008). MYH9 is a major-effect risk gene for focal segmental glomerulosclerosis. *Nat. Genet.* 40 1175–1184.

39. Kraepelin, E. (1899). *Psychiatrie: Ein Lehrbuch fur Studirende und Aerzte.* Barth, Leipzig.

40. Kruglyak, L., Daly, M. J., Reeve-Daly, M. P. and Lander, E. S. (1996). Parametric and nonparametric linkage analysis: A unified multipoint approach. *Am. J. Hum. Genet.* 58 1347–1363.

41. Lander, E. S. and Green, P. (1987). Construction of multilocus genetic linkage maps in humans. *Proc. Natl. Acad. Sci. USA* 84 2363–2367.

42. Lange, C., Silverman, E. K., Xu, X., Weiss, S. T. and Laird, N. M. (2003). A multivariate family-based association test using generalized estimating equations: FBAT-GEE. *Biostatistics* 4 195–306.

43. Li, H. Z. and Thompson, E. (1997). Semiparametric estimation of major gene and family-specific random effects for age of onset. *Biometrics* 53 282–293.

44. Li, M. D. and Burmeister, M. (2009). New insights into the genetics of addiction. *Nat. Rev. Genet.* 10 225–231.

45. Liang, K.-Y. and Rathouz, P. J. (1999). Hypothesis testing under mixture models: Application to genetic linkage analysis. *Biometrics* 5565–74.

46. Liu, H., Tang, Y. and Zhang, H. H. (2009). A new chi-square approximation to the distribution of non-negative definite quadratic forms in non-central normal variables. *Comput. Statist. Data Anal.* 53 853–856.

47. Liu, Y., Tritchler, D. and Bull, S. B. (2002). A unified framework for transmission-disequilibrium test analysis of discrete and continuous traits. *Genet. Epidemiol.* 22 26–40.

48. Lunetta, K. L., Farone, S. V., Biederman, J. and Laird, N. M. (2000). Family based tests of association and linkage that used unaffected sibs, covariates, and interactions. *Am. J. Hum. Genet.* 66 605–614.

49. Martin, E. R., Monks, S. A., Warren, L. L. and Kaplan, N. L. (2000). A test for linkage and association in general pedigrees: The pedigree disequilibrium test. *Am. J. Hum. Genet.* 67 146–154.

50. Merikangas, K. R., Stolar, M., Stevens, D. E., Goulet, J., Preisig, M. A., Fenton, B., Zhang, H., O'Malley, S. S. and Rounsaville, B. J.

(1998). Familial transmission of substance use disorders. *Arch. Gen. Psychiatry* 55 973–979.

51. Morton, N. E. (1955). Sequential tests for the detection of linkage. *Am. J. Hum. Genet.* 7 277–318.

52. Mulligan, M. K., Ponomarev, I., Hitzemann, R. J., Belknap, J. K., Tabakoff, B., Harris, R. A., Crabbe, J. C., Blednov, Y. A., Grahame, N. J., Phillips, T. J., Finn, D. A., Hoffman, P. L., Iyer, V. R., Koob, G. F. and Bergeson, S. E. (2006). Toward understanding the genetics of alcohol drinking through transcriptome meta-analysis. *Proc. Natl. Acad. Sci. USA* 103 6368–6373.

53. Ott, J. (1974). Estimation of the recombination fraction in human pedigrees: Efficient computation of the likelihood for human linkage studies. *Am. J. Hum. Genet.* 26 588–597.

54. Ott, J. (1999). *Analysis of Human Genetic Linkage*, 3rd ed. Johns Hopkins Univ. Press, Baltimore, MD.

55. Pauls, D. L. and Leckman, J. F. (1986). The inheritance of Gilles de la Tourette's syndrome and associated behaviors. Evidence for autosomal dominant transmission. *New Eng. J. Med.* 315 993–997.

56. Pearson, E. S. (1959). Note on an approximation to the distribution of non-central ^{2}. *Biometrika* 46 364.

57. Plomin, R., DeFries, J. C., McClearn, G. E. and Rutter, M. (1997). *Behavioral Genetics*, 3rd ed. Freeman, New York.

58. Purcell, S., Neale, B., Todd-Brown, K., Thomas, L., Ferreira, M. A. R., Bender, D., Maller, J., Sklar, P., de Bakker, P. I. W., Daly, M. J. and Sham, P. C. (2007). PLINK: A toolset for whole-genome association and population-based linkage analysis. *Am. J. Hum. Genet.* 81 559–575.

59. Rabinowitz, D. (1997). A transmission disequilibrium test for quantitative trait loci. *Hum. Hered.* 47 342–350.

60. Rabinowitz, D. and Laird, N. M. (2000). A unified approach to adjusting association tests for population admixture with arbitrary pedigree structure and arbitrary missing marker information. *Hum. Hered.* 50 211–223.

61. Rao, S. and Xu, S. (1998). Mapping quantitative trait loci for ordered categorical traits in four-way crosses. *Heredity* 81 214–224.

62. Risch, N. R. and Zhang, H. P. (1995). Extreme discordant sib pairs for mapping quantitative trait loci in humans. *Science* 268 1584–1589.

63. Rosenthal, D. (1972). Three adoption studies of heredity in the schizophrenic disorders. *Internat. J. Mental Health* 1 63–75.

64. Scharf, J. M., Moorjani, P., Fagerness, J., Platko, J. V., Illmann, C., Galloway, B., Jenike, E., Stewart, S. E., Pauls, D. L. and The Tourette Syndrome International Consortium for Genetics (2008). Lack of association between SLITRK1var321 and Tourette syndrome in a large family-based sample. *Neurology* 70 1495–1496.

65. Schork, N. J. (1993). Extended multipoint identity-by-descent analysis of human quantitative traits: Efficiency, power, and modeling considerations. *Am. J. Hum. Genet.* 53 1306–1319.

66. Siegmund, K. and McKnight, B. (1998). Modeling hazard functions in families. *Genet. Epidemiol.* 15 147–171.

67. Solomon, H. and Stephens, M. A. (1977). Distribution of a sum of weighted chi-square variables. *J. Amer. Statist. Assoc.* 72 881–885.

68. Spielman, R. S. and Ewens, W. J. (1998). A sibship rest for linkage in the presence of association: The sib transmission/disequilibrium test. *Am. J. Hum. Genet.* 62 450–458.

69. Spielman, R. S., McGinnis, R. E. and Ewens, W. J. (1993). Transmission test for linkage disequilibrium: The insulin gene region and insulin-dependent diabetes mellitus (IDDM). *Am. J. Hum. Genet.* 52 506–516.

70. Steinke, J. W., Borish, L. and Rosenwasser, L. J. (2003). Genetics of hypersensitivity. *J. Allergy Clin. Immunol.* 111 S495–S501.

71. Tienari, P. (1991). Interaction between genetic vulnerability and family environment: The Finnish adoptive family study of schizophrenia.*Acta Psychiatr. Scand.* 84 460–465.

72. True, W. R., Heath, A. C., Scherrer, J. F., Xian, H., Lin, N., Eisen, S. A., Lyons, M. J., Goldberg, J. and Tsuang, M. T. (1999). Interrelationship of genetic and environmental influences on conduct disorder and alcohol and marijuana dependence symptoms. *Am. J. Med. Genet.* 88 391–397.

73. Vergne, L., Bourgeois, A., Mpoudi-Ngole, E., Mougnutou, R., Mbuagbaw, J., Liegeois, F., Laurent, C., Butel, C., Zekeng, L., Delaporte, E. and Peeters, M. (2003). Biological and genetic characteristics of HIV infections in Cameroon reveals dual group M and O infections and a correlation between SI-inducing phenotype of the predominant CRF02_ AG variant and disease stage. *Virology* 310 254–266.

74. Wang, X. Q., Ye, Y. Q. and Zhang, H. P. (2006). Family-based association tests for ordinal traits adjusting for covariates. *Genet. Epidemiol.* 30 728–736.

75. Weinberg, C. R. (1999). Allowing for missing parents in genetic studies of case-parental triads. *Am. J. Hum. Genet.* 64 1186–1193.

76. Whittemore, A. S. (1996). Genome scanning for linkage: An overview. *Am. J. Hum. Genet.* 59 704–716.

77. Xu, S. and Xu, C. (2006). A multivariate model for ordinal trait analysis. *Heredity* 97 409–417.

78. Zhang, H. P., Feng, R. and Zhu, H. (2003). A latent variable model of segregation analysis for ordinal traits. *J. Amer. Statist. Assoc.* 98 1023–1034.

79. Zhang, H. P., Liu, C.-T. and Wang, X. (2010). An association test for multiple traits based on the generalized Kendall's tau. *J. Amer. Statist. Assoc.* 105 473–481.

80. Zhang, H. P. and Merikangas, K. (2000). A frailty model of segregation analysis: Understanding the familial transmission of alcoholism. *Biometrics* 56 815–823.

81. Zhang, H. P., Wang, X. Q. and Ye, Y. Q. (2006). Detection of genes for ordinal traits in nuclear families and a unified approach for association studies. *Genetics* 172 693–699.

82. Zhang, M., Feng, R., Chen, X., Hu, B. and Zhang, H. (2008). LOT: A tool for linkage analysis of ordinal traits for pedigree data. *Bioinformatics* 24 1737–1739.

83. Zheng, G., Joo, J., Zaykin, D., Wu, C. and Geller, N. (2009). Robust tests in genome-wide scans under incomplete linkage disequilibrium. *Statist. Sci.* 24 503–516.

84. Zhu, W. S., Jiang, Y. and Zhang, H. P. (2010). Covariate-adjusted association tests and power calculations based on the generalized Kendall's tau. Technical report.

85. Zhu, W. and Zhang, H. (2009). Why do we test multiple traits in genetic association studies? *J. Korean Statist. Soc.* 38 1–10.

86. Zhu, X., Cooper, R., Kan, D., Cao, G. and Wu, X. (2005). A genome-wide linkage and association study using COGA data. *BMC Genet.* 6 S128.

Chapter 12

USING EQTL WEIGHTS TO IMPROVE POWER FOR GENOME-WIDE ASSOCIATION STUDIES: A GENETIC STUDY OF CHILDHOOD ASTHMA

Lin Li[1], Michael Kabesch[2], Emmanuelle Bouzigon[3,4], Florence Demenais[3,4], Martin Farrall[5], Miriam F. Moffatt[6], Xihong Lin[1] and Liming Liang[1,7]

[1]Department of Biostatistics, Harvard School of Public Health, Boston, MA, USA

[2]Department of Pediatric Pneumology and Allergy, KUNO University Children's Hospital Regensburg, Regensburg, Germany

[3]INSERM, Genetic Variation and Human Diseases Unit, U946, Paris, France

[4]Sorbonne Paris Cité, Institut Universitaire d'Hématologie, Université Paris Diderot, Paris, France

[5]Wellcome Trust Centre for Human Genetics, Oxford, UK

[6]Molecular Genetics and Genomics Section, National Heart and Lung Institute, Imperial College London, London, UK

[7]Department of Epidemiology, Harvard School of Public Health, Boston, MA, USA

ABSTRACT

Increasing evidence suggests that single nucleotide polymorphisms (SNPs) associated with complex traits are more likely to be expression quantitative trait loci (eQTLs). Incorporating eQTL information hence has potential to increase power of genome-wide association studies (GWAS). In this paper, we propose using eQTL weights as prior information in SNP based association tests to improve test power while maintaining control of the family-wise error rate (FWER) or the false discovery rate (FDR). We apply the proposed methods to the analysis of a GWAS for childhood asthma consisting of 1296 unrelated individuals with German ancestry. The results confirm that eQTLs are enriched for previously reported asthma SNPs. We also find that some SNPs are insignificant using procedures without eQTL weighting, but become significant using eQTL-weighted Bonferroni or Benjamini–Hochberg procedures, while controlling the same FWER or FDR level. Some of these SNPs have been reported by independent studies in recent literature. The results suggest that the eQTL-weighted procedures provide a promising approach for improving power of GWAS. We also report the results of our methods applied to the large-scale European GABRIEL consortium data.

INTRODUCTION

Asthma is a disorder characterized by inflamed mucosa of small airways of lung, causing wheezing and shortness of breath (Moffatt et al., 2010). Among the most common chronic diseases of childhood, asthma has been reported to affect more than 10% of children in many westernized societies (Cookson, 2004). It is caused by a combination of genetic and environmental factors (Cookson, 2004; Moffatt et al., 2007), and several genome-wide association studies (GWAS) have been conducted to study the genetic basis underlying the complex disorder. More than 50 single nucleotide polymorphisms (SNPs) have been reported to be associated with asthma, according to the GWAS catalog (www.genome.gov/gwastudies, accessed on January 15, 2013). Remarkably, the recent report (Moffatt et al., 2010) from the GABRIEL (A Multidisciplinary Study to Identify the Genetic and Environmental Causes of Asthma in the European Community) consortium identified several SNPs reaching genome-wide significance through a large-scale meta-analysis.

Prior biological information, often available in practice, has potential to increase power of GWAS. The common practice of GWAS, "agnostic" in some sense, assumes no prior information about any of the SNPs under investigation, meaning that all the SNPs have an equal likelihood of being causal. Some recent studies have taken advantage of information from linkage analysis (Roeder et al., 2006) and gene expression (Xiong et al., 2012) in genome-wide association scans. In genetic studies of etiology of asthma, it is of our particular interest to employ similar approaches and explore potentials of power gain in identifying asthma-associated SNPs by incorporating expression quantitative trait loci (eQTL) information.

Catalogs of eQTLs in multiple tissues have been made publicly available, resulting from recent efforts of GWAS of gene expressions (Stranger et al., 2005, 2007, 2012; Dixon et al., 2007; Dimas et al., 2009; Yang et al., 2010). eQTLs provide insight into biology of transcription regulation. It has been shown that eQTLs are enriched for SNPs associated with complex diseases and traits using GWAS (Cookson et al., 2009; Nicolae et al., 2010). eQTL results can be used to provide functional interpretation for findings from GWAS (Moffatt et al., 2007; Heid et al., 2010; Hsu et al., 2010; Lango Allen et al., 2010; Speliotes et al., 2010; Chu et al., 2011; Wu et al., 2012) and prioritize genes in an association region for carrying out functional experiments using animal models (Teslovich et al., 2010). Focusing on eQTLs may also be useful to identify genetic pathways associated with the risk of complex diseases and traits, such as basal cell carcinoma in a skin cancer GWAS (Zhang et al., 2012) and type 2 diabetes (Zhong et al., 2010). Other results show that many cis eQTLs are shared across tissues (Ding et al., 2010) and that a comprehensive eQTL catalog in one tissue might be used to increase the power of capturing relevant transcripts for other diseases (including those that are only weakly or incidentally expressed in tissues where eQTL information was collected).

As single-SNP analysis still remains the most popular in GWAS, we focus on those methods designed for this type of analysis. Single-SNP analysis tests one SNP at a time for association by scanning across the whole genome, and hence involves a large number of hypotheses. To correct for multiple comparisons, statistical methods have been proposed and applied to control for the family-wise error rate (FWER) (Bonferroni, 1936; Holm, 1979) or the false discovery rate (FDR) (Benjamini and Hochberg, 1995; Storey and Tibshirani, 2003). Recent advances in statistical methodology make it possible to incorporate prior information through weighted hypothesis testing. In several of such methods (Genovese et al., 2006; Roeder et al., 2006, 2007), hypotheses are up-weighted or down-weighted based on prior likelihood of association with the trait of interest. While keeping the FWER or FDR under control, the procedures can improve power with informative weights and suffer small loss in power with uninformative weights (Genovese et al., 2006; Roeder and Wasserman, 2009). This feature is appealing as compared to prescreening SNPs based on prior information (e.g., to consider only eQTLs for association testing). In this paper, we propose to use eQTLs as prior information, and apply these weighted hypothesis testing methods to reanalyze the MAGICS (Multicentre Asthma Genetics in Childhood Study) data of asthma GWAS (Moffatt et al., 2007) as well as the GABRIEL meta-study of asthma (Moffatt et al., 2010).

RESULTS

Published Asthma Associations are Enriched with eQTLs

We extracted published asthma associations from the GWAS catalog maintained by the National Human Genome Research Institute. As of January 15, 2013, 52 distinct reference SNPs in or near more than 40 genes have been reported to be associated with asthma (Table A1). According to the eQTL database (described in Materials and Methods), 20 of these 52 SNPs (38.5%) are eQTLs. Using the proxy SNP search tool SNAP (Johnson et al., 2008), we then obtained an extended list of 506 SNPs that were either in the GWAS catalog or in strong linkage disequilibrium (LD) with the 52 SNPs ($r^2 \geq 0.8$). We called all these 506 SNPs the extended set of asthma-associated SNPs.

We calculated an eQTL enrichment p-value (Hosack et al., 2003) using the MAGICS data. There are 300,821 SNPs that passed quality control in the MAGICS data. Among these SNPs, 29 SNPs are in the GWAS catalog, and 64 SNPs are among the 506 extended asthma-associated SNPs defined previously. To account for the LD between SNPs in the calculation of enrichment p-value, we conducted LD pruning with the r^2 threshold of 0.8 on the 300,821 SNPs. This resulted in 251,826 SNPs and 38 of them are extended asthma-associated SNPs according to the GWAS catalog. According to the eQTL database, 22,922 SNPs (9.1% of 251,826 SNPs) are eQTLs, and 13 asthma associated SNPs (34.2% of 38 SNPs) are eQTLs. The corresponding enrichment p-value is 6.78×10^{-5}, suggesting the asthma associations are enriched with eQTLs in the MAGICS data. Note that other analyses considered all the SNPs rather than the pruned set of SNPs.

These results are in line with the previous findings (Nicolae et al., 2010), which studied the eQTLs in lymphoblastoid cell lines (LCL) from the HapMap samples and the GWAS catalog. Their results suggest that SNPs associated with complex traits are more likely to be eQTLs.

Weights Using eQTL Information

We calculated two kinds of weights using eQTL information for the MAGICS data. All the 300,821 SNPs passing quality control were considered. First, we defined a SNP as an eQTL SNP if it was labeled as an eQTL in the eQTL database (see details in Materials and Methods). There are 31,781cis eQTL SNPs (10.6% of 300,821 SNPs) according to the definition, and for each of them, we retrieved an eQTL p-value p_{eQTL}. Next, we considered two choices of weights, the general weight and the binary weight. The general weight is $w_g = \sqrt{-\log_{10} p_{eQTL}}$ for an eQTL SNP, and $w_g = 1$ otherwise. The binary weight takes only two possible values, $w_b = 3.70$ for any eQTL SNP and $w_b = 0.68$ otherwise. The two values of the binary weight were chosen to maximize the minimum power while keeping at least 10.6% (also the percentage of eQTL SNPs) of all the hypotheses with a power of 60%. The parameters for calculating the binary weight are $\epsilon = 0.106$, $\alpha = 0.05$, and $\beta = 0.4$ (see details in Materials and Methods). Last, both weights were normalized to have the mean equal to 1 which is necessary for the weighted hypothesis testing methods to maintain the correct FWER or FDR (Genovese et al., 2006). After normalization, the general weight w_g has a mean of 2.44 and a median of 2.21 among eQTL SNPs, while the binary weight w_b is 3.70 for all eQTL SNPs (Figure 1).

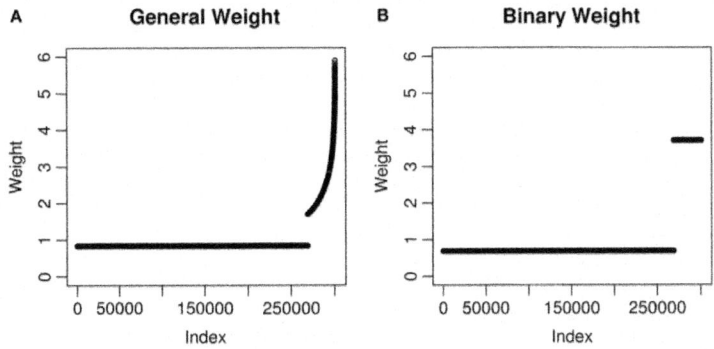

Figure 1: Weights used in the MAGICS analysis. Each weight corresponds to a SNP and a hypothesis. The weights have been normalized to have mean 1 and shown in the ascending order. **(A)** The weights are based on the square root of $-\log_{10} p_{eQTL}$ where p_{eQTL} is the eQTL p-value; **(B)** the weights take only two possible values, which are decided using the method described in Materials and Methods.

Weighted Hypothesis Testing

We applied the weighted hypothesis testing methods (Genovese et al., 2006; Roeder and Wasserman, 2009) using the general weight w_g and the binary weight w_b to the MAGICS data. For each of the 300,821 SNPs, we calculated the trait association p-value, p, from the single-SNP association test on the phenotypes of asthma status, as well as the weighted p-values $Q_g = p/w_g$ and $Q_b = p/w_b$. Multiple testing adjustments were done for both the original p-values (p) and the weighted p-values (Q_g and Q_b). Bonferroni (1936) and Holm's (1979) methods were considered to control for FWER, and Benjamini and Hochberg's (1995) method was used to control for FDR.

We first ranked the SNPs using their p-values in the ascending order, and compared the ranks based on the weighted p-values with those based on the original p-values (Figure 2). Since only 10.6% SNPs are eQTL SNPs according to the eQTL database, and hypotheses for eQTL SNPs are up-weighted, eQTLs generally have higher ranks after weighting, and non-eQTLs' ranks are lower but the magnitude of changes is small. This is true for both the general and binary weights. When restricting to the 29 asthma-associated SNPs reported in the GWAS catalog, we also observed similar behaviors, suggesting that weighting hypotheses may improve power using informative weights, and sacrifice a little power using uninformative weights (Roeder and Wasserman, 2009).

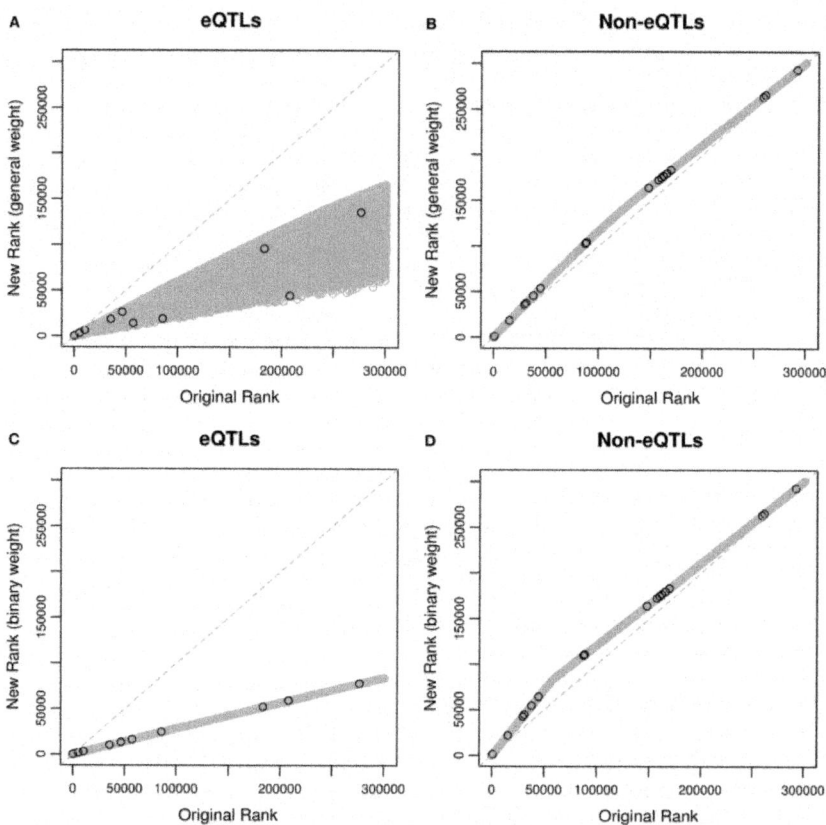

Figure 2: Rankings of the SNPs based on original p-values and weighted p-values in the MAGICS analysis. (A) Original ranks of eQTLs compared to their new ranks based on the general weight; (B) original ranks of non-eQTLs compared to their new ranks based on the general weight; (C) original ranks of eQTLs compared to their new ranks based on the binary weight; (D) original ranks of non-eQTLs compared to their new ranks based on the binary weight. The black circles represent the reported asthma-associated SNPs in the GWAS catalog, and the gray circles represent the rest of the SNPs in the data.

We then looked at the Q–Q plots of the original and weighted p-values. For p-values greater than 0.0001, the Q–Q curves (Figure 3) are similar between the original and weighted p-values, regardless of the weights used. For those p-values less than 0.0001, some weighted p-values are smaller than original ones, and the difference is larger using the binary weight. We also observed that 3 asthma-associated SNPs in the GWAS catalog are among the top SNPs with original p-values less than 10^{-6}. The weighted p-values for all the 3 asthma-associated SNPs are smaller than original ones.

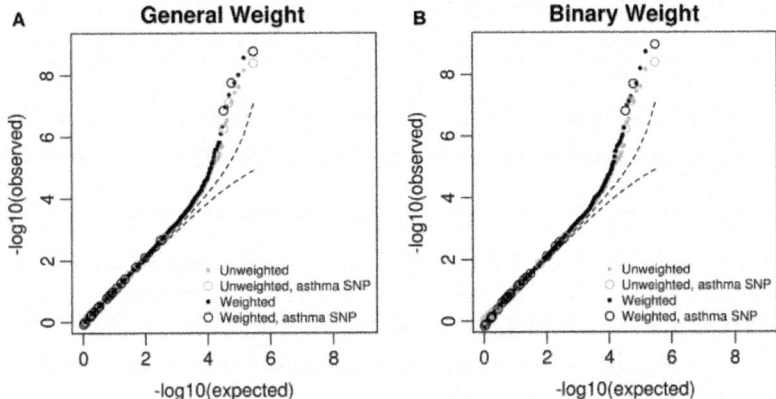

Figure 3: Q–Q plots of original p-values and weighted p-values in the MAGICS analysis. The weighted p-values are based on (A) the general weight, or (B) the binary weight. The reported asthma-associated SNPs in the GWAS catalog are shown in circles.

Next, we applied the methods to control for the FWER. An effective ratio of 0.791 (Li et al., 2012) was used to calculate the effective number of SNPs (300,821 × 0.791). Controlling for an FWER level of 0.05, we obtained significant SNPs using both the original and weighted p-values. Both Bonferroni and Holm's methods gave the same results, and both weights (binary and general) also gave the same results (Tables A2, A3). The unweighted hypothesis testing claimed 6 SNPs to be significant, all on chromosome 17, including 2 asthma-associated SNPs (rs3894194 with GSDMA, and rs7216389 with ORDML3) that have been reported previously (Moffatt et al., 2007, 2010). After applying the weighted hypothesis testing, we obtained 9 significant SNPs including all the 6 SNPs identified by the unweighted method, although the ranks are not exactly the same. The 3 SNPs additionally identified by eQTL weighting were rs3902025, rs4795405, and rs2305480. The SNP rs2305480, a missense SNP in the gene GSDMB, was not reported in the previous GWAS study (Moffatt et al., 2007) but has been reported as an asthma-associated SNP in a later larger scale study by the GABRIEL consortium (including the MAGIC data, Moffatt et al., 2010) and was found to be strongly interacting with exposure to tobacco smoke in early life (Bouzigon et al., 2008). We found that rs2305480 is actually in LD ($r^2 = 0.702$, D' = 0.926) with rs7216389 that was identified by the unweighted methods, suggesting that rs2305480 may not represent a new association. Using a stringent r^2 threshold of 0.4, we found that the other two SNPs, rs3902025 and rs4795405, are also in LD with at least one SNP identified by the unweighted methods. So there is no new association identified by the weighted methods in this particular analysis.

Besides controlling for FWER, we also used Benjamini and Hochberg's (BH) procedure (Benjamini and Hochberg, 1995) to control for a FDR level of 0.05 (Table A4). Based on the original p-values without weighting, the BH procedure gave 11 positive results (SNPs). The weighted BH procedures based on the general weight and the binary weight resulted in 7 and 8 additional positive results (SNPs), respectively. Using a stringent r^2 threshold of 0.4, we found that 5 SNPs (Table 1) are not in LD with any of the SNPs identified without weighting. Although none of the 5 SNPs are, or in LD with, any asthma-associated SNPs according to the GWAS catalog, there are some SNPs that seem interesting. Some of the SNPs are in or close to the genes PGAP3 and STARD3 on chromosome 17, and interestingly, rs2941504 has been reported in a recent independent study (Anantharaman et al., 2011) to be associated with asthma, although it does not meet the criteria for inclusion in the GWAS catalog. This suggests that the reanalysis using the eQTL weighting approaches is promising and potentially useful.

Table 1: Additional significant SNPs or positive results identified by eQTL weighting methods after accounting for linkage disequilibrium in the MAGICS analysis

Chr	SNP	Gene	p-value	Q_g	Q_b	Method
17	rs1877031	STARD3	4.32×10^{-6}	1.54×10^{-6}	1.17×10^{-6}	BH
17	rs931992	TCAP STARD3	5.61×10^{-6}	1.96×10^{-6}	1.52×10^{-6}	BH
17	rs1565922	PGAP3	7.28×10^{-6}	2.55×10^{-6}	1.97×10^{-6}	BH
17	rs2941504	PGAP3	7.64×10^{-6}	2.68×10^{-6}	2.06×10^{-6}	BH
10	rs11191325	SUFU	8.49×10^{-6}	3.37×10^{-6}	2.29×10^{-6}	BH

Q_g and Q_b are the eQTL weighted p-values based on the general weight and the binary weight, respectively. Using both weights result in the same set of SNPs.

Reanalysis of the GABRIEL Data

As another application, we reanalyzed the GABRIEL data using the eQTL weighted approaches. Since only the p-values are necessary for the use of eQTL weighting, we took the p-values of the meta analysis of 37 studies that were calculated based on imputed data. In total, there are 2,473,850 SNPs and their p-values available in the GABRIEL study, which include 267,350 out of 268,204 eQTL SNPs in the eQTL database. The weights based on eQTL information were calculated in the similar way to the MAGICS data analysis.

We applied Bonferroni and Holm's methods with an FWER level of 0.05, as well as the BH procedure with an FDR level of 0.05. An effective ratio of 0.30 (Li et al., 2012) was used to calculate the effective number of SNPs ($2,473,850 \times 0.30$). After obtaining the lists of significant SNPs using different methods, we report any SNPs identified by eQTL weighting that are not in LD with any SNPs identified by unweighted methods using an r^2 threshold of 0.4. Such SNPs may be informative and suggest new associations. Tables 2 and 3 show the SNPs that were identified based on the general weight and the binary weight, respectively.

Table 2: Additional significant SNPs or positive results identified by eQTL weighting methods after accounting for linkage disequilibrium in the GABRIEL analysis

Chr	SNP	Gene	p-value	Q_g	Q_b	Method
5	rs244749	MEF2C	6.25×10^{-5}	2.51×10^{-5}	1.54×10^{-5}	BH
5	rs10075941		6.13×10^{-5}	3.03×10^{-5}	1.51×10^{-5}	BH
17	rs7503195		7.66×10^{-5}	3.07×10^{-5}	1.88×10^{-5}	BH
17	rs17637472		6.13×10^{-5}	3.56×10^{-5}	1.51×10^{-5}	BH
9	rs7047575		6.87×10^{-5}	3.92×10^{-5}	1.69×10^{-5}	BH

The results are based on the general weight.

Table 3: Additional significant SNPs or positive results identified by eQTL weighting methods after accounting for linkage disequilibrium in the GABRIEL analysis

Chr	SNP	Gene	p-value	Q_g	Q_b	Method
5	rs736801		2.15×10^{-7}	8.81×10^{-8}	5.29×10^{-8}	Bonferroni, Holm
6	rs2596450	HCG26	2.72×10^{-7}	1.16×10^{-7}	6.69×10^{-8}	Bonferroni, Holm
5	rs10075941		6.13×10^{-5}	3.03×10^{-5}	1.51×10^{-5}	BH
17	rs17637472		6.13×10^{-5}	3.56×10^{-5}	1.51×10^{-5}	BH
5	rs244749	MEF2C	6.25×10^{-5}	2.51×10^{-5}	1.54×10^{-5}	BH
9	rs7047575		6.87×10^{-5}	3.92×10^{-5}	1.69×10^{-5}	BH
17	rs7503195		7.66×10^{-5}	3.07×10^{-5}	1.88×10^{-5}	BH
6	rs9273363	HLA_DQB1	8.38×10^{-5}	4.41×10^{-5}	2.06×10^{-5}	BH
5	rs4351182		9.98×10^{-5}	4.91×10^{-5}	2.45×10^{-5}	BH
2	rs13391794		1.01×10^{-4}	5.62×10^{-5}	2.49×10^{-5}	BH
5	rs10044342	MEF2C	1.10×10^{-4}	4.59×10^{-5}	2.71×10^{-5}	BH
2	rs6751196		1.14×10^{-4}	6.22×10^{-5}	2.80×10^{-5}	BH
6	rs176095	PBX2, GPSM3	1.25×10^{-4}	5.12×10^{-5}	3.08×10^{-5}	BH
2	rs2675073		1.34×10^{-4}	5.40×10^{-5}	3.29×10^{-5}	BH
2	rs1913621		1.46×10^{-4}	7.92×10^{-5}	3.60×10^{-5}	BH
2	rs10497621	NUP35	1.59×10^{-4}	7.55×10^{-5}	3.91×10^{-5}	BH
5	rs244750	MEF2C	1.70×10^{-4}	7.08×10^{-5}	4.17×10^{-5}	BH
6	rs9366689	POM121L2	1.72×10^{-4}	8.73×10^{-5}	4.23×10^{-5}	BH
6	rs7775759		1.81×10^{-4}	6.43×10^{-5}	4.46×10^{-5}	BH
6	rs7741091		1.81×10^{-4}	6.43×10^{-5}	4.46×10^{-5}	BH
8	rs6601649		1.90×10^{-4}	9.68×10^{-5}	4.68×10^{-5}	BH
6	rs204994		1.94×10^{-4}	8.02×10^{-5}	4.76×10^{-5}	BH

The results are based on the binary weight.

Size Simulations on FWER

We conducted simulations using 5000 permutations based on the MAGICS data, and calculated the percentage of having at least one false positive claimed by Bonferroni and Holm's methods ($\alpha = 0.05$, with an effective ratio of 0.791). In fact, any SNPs claimed significant using the two methods would be a false positive. The calculated percentages (Table 4) provide estimates of the FWER. Bonferroni and Holm's methods give the same results. The results suggest that, under the null hypothesis, the FWER level is controlled for the methods based on both the original (unweighted) and the weighted p-values. The simulations confirm the validity of the weighted hypothesis method (Genovese et al., 2006).

Table 4: Family-wise error rate estimates in 5000 permutations

P-value	FWER
Original	0.0454
Weighted by general weight	0.0458
Weighted by binary weight	0.0460

Both Bonferroni and Holm's methods gave the same results in the same scenarios.

DISCUSSION

It is of substantial interest to enhance the power for identifying associations in the era of post-GWAS. Besides meta-analysis that has been proved successful in power gain (Moffatt et al., 2010), incorporating prior information has also received increasing attention. Such information can be obtained from various sources and levels, such as linkage analysis (Roeder et al., 2006), gene expression (Yang et al., 2010), and annotation information of variants (Adzhubei et al., 2010), genes (Saccone et al., 2007), and pathways (Wang et al., 2007). The so-called "agnostic" GWAS may benefit from incorporating useful prior information. In our study of asthma, gene expression information is of particular interest, as a recent study (Moffatt et al., 2007) identified several eQTLs associated with asthma.

In the reanalysis of the MAGICS data (Moffatt et al., 2007), we applied recently developed statistical methods that can improve power by weighting hypothesis (Genovese et al., 2006; Roeder and Wasserman, 2009). Using eQTL information obtained from an independent dataset, we employed weighted procedures that up-weighted eQTL SNPs and down-weighted non-eQTL SNPs while controlling for the FWER or the FDR. It has been proved (Genovese et al., 2006) that any set of nonnegative weights can guarantee substantial power gain given informative weights and little power loss for uninformative weights. The property implies that the weighted procedures are robust to informativeness of weights and to the uneven coverage of genes and expression targets on the genome. We took advantage of this robustness and applied the procedures to an asthma study. We found additional SNPs that were significantly associated with asthma according to the weighting hypothesis methods. Some of them were interesting after we accounted for LD and compared them to literature. Our analysis was the first application of this approach to asthma GWAS studies, and the results successfully illustrated the use of eQTL weighting in the context of asthma studies. As another application, we also reanalyzed the GABRIEL meta-analysis p-values and reported corresponding results.

It is noted that the weighted procedures can utilize eQTL information from a reference database. Multiple choices of eQTL databases have already been made available (e.g., Yang et al., 2010; Liang et al., 2013), and future efforts may provide even better reference of eQTL information. For example, the eQTL information considered in our reanalysis was obtained through a single platform (Affymetrix HG-U133 Plus 2.0), and better coverage of gene expression profiling may be achieved through RNA-Seq technologies or by combining information from various platforms.

Besides the weighted procedures, an alternative method of using eQTL information is to simply test association between eQTL SNPs and the trait of interest. Such a method is not recommended in GWAS as it excludes non-eQTL completely and relies on the prior information too heavily. By contrast, weighted procedures make it possible to consider eQTLs and non-eQTLs simultaneously. More importantly, they can possibly increase power if the prior information is useful and are able to maintain the type I error under the null.

Applying the weighted procedures in our reanalysis only requires p-value of eQTL SNPs. This flexibility means that such analyses can be applied to any existing GWAS data, even if they do not have accompanying gene expression data. Although gene expression may have tissue-specific patterns, a substantial fraction of eQTLs may be shared across tissues (Ding et al., 2010). Hence eQTLs developed from tissues that are not directly relevant to the outcome of interest, such as those from publicly available eQTL databases based on LCL, can be used to improve power on GWAS. It is possible that using eQTL information from relevant tissues may result in even more power gain, if such information is available.

Besides the particular weighting hypothesis method (Genovese et al., 2006) we adopted, Bayesian methods are potentially alternative strategies to incorporate eQTL information. The use of Bayes factors has been applied to genetic association studies (Wellcome Trust Case Control Consortium, 2007; Stephens and Balding, 2009). In single SNP analysis, a prior is assumed for each SNP effect [e.g., $N(0, 0.2^2)$ under a model of association in Wellcome Trust Case Control Consortium (2007)]. eQTL information can be naturally incorporated into the prior, although it may be challenging to choose a realistic yet tractable alternative model and to assess error rates (Hoggart et al., 2008), especially with the eQTL weight. One possible choice is through modifying the variance of the prior, for example assuming a prior $N(0.2^2 w)$, where w is a weight of eQTL signal. Another possible choice is to keep the variance the same and increase the probability of association for eQTLs a priori. It is of interest in future research to explore these possibilities and consider the extension of Bayesian methods to incorporate eQTL information.

In our analysis, we took into account the LD between SNPs by considering the effective number of SNPs (Li et al., 2012). As an alternative, testing SNP sets for association has potential of improving power and reducing the correlation between tests. Since the focus of this paper is to demonstrate the use of eQTL information in association testing, we will consider the weighted correlated hypothesis in future research.

Two choices of weights were applied in our analysis including a binary weight and a weight using strength of eQTLs, and the results using the two weights were similar in our analysis. Theoretical results exist (Roeder and Wasserman, 2009) for the optimal binary weight, which provide guidance in choosing the values of the weight. The weight taking advantage of the eQTL strength may possibly provide more useful information, and what is the best choice of weights is still under research.

Through an application to an asthma GWAS, we demonstrated the usefulness of eQTL weights in GWAS. Although results may vary depending on the traits of interest and the underlying biological mechanism, the potentials of increasing power and little investment required for reanalysis make the eQTL-weighted procedures desirable for reanalysis of existing GWAS data and useful for design and analysis of future studies.

MATERIALS AND METHODS

The MAGICS Asthma GWAS Samples and Data

The MAGICS (Multicentre Asthma Genetics in Childhood Study) study data (Moffatt et al., 2007), part of the GABRIEL consortium, were reanalyzed by incorporating eQTL information. Quality control procedures were conducted similarly to a published protocol (Anderson et al., 2010). Individuals with missing phenotypes, elevated missing rates ($\geq 5\%$), or outlying heterozygosity rate were removed. Markers with an excessive missing rate ($\geq 5\%$), low MAF ($<5\%$), or failing in the HWE test (p-value $< 10^{-5}$) were all excluded as well. The remaining dataset contains 1296 individuals (647 affected and 649 unaffected) genotyped across 300,821 SNPs.

To account for possible divergent ancestry and population stratification, principal component analysis (PCA) was conducted using EIGENSOFT 4.2 (Patterson et al., 2006; Price et al., 2006). The genotype data were pruned for LD prior to the PCA. The PCA result (Figure A1) suggests that no obvious stratification exists, and the signal of the first principal component is very weak.

In the subsequent analysis, we still included the first principal component as a covariate.

LD pruning was considered only in the calculation of enrichment p-value. It was conducted using PLINK (v1.07, downloaded from http://pngu.mgh. harvard.edu/purcell/plink/) (Purcell et al., 2007). A moving window with a width of 50 SNPs and a step size of 5 SNPs was considered, and pairwise LDs were calculated and pruned if $r^2 > 0.8$ (corresponding PLINK arguments: "–indep-pairwise 50 5 0.8").

The GABRIEL Meta-Analysis p-Values

Association testing results, including SNP ID and p-values, were obtain from a reanalysis of the GABRIEL consortium data using imputed SNPs (Bouzigon et al., personal communication). The meta-analysis considered imputation of SNP genotypes using the HapMap 2 reference data for 37 studies, and calculated a meta-analysis p-value for each SNP using available data. Imputed SNPs were kept for analysis if their imputation scores (Rsq) were ≥ 0.5 and if their minor allele frequencies were $\geq 1\%$. In total there were 2,473,850 SNPs that passed the quality control. Only the SNP ID and the p-values of these SNPs were obtained and used for the reanalysis described in this paper.

Expression Quantitative Trait Loci Data

An eQTL database (http://www.hsph.harvard.edu/liming-liang/software/eqtl/) resulting from an independent dataset was used as prior information to be incorporated in the GWAS. The sample contains 405 siblings from a panel of families of British descent (MRC-A) (Dixon et al., 2007). Global gene expression in LCLs was measured using Affymetrix HG-U133 Plus 2.0 chips. All siblings were genotyped using the Illumina Sentrix HumanHap300 BeadChip (ILMN300K) and/or the Illumina Sentrix Human-1 Genotyping BeadChip (ILMN100K). The SNP genotype data were further imputed using the MaCH program, and each SNP was tested for association with probes in the gene expression data. Restricting to cis eQTLs (1 Mb region) and controlling for the FDR of 1%, there are 515,947 tests with logarithm of odds (LOD) scores greater than 3.172, corresponding to 268,204 unique SNPs. In case a SNP has multiple p-values reported for associations with different probes, the minimal p-value was used for that SNP. These 268,204 SNPs are considered as eQTLs, and the database contains information of their physical positions, LOD scores, p-values, and residing or nearby genes. Details of the database are described by Liang et al. (2013).

Genetic Association Analysis

Genetic association analysis of the MAGICS data was conducted in PLINK. Logistic regression was used to test for disease-trait SNP association while adjusted for gender and the first principal component. Meta-analysis on GABRIEL data was carried out by combining association results from 37 studies using a random effect model, and all computations were done using Stata software.

p-Value Weighting Methods

Consider m hypotheses $H_1, \ldots, H_m$ and their test statistic p-values, $P_1, \ldots, P_m$. Suppose there are weights $W_1, \ldots, W_m$ available for the m tests, respectively, satisfying $W_i > 0$ and $\sum_{i=1}^{m} W_i = m$. Define $Q_i = P_i/W_i$ and let $Q_{(1)} \leq \ldots \leq Q_{(m)}$ be the sorted values. Let $P_{(1)}, \ldots, P_{(m)}$ and $W_{(1)}, \ldots, W_{(m)}$ be the values in the corresponding order. Q_i is sometimes referred to a "weighted p-value" (e.g.,Roeder and Wasserman, 2009), although it is not a p-value.

The weighted Bonferroni procedure is to reject any hypothesis H_j ($1 \leq j \leq$ m) that satisfies $Q_j \leq \alpha/m$, where α is the desired level of FWER. Genovese et al. (2006) showed that this procedure controls FWER at level no greater than α.

Holm's weighted procedure (1979) is carried out as follows: given the desired α level of FWER, if $Q_{(1)} \geq \alpha/m$, no hypothesis is rejected; otherwise, find the largest j that satisfies $Q_{(i)} \leq \alpha/\sum_{k=i}^{m} W_{(k)}$ for all $i \leq j$, and reject the hypotheses corresponding to the j smallest Q_j's. Genovese et al. (2006)also prove that this procedure can work for a general setting of weights.

We also consider Benjamini and Hochberg's procedure (1995) for controlling FDR. Given the desired level α, find the largest j such that $Q_{(j)} \leq \alpha \cdot j/m$, and reject the hypotheses corresponding to the j smallest Q_j's. Genovese et al. (2006) prove that this procedure controls FDR at level α.

eQTL Information as Weights

The eQTL p-values were used to construct weights for the SNPs in the asthma GWAS reanalysis. We considered two kinds of weights, the binary weight w_b and the general weight w_g. The binary weight takes only two possible values that are predefined, denoted by w_{eQTL} and $w_{non-eQTL}$. For mhypotheses, a binary weight is defined as $w_b = (w_{b,1}, \ldots, w_{b,m})$ where $w_{b,j} = w_{eQTL}$ if the jth SNP is an eQTL SNP, and $w_{b,j} = w_{non-eQTL}$ if it is not an eQTL SNP. Given the values of α, β, and ϵ, the optimal values of w_{eQTL} and $w_{non-eQTL}$ were chosen (Roeder and Wasserman, 2009) to maximize the minimum power among all the hypotheses while having at least a fraction ϵ with high power $1-\beta$. Here α is either the level of FWER or FDR. We also considered a general weight, where the weight w_g = $(w_{g,1}, ..., w_{g,m})$ has wg, $j = \sqrt{-\log_{10} p_{eQTL}}$ if the jth SNP is an eQTL SNP with the eQTL p-value p_{eQTL}, and $w_{g,j}$ = 1 otherwise. The particular form was intuitively chosen prior to the reanalysis of the GWAS data in consideration of avoiding up-weighting top eQTL SNPs too much. Both w_h and w_g were then normalized such that the means equal to 1, i.e., $\overline{w}_B = 1$ and $\overline{w}_G = 1$.

Reported Associations in the GWAS Catalog

Asthma-associated SNPs and genes reported in publications were retrieved from the online catalog of published GWAS on January 15, 2013. The catalog limits the associations to those with p-values less than 1.0×10^{-5} and records only one SNP with a gene or region of high LD unless there was evidence of independent association. The reported associations were compared against the findings in the asthma GWAS data we reanalyzed.

Linkage Disequilibrium Information

To account for LD between SNPs, LD information based on HapMap 2 was obtained. The SNAP proxy search tool (http://www.broadinstitute.org/mpg/snap/ldsearch.php) was used to obtain the information, based on the HapMap 2 (rel22) reference and a distance limit of 500kb.

Size Simulation Using the Asthma GWAS Data

Besides analyzing the MAGICS asthma GWAS data, we also conducted size simulations by permuting the disease status in the data. Logistic regression was considered where the dependent variable was the disease status (affected or unaffected) and the independent variables included a single SNP effect, gender, and the first principal component. The regression was applied to all the ~300,000 SNPs across the whole genome. Five thousand permutations were done by permuting the disease status among all the individuals, and then the model was refitted for each SNP. In the end of simulations, 5000 permutation p-values were obtained for each of the ~300,000 SNPs.

Conflict of Interest Statement

The authors declare that the research was conducted in the absence of any commercial or financial relationships that could be construed as a potential conflict of interest.

ACKNOWLEDGMENTS

Lin Li and Xihong Lin's research was supported by grants from National Cancer Institute (R37 CA076404 and P01 CA134294). Martin Farrall is supported by the British Heart Foundation Centre for Research Excellence in Oxford and the Wellcome Trust core award [090532/Z/09/Z]. We thank the members of the GABRIEL consortium groups for providing data and summary results (a full list of the GABRIEL groups can be found in the Supplement of Moffatt et al., 2010).

RESOURCES

R functions have been made available for users' convenience to compute the eQTL weighted p-values in order to conduct weighted procedures such as Bonferroni, Holm's and Benjamini–Hochberg procedures. The functions can be found at http://www.hsph.harvard.edu/liming-liang/eqtl-weighted-gwas.

REFERENCES

1. Adzhubei, I. A., Schmidt, S., Peshkin, L., Ramensky, V. E., Gerasimova, A., Bork, P., et al. (2010). A method and server for predicting damaging missense mutations. Nat. Methods 7, 248–249. doi: 10.1038/nmeth0410-248

2. Anantharaman, R., Andiappan, A. K., Nilkanth, P. P., Suri, B. K., Wang, D. Y., and Chew, F. T. (2011). Genome-wide association study identifies PERLD1 as asthma candidate gene. BMC Med. Genet. 12:170. doi: 10.1186/1471-2350-12-170

3. Anderson, C. A., Pettersson, F. H., Clarke, G. M., Cardon, L. R., Morris, A. P., and Zondervan, K. T. (2010). Data quality control in genetic case-control association studies. Nat. Protoc. 5, 1564–1573. doi: 10.1038/nprot.2010.116

4. Benjamini, Y., and Hochberg, Y. (1995). Controlling the false discovery rate: a practical and powerful approach to multiple testing. J. R. Stat. Soc. Ser. B 57, 289–300. doi: 10.2307/2346101

5. Bonferroni, C. E. (1936). Teoria statistica delle classi e calcolo delle probabilità. Pubblicazioni del R Istituto Superiore di Scienze Economiche e Commerciali di Firenze 8, 3–62.

6. Bouzigon, E., Corda, E., Aschard, H., Dizier, M. H., Boland, A., Bousquet, J., et al. (2008). Effect of 17q21 variants and smoking exposure in early-onset asthma. N. Engl. J. Med. 359, 1985–1994. doi: 10.1056/NEJMoa0806604

7. Chu, X., Pan, C.-M., Zhao, S.-X., Liang, J., Gao, G.-Q., Zhang, X.-M., et al. (2011). A genome-wide association study identifies two new risk loci for Graves' disease. Nat. Genet. 43, 897–901. doi: 10.1038/ng.898

8. Cookson, W., Liang, L., Abecasis, G., Moffatt, M., and Lathrop, M. (2009). Mapping complex disease traits with global gene expression. Nat. Rev. Genet. 10, 184–194. doi: 10.1038/nrg2537

9. Cookson, W. (2004). The immunogenetics of asthma and eczema: a new focus on the epithelium. Nat. Rev. Immunol. 4, 978–988. doi: 10.1038/nri1500

10. Dimas, A. S., Deutsch, S., Stranger, B. E., Montgomery, S. B., Borel, C., Attar-Cohen, H., et al. (2009). Common regulatory variation impacts gene expression in a cell type-dependent manner. Science 325, 1246–1250. doi: 10.1126/science.1174148

11. Ding, J., Gudjonsson, J. E., Liang, L., Stuart, P. E., Li, Y., Chen, W., et al. (2010). Gene expression in skin and lymphoblastoid cells: refined statistical method reveals extensive overlap in cis-eQTL signals. Am. J. Hum. Genet. 87, 779–789. doi: 10.1016/j.ajhg.2010.10.024

12. Dixon, A. L., Liang, L., Moffatt, M. F., Chen, W., Heath, S., Wong, K. C. C., et al. (2007). A genome-wide association study of global gene expression. Nat. Genet. 39, 1202–1207. doi: 10.1038/ng2109

13. Genovese, C. R., Roeder, K., and Wasserman, L. (2006). False discovery control with p-value weighting. Biometrika 93, 509–524. doi: 10.1093/biomet/93.3.509

14. Heid, I. M., Jackson, A. U., Randall, J. C., Winkler, T. W., Qi, L., Steinthorsdottir, V., et al. (2010). Meta-analysis identifies 13 new loci associated with waist-hip ratio and reveals sexual dimorphism in the genetic basis of fat distribution. Nat. Genet. 42, 949–960. doi: 10.1038/ng.685

15. Hoggart, C. J., Clark, T. G., De Iorio, M., Whittaker, J. C., and Balding, D. J. (2008). Genome-wide significance for dense SNP and resequencing data. Genet. Epidemiol. 32, 179–185. doi: 10.1002/gepi.20292

16. Holm, S. (1979). A simple sequentially rejective multiple test procedure. Scand. J. Stat. 6, 65–70. doi: 10.2307/4615733

17. Hosack, D. A., Dennis, G. Jr. Sherman, B. T., Lane, H. C., and Lempicki, R. A. (2003). Identifying biological themes within lists of genes with EASE. Genome Biol. 4:R70. doi: 10.1186/gb-2003-4-10-r70

18. Hsu, Y.-H., Zillikens, M. C., Wilson, S. G., Farber, C. R., Demissie, S., Soranzo, N., et al. (2010). An integration of genome-wide association study and gene expression profiling to prioritize the discovery of

novel susceptibility Loci for osteoporosis-related traits. PLoS Genet. 6:e1000977. doi: 10.1371/journal.pgen.1000977

19. Johnson, A. D., Handsaker, R. E., Puilt, S., Nizzari, M. M., O'Donnell, C. J., and de Bakker, P. I. W. (2008). SNAP: a web-based tool for identification and annotation of proxy SNPs using HapMap. Bioinformatics 24, 2938–2939. doi: 10.1093/bioinformatics/btn564

20. Lango Allen, H., Estrada, K., Lettre, G., Berndt, S. I., Weedon, M. N., Rivadeneira, F., et al. (2010). Hundreds of variants clustered in genomic loci and biological pathways affect human height. Nature 467, 832–838. doi: 10.1038/nature09410

21. Li, M. X., Yeung, J. M. Y., Cherny, S. S., and Sham, P. C. (2012). Evaluating the effective numbers of independent tests and significant p-value thresholds in commercial genotyping arrays and public imputation reference datasets. Hum. Genet. 131, 747–756. doi: 10.1007/s00439-011-1118-2

22. Liang, L., Morar, N., Dixon, A. L., Lathrop, G. M., Abecasis, G. R., Moffatt, M. F., et al. (2013). A cross-platform catalogue of 14, 177 expression quantitative trait loci derived from lymphoblastoid cell lines. Genome Res. 23, 716–726. doi: 10.1101/gr.142521.112

23. Moffatt, M. F., Kabesch, M., Liang, L., Dixon, A. L., Strachan, D., Heath, S., et al. (2007). Genetic variants regulating ORMDL3 expression contribute to the risk of childhood asthma. Nature 448, 470–473. doi: 10.1038/nature06014

24. Moffatt, M. F., Gut, I. G., Demenais, F., Strachan, D. P., Bouzigon, E., Heath, S., et al. (2010). A large-scale, consortium-based genomewide association study of asthma. N. Engl. J. Med. 363, 1211–1221. doi: 10.1056/NEJMoa0906312

25. Nica, A. C., Parts, L., Glass, D., Nisbet, J., Barrett, A., Sekowska, M., et al. (2011). The architecture of gene regulatory variation across multiple human tissues: the MuTHER study. PLoS Genet. 7:e1002003. doi: 10.1371/journal.pgen.1002003

26. Nicolae, D. L., Gamazon, E., Zhang, W., Duan, S., Dolan, M. E., and Cox, N. J. (2010). Trait-associated SNPs are more likely to be eQTLs: annotation to enhance discovery from GWAS. PLoS Genet. 6:e1000888. doi: 10.1371/journal.pgen.1000888

27. Patterson, N., Price, A. L., and Reich, D. (2006). Population structure and eigenanalysis. PLoS Genet. 2:e190. doi: 10.1371/journal.pgen.0020190

28. Price, A. L., Patterson, N. J., Plenge, R. M., Weinblatt, M. E., Shadick, N. A., and Reich, D. (2006). Principal components analysis corrects for

stratification in genome-wide association studies. Nat. Genet. 38, 904–909. doi: 10.1038/ng1847

29. Purcell, S., Neale, B., Todd-Brown, K., Thomas, L., Ferreira, M. A. R., Bender, D., et al. (2007). PLINK: a tool set for whole-genome association and population-based linkage analyses. Am. J. Hum. Genet. 81, 559–575. doi: 10.1086/519795

30. Roeder, K., and Wasserman, L. (2009). Genome-wide significance levels and weighted hypothesis testing. Stat. Sci. 24, 398–413. doi: 10.1214/09-STS289

31. Roeder, K., Bacanu, S.-A., Wasserman, L., and Devlin, B. (2006). Using linkage genome scans to improve power of association in genome scans. Am. J. Hum. Genet. 78, 243–252. doi: 10.1086/500026

32. Roeder, K., Devlin, B., and Wasserman, L. (2007). Improving power in genome-wide association studies: weights tip the scale.Genet. Epidemiol. 31, 741–747. doi: 10.1002/gepi.20237

33. Saccone, S. F., Hinrichs, A. L., Saccone, N. L., Chase, G. A., Konvicka, K., Madden, P., et al. (2007). Cholinergic nicotinic receptor genes implicated in a nicotine dependence association study targeting 348 candidate genes with 3713 SNPs. Hum. Mol. Genet. 16, 36–49. doi: 10.1093/hmg/ddl438

34. Speliotes, E. K., Willer, C. J., Berndt, S. I., Monda, K. L., Thorleifsson, G., Jackson, A. U., et al. (2010). Association analyses of 249, 796 individuals reveal 18 new loci associated with body mass index. Nat. Genet. 42, 937–948. doi: 10.1038/ng.686

35. Stephens, M., and Balding, D. (2009). Bayesian statistical methods for genetic association studies. Nat. Rev. Genet. 10, 681–690. doi: 10.1038/nrg2615

36. Storey, J. D., and Tibshirani, R. (2003). Statistical significance for genomewide studies. Proc. Natl. Acad. Sci. U.S.A. 100, 9440–9445. doi: 10.1073/pnas.1530509100

37. Stranger, B. E., Forrest, M. S., Clark, A. G., Minichiello, M. J., Deutsch, S., Lyle, R., et al. (2005). Genome-wide associations of gene expression variation in humans. PLoS Genet. 1:e78. doi: 10.1371/journal.pgen.0010078

38. Stranger, B. E., Nica, A. C., Forrest, M. S., Dimas, A., Bird, C. P., Beazley, C., et al. (2007). Population genomics of human gene expression. Nat. Genet. 39, 1217–1224. doi: 10.1038/ng2142

39. Stranger, B. E., Montgomery, S. B., Dimas, A. S., Parts, L., Stegle, O., Ingle, C. E., et al. (2012). Patterns of cis regulatory variation in diverse

human populations. PLoS Genet. 8:e1002639. doi: 10.1371/journal.pgen.1002639

40. Teslovich, T. M., Musunuru, K., Smith, A. V., Edmondson, A. C., Stylianou, I. M., Koseki, M., et al. (2010). Biological, clinical and population relevance of 95 loci for blood lipids. Nature 466, 707–713. doi: 10.1038/nature09270

41. Wang, K., Li, M., and Bucan, M. (2007). Pathway-based approaches for analysis of genomewide association studies. Am. J. Hum. Genet. 81, 1278–1283. doi: 10.1086/522374

42. Wellcome Trust Case Control Consortium. (2007). Genome-wide association study of 14, 000 cases of seven common diseases and 3, 000 shared controls. Nature 447, 661–678. doi: 10.1038/nature05911

43. Wu, C., Miao, X., Huang, L., Che, X., Jiang, G., Yu, D., et al. (2012). Genome-wide association study identifies five loci associated with susceptibility to pancreatic cancer in Chinese populations. Nat. Genet. 44, 62–66. doi: 10.1038/ng.1020

44. Xiong, Q., Ancona, N., Hauser, E. R., Mukherjee, S., and Furey, T. S. (2012). Integrating genetic and gene expression evidence into genome-wide association analysis of gene sets. Genome Res. 22, 386–397. doi: 10.1101/gr.124370.111

45. Yang, T.-P., Beazley, C., Montgomery, S. B., Dimas, A. S., Gutierrez-Arcelus, M., Stranger, B. E., et al. (2010). Genevar: a database and Java application for the analysis and visualization of SNP-gene associations in eQTL studies. Bioinformatics 26, 2474–2476. doi: 10.1093/bioinformatics/btq452

46. Zhang, M., Liang, L., Morar, N., Dixon, A. L., Lathrop, G. M., Ding, J., et al. (2012). Integrating pathway analysis and genetics of gene expression for genome-wide association study of basal cell carcinoma. Hum. Genet. 131, 615–623. doi: 10.1007/S00439-011-11047-8

47. Zhong, H., Yang, X., Kaplan, L. M., Molony, C., and Schadt, E. E. (2010). Integrating pathway analysis and genetics of gene expression for genome-wide association studies. Am. J. Hum. Genet. 86, 581–591. doi: 10.1016/j.ajhg.2010.02.020

CITATION

CHAPTER 1

He T, Sa J, Zhong P-S, Cui Y (2014) Statistical Dissection of Cyto-Nuclear Epistasis Subject to Genomic Imprinting in Line Crosses. PLoS ONE 9(3): e91702. doi:10.1371/journal.pone.0091702.

CHAPTER 2

Bernd Genser, Philip J Cooper, Maria Yazdanbakhsh, Mauricio L Barreto and Laura C Rodrigues, A guide to modern statistical analysis of immunological data, DOI: 10.1186/1471-2172-8-27.

CHAPTER 3

Fenger M (2014) Next generation genetics. *Front. Genet.* **5**:322. doi: 10.3389/fgene.2014.00322.

CHAPTER 4

Li Y, Guo Y, Wang J, Hou W, Chang MN, Liao D, et al. (2011) A Statistical Design for Testing Transgenerational Genomic Imprinting in Natural Human Populations. PLoS ONE 6(2): e16858. doi:10.1371/journal.pone.0016858.

CHAPTER 5

Shen M, Broeckling CD, Chu EY, Ziegler G, Baxter IR, Prenni JE, et al. (2013) Leveraging Non-Targeted Metabolite Profiling via Statistical Genomics. PLoS ONE 8(2): e57667. doi:10.1371/journal.pone.0057667.

CHAPTER 6

A. N. Diaz-Lacava, M. Walier, D. Holler, et al., "Genetic Geostatistical Framework for Spatial Analysis of Fine-Scale Genetic Heterogeneity in Modern Populations: Results from the KORA Study," International Journal of Genomics, vol. 2015, Article ID 693193, 15 pages, 2015. doi:10.1155/2015/693193.

CHAPTER 7

C. O. Aremu, Exploring Statistical Tools in Measuring Genetic Diversity for Crop Improvement, http://cdn.intechopen.com/pdfs/31485.pdf.

CHAPTER 8

Alex Clarke and Timothy J Vyse, Genetics of rheumatic disease, DOI: 10.1186/ar2781.

CHAPTER 9

Benjamin FrenchEmail author, Jungnam JooEmail author, Nancy L Geller, Stephen E Kimmel, Yves Rosenberg, Jeffrey L Anderson, Brian F Gage, Julie A Johnson, Jonas H Ellenberg and the COAG (Clarification of Optimal Anticoagulation through Genetics) Investigators, Statistical design of personalized medicine interventions: The Clarification of Optimal Anticoagulation through Genetics (COAG) trial, DOI: 10.1186/1745-6215-11-108.

CHAPTER 10

Michael Lässig, From biophysics to evolutionary genetics: statistical aspects of gene regulation, DOI: 10.1186/1471-2105-8-S6-S7.

CHAPTER 11

Zhang, Heping. Statistical Analysis in Genetic Studies of Mental Illnesses. Statist. Sci. 26 (2011), no. 1, 116--129. doi:10.1214/11-STS353. http://projecteuclid.org/euclid.ss/1307626569.

CHAPTER 12

Li L, Kabesch M, Bouzigon E, Demenais F, Farrall M, Moffatt MF, Lin X and Liang L (2013) Using eQTL weights to improve power for genome-wide association studies: a genetic study of childhood asthma. Front. Genet. 4:103. doi: 10.3389/fgene.2013.00103.

INDEX